British Tertiary Volcanic Province

THE GEOLOGICAL CONSERVATION REVIEW SERIES

The comparatively small land area of Great Britain contains an unrivalled sequence of rocks, mineral and fossil deposits, and a variety of landforms that span much of the earth's long history. Well-documented ancient volcanic episodes, famous fossil sites, and sedimentary rock sections used internationally as comparative standards, have given these islands an importance out of all proportion to their size. These long sequences of strata and their organic and inorganic contents, have been studied by generations of leading geologists thus giving Britain a unique status in the development of the science. Many of the divisions of geological time used throughout the world are named after British sites or areas, for instance the Cambrian, Ordovician and Devonian systems, the Ludlow Series and the Kimmeridgian and Portlandian stages.

The Geological Conservation Review (GCR) was initiated by the Nature Conservancy Council in 1977 to assess, document, and ultimately publish accounts of the most important parts of this rich heritage. Since 1991 the task of publication has been assumed by the Joint Nature Conservation Committee on behalf of the three country conservation agencies, English Nature, Scottish Natural Heritage and the Countryside Council for Wales. The GCR series of volumes will review the current state of knowledge of the key earth-science sites in Great Britain and provide a firm basis on which site conservation can be founded in years to come. Each GCR volume will describe and assess networks of sites in the context of a portion of the geological column, or a geological, palaeontological, or mineralogical topic. The full series of approximately 50 volumes will be published by the year 2000.

Within each individual volume, every GCR locality is described in detail in a self-contained account, consisting of highlights (a précis of the special interest of the site), an introduction (with a concise history of previous work), a description, an interpretation (assessing the fundamentals of the site's scientific interest and importance), and a conclusion (written in simpler terms for the non-specialist). Each site report is a justification of a particular scientific interest at a locality, of its importance in a British or international setting, and ultimately of its worthiness for conservation.

The aim of the Geological Conservation Review series is to provide a public record of the features of interest in sites being considered for notification as Sites of Special Scientific Interest (SSSIs). It is written to the highest scientific standards but in such a way that the assessment and conservation value of the site is clear. It is a public statement of the value set on our geological and geomorphological heritage by the earth-science community which has participated in its production, and it will be used by the Joint Nature Conservation Committee, English Nature, the Countryside Council for Wales, and Scottish Natural Heritage in carrying out their conservation functions.

All the sites in this volume have been proposed for notification as SSSIs, the final decision to notify or renotify lies with the governing Councils of the appropriate country conservation agency.

Information about the GCR publication programme may be obtained from:

Earth Science Branch,
Joint Nature Conservation Committee,
Monkstone House,
City Road,
Peterborough PE1 1JY.

Titles in the series

1. **Geological Conservation Review**
   An introduction
2. **Quaternary of Wales**
   S. Campbell and D. Q. Bowen
3. **Caledonian Structures in Britain South of the Midland Valley**
   Edited by J. E. Treagus
4. **British Tertiary Volcanic Province**
   C. H. Emeleus and M. C. Gyopari
5. **Igneous Rocks of South–West England**
   P. A. Floyd, C. S. Exley and M. T. Styles
6. **Quaternary of Scotland**
   Edited by J. E. Gordon and D. G. Sutherland

British Tertiary Volcanic Province

C. H. Emeleus

Reader in Geology,
University of Durham

and

M. C. Gyopari

Senior Hydrogeologist,
Groundwater Consulting Services

(with contributions from G. P. Black and I. Williamson)

GCR editors: W. A. Wimbledon and P. H. Banham

CHAPMAN & HALL

London · Glasgow · New York · Tokyo · Melbourne · Madras

Published by Chapman & Hall, 2–6 Boundary Row, London SE1 8HN

Chapman & Hall, 2–6 Boundary Row, London SE1 8HN, UK

Blackie Academic & Professional, Wester Cleddens Road, Bishopbriggs, Glasgow G64 2NZ, UK

Chapman & Hall, 29 West 35th Street, New York NY10001, USA

Chapman & Hall Japan, Thomson Publishing Japan, Hirakawacho Nemoto Building, 6F, 1–7–11 Hirakawa-cho, Chiyoda-ku, Tokyo 102, Japan

Chapman & Hall Australia, Thomas Nelson Australia, 102 Dodds Street, South Melbourne, Victoria 3205, Australia

Chapman & Hall India, R. Seshadri, 32 Second Main Road, CIT East, Madras 600 035, India

First edition 1992

Typeset in 10/12 Garamond by Columns Design & Production Services Ltd, Reading
Printed in Great Britain at the University Press, Cambridge

ISBN 0 412 47980 X

A catalogue record for this book is available from the British Library

Library of Congress Cataloging-in-Publication data available

Contents

Contents

Acknowledgements

Work on this volume was initiated by the Nature Conservancy Council and has been seen to completion by the Joint Nature Conservation Committee on behalf of the three country agencies, English Nature, Scottish Natural Heritage and the Countryside Council for Wales. Work began on the *British Tertiary Volcanic Province* volume shortly after the Geological Conservation Review (GCR) had been initiated in 1977 by Dr G. P. Black, then Head of the Geology and Physiography Section of the Nature Conservancy Council. With the help of Mr A. P. McKirdy, now with Scottish Natural Heritage, sites were identified, assessed and selected for initial writing-up by Dr Black (Ardnamurchan and Mull), Dr I. Williamson (Skye) and Dr C. H. Emeleus (Rum). The major early contributions from these specialists is very gratefully acknowledged.

In 1987, Dr Emeleus and Dr M. C. Gyopari of the University of Durham were commissioned to write this volume for publication according to the format approved by the GCR Publications Management (now Advisory) Committee. Since that time, many others have contributed in various ways to the production of the volume; in particular, the editors wish to acknowledge the following: Professor Sir Malcolm Brown, FRS, who suggested several important improvements while refereeing the volume as a whole; Dr G. F. Marriner, Dr M. F. Thirlwall and Dr J. N. Walsh of Royal Holloway and Bedford New College, University of London, who commented helpfully on portions of the text; past and present members of the Publications Management and Advisory Committees for their support and advice; Dr D. O'Halloran, who has calmly managed this project on behalf of the Joint Nature Conservation Committee, and the GCR production team of Valerie Wyld (Sub-editor), Nicholas D. W. Davey (Scientific Officer) and Caroline Mee (Administrative Officer); the several Chief Wardens of the Rum NNR who over the years have greatly assisted Dr Emeleus in the field; Chapman & Hall for their help and advice at the final stages of publication; Lovell Johns Limited, Colwyn Bay, for cartographic drafting and David C. Davies for cartographic editing.

W. A. Wimbledon and P. H. Banham

Access to the countryside

This volume is not intended for use as a field guide. The description or mention of any site should not be taken as an indication that access to a site is open or that a right of way exists. Most sites described are in private ownership, and their inclusion herein is solely for the purpose of justifying their conservation. Their description or appearance on a map in this work should in no way be construed as an invitation to visit. Prior consent for visits should always be obtained from the landowner and/or occupier.

Information on conservation matters, including site ownership, relating to Sites of Special Scientific Interest (SSSIs) or National Nature Reserves (NNRs) in particular counties or districts may be obtained from the relevant country conservation agency headquarters listed below:

Scottish Natural Heritage,
12 Hope Terrace,
Edinburgh EH9 2AS.

Countryside Council for Wales,
Plas Penrhos,
Ffordd Penrhos,
Bangor,
Gwynedd LL57 2LQ.

English Nature,
Northminster House,
Peterborough PE1 1UA.

Foreword

When setting out to produce the Geological Conservation Review series the then Nature Conservancy Council rightly selected the British Tertiary Volcanic Province as one of its front-running topics. By any standards the Province is one of the outstanding features of British geology. It has contributed to the development of geological ideas, applied world-wide, over about 200 years, and holds a continuing position as a focus for international research. A prime earth science concern of any conservation agency must be to preserve the evidence on which scientific advances have been based, and to ensure that future generations have an opportunity to study the problems that remain. Of course, geological features are generally speaking pretty robust. Casual and thoughtless destruction on a grand scale, to which biological assemblages are so vulnerable, is not generally a serious problem. But geological knowledge depends on seeing the relationships between rock masses, the critical areas of outcrop may be few and far between, both in Highland Scotland with its peat moors and forests and in the agricultural lands to the south. It is documentation of these sites showing critical inter-relationships that is the business of the Geological Conservation Review series. This volume provides the NCC's successor body, Scottish Natural Heritage, with the scientific justification to safeguard the described sites, which are the highlights of the Province.

Reading this volume brought home to me the scale of the scientific resource provided by the volcanic rocks and associated intrusions which make up the British Tertiary province. To a geologist many of the place names ring out like the names of great battles: Waterloo, Trafalgar – Skye, Mull, Rum, Ardnamurchan! And lesser places have lent their names to rock types, used all over the world: benmoreite, allivalite, mugearite - to mention but a few. You can go to a scientific conference in California and hear American geologists talking about rocks named after tiny Hebridean hamlets, but also describing research, at the forefront of the international scene, which they are carrying out now on rocks from the Scottish Tertiary.

The province developed when the opening of the North Atlantic reached British latitudes about 65 million years ago. Truly vast outpourings of lava occurred, particularly in then-adjacent East Greenland, very probably associated with a 'hot spot' in the Earth's mantle which lives on to this day under Iceland. While the immense basalt fields of East Greenland are fearsomely inaccessible, the west of Scotland provides relatively easy access to the deeply eroded relics of the basalt pile and the frozen equivalents of the magma chambers (the central complexes) which lay beneath. Research on igneous rocks can take place at various scales, all admirably served by the Scottish Tertiary: in a regional setting, to understand problems such as the mechanisms and driving force behind ocean-opening, and the relationship between 'hot spots' and ocean formation; on the scale of a single volcano, to enable us to understand the processes of igneous intrusion, the evolution of magmas and the controls on episodic volcanic activity; and on the scale of the outcrop, where, for example, igneous layering, nowhere better shown than on Rum, still presents many totally enigmatic features.

Scotland's great natural laboratory is splendidly documented in this volume of the Geological Conservation Review. Henry Emeleus is an outstanding expert on the Tertiary igneous province, particularly with respect to the field relationships, at once the most fundamental and also the most difficult type of geological observation. He and his younger co-author, Mark Gyopari, have provided clear, crisp, beautifully illustrated accounts of sites, which will be of outstanding value to students, researchers, amateur geologists and professional conservationists. Because of its thematic character and style of presentation, placing local detail into a regional context, it provides an unmatched teaching resource. Finally, and most important of all, it sets out and values, for the first time, publicly and clearly, those sites which have contributed most in the past, and most probably will contribute in the future, to our enjoyment and understanding of one of the grandest events in the geological growth of Britain.

Ian Parsons FRSE
Professor of Mineralogy, The University of Edinburgh

Chapter 1

Introduction to the British Tertiary Volcanic Province

THE GENERAL SETTING

The north-western British Isles was the site of intense igneous activity during the Palaeocene and early Eocene (*c.* 63–52 Ma) which accompanied continental separation and lithospheric attenuation during the early stages of the opening of the North Atlantic. In Great Britain, volcanism was most vigorous in the Inner Hebrides and adjoining north-west Scotland but also extended to southern Scotland, north-east England and the Outer Hebrides. Intrusions of similar age in North Wales and the English Midlands are probably outlying representatives of contemporaneous activity in north-east Ireland; additional activity in the Bristol Channel centred on Lundy. The region encompassing this activity is known as the British Tertiary Volcanic Province (BTVP) or the British Tertiary Igneous Province (BTIP) (Fig. 1.1).

The igneous activity took many forms and involved a wide variety of magmas and rock types. The remains of large accumulations of dominantly basaltic lava flows cover extensive areas in Skye and Mull. Laterally extensive swarms of basaltic dykes are most intense near Skye, Rum, Mull, and Arran but extend to the Outer Hebrides, southern Scotland, Cleveland in north Yorkshire and parts of North Wales and central England. Central intrusive complexes consisting of granite, gabbro, peridotite and other rock types occur on a line from Skye to the Bristol Channel and at several places in the north-east Atlantic (Fig. 1.1); these are the deeply-dissected roots of major volcanoes.

The varied igneous rocks, together with associated sediments and metamorphic rocks have responded in different ways to the profound erosion of the last fifty million years. Gabbro and peridotite have given rise to the rugged mountain scenery of St Kilda, the Skye Cuillins and Rum. The considerable, but generally less-rugged, mountains of northern Arran and the Skye Red Hills are composed of granite, while the piles of flat-lying lavas form tabular, 'trap-featured' hills in northern Skye, rising sometimes to form high mountains, as at Ben More, Mull.

Geologists, mineralogists, mountaineers and others have been attracted to the Province for over two centuries by the spectacular mountain and coastal scenery, and by the abundance of fresh rock exposures from sea-level to over 1000 metres altitude. Pennant, Necker de Saussure, Ami Boué and MacCulloch were among early visitors; notable scientists who worked on the Province in later years included Judd, Geikie, Harker, Bowen, Holmes and Wager; and the area is one where at the present time research into fundamental problems of igneous geology is actively pursued. The intensity of scientific investigation has made the BTVP one of the most historically important and deeply studied igneous provinces in the world. Furthermore, it is acknowledged as one of the best areas in the British Isles in which to demonstrate igneous rocks in the field, and consequently it is visited annually by numerous groups from universities, schools and scientific societies from Britain and abroad.

GCR sites have been selected to cover the features of the BTVP within Great Britain. Additional sites have been identified in Northern Ireland, but these are outside the terms of reference of this published review. The sites vary considerably in importance, size and scope. Some are whole mountain groups (for example, the Skye Cuillin Hills); others may be merely stream sections or small quarries but all have been selected as scientifically significant examples of their kind, where the features described can be observed and appreciated at present and which should be preserved for future study and research.

The Tertiary igneous activity occurred in geographically well-defined areas which usually include a central volcano and surrounding lavas. These areas have been made the basis of individual chapters.

THE IGNEOUS SEQUENCE

Volcanism in the BTVP extended over a period of about twelve million years, largely within the Palaeocene Epoch, apart from the later intrusions within the Eastern Red Hills Centre of Skye (Dickin, 1981), the Sgurr of Eigg pitchstone (Dickin and Jones, 1983), the Lundy granite complex (Hampton and Taylor, 1983) and the youngest intrusion in the Mourne Mountains Western Centre, Ireland (Meighan *et al.*, 1988). The Palaeocene volcanism was preceded by Cretaceous igneous activity in the eastern Atlantic (Harrison, 1982) and it is also possible that the submarine central complex at the Blackstones, south-west of Mull, is Cretaceous in age (Durant *et al.*, 1982). Within the BTVP, the life span of individual central complexes was short, of the order of two or three million years (or even less) and the thick lava accumulations built up over even shorter periods (Table 1.1).

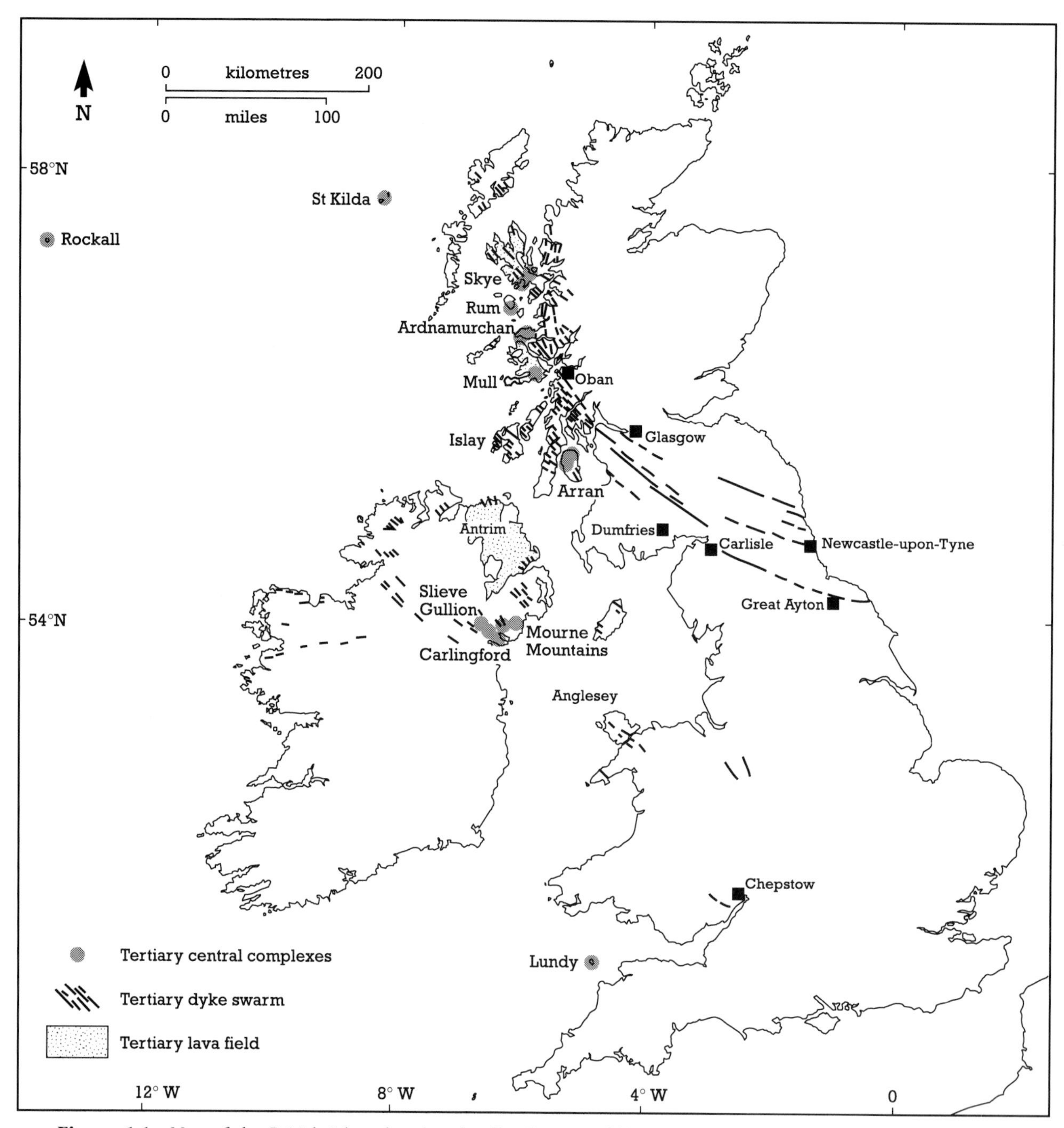

Figure 1.1 Map of the British Isles, showing the distribution of Tertiary central complexes, dyke swarms and lavas (submarine occurrences not shown). Modified from Emeleus, in Sutherland (1982, figure 29.1).

Volcanic activity in a given area often started with the formation of small amounts of basaltic ash and other volcaniclastic accumulations. These were quickly followed by voluminous subaerial eruptions of basaltic lavas which covered the peneplaned surfaces of older sedimentary and metamorphic rocks ranging in age from the Precambrian to the Cretaceous. Occasionally, the lavas covered landscapes of considerable relief, filling valleys, burying hills and sometimes flowing into shallow lakes, where pillow lavas and hyaloclastites formed. Sedimentary horizons are not common in the lavas, but fluviatile conglomerates, sandstones and fine-grained plant-bearing horizons do occur and provide valuable stratigraphic and palaeogeographical information. The lavas were often subjected to intense weathering between flow extrusion, with the

Table 1.1 British Tertiary Volcanic Province: summary of the geological successions, radiometric ages and magnetic polarities (after Mussett *et al.*, 1988, figure 2)

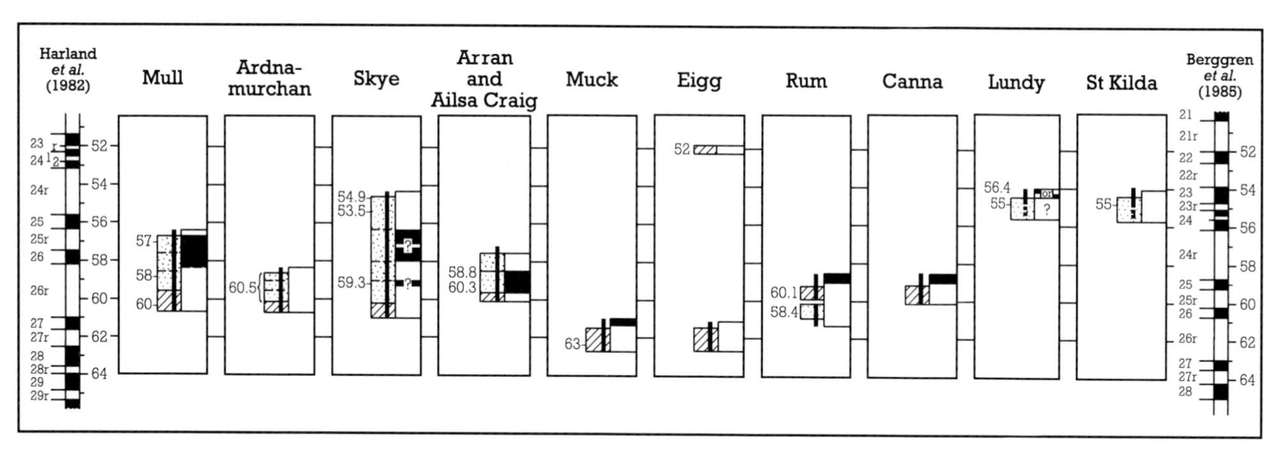

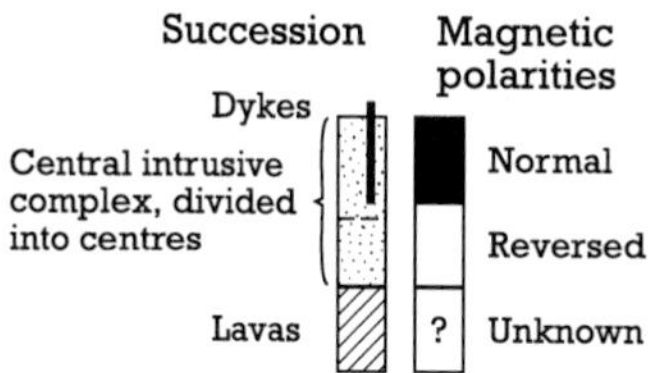

formation of bright red lateritic deposits. With increasing thickness of lavas, heated waters circulating through the flows altered the basalts and deposited distinctive suites of zeolite minerals, for which Skye and Mull are particularly noted.

The lavas were principally fed from fissure eruptions similar to those of present day Iceland. The actual feeders are among the multitude of dykes forming the swarms which extend across the Province; dykes intrude virtually all intrusions and extrusions in the BTVP, so it is likely that lava effusion also occurred throughout the life span of the Province. However, the thick sequences of lavas now preserved in Mull, Skye, and the Small Isles built up between about 63 Ma and 60 Ma, early in the life of the BTVP. Occasionally, lavas must have been erupted as the central complexes developed; there is good evidence from several centres that silicic and intermediate lavas were closely associated with central complexes but there are few substantiated examples of basaltic lavas, with the exception of pillow lavas within the Mull centre. Not all the (predominantly basaltic) magma reached the surface to form flows, some froze in conduits as dykes and plugs and quite large amounts spread laterally through the Mesozoic sediments beneath the lavas to form the prominent dolerite sills of northern Skye and Arran.

The central complexes generally post-date the adjoining lavas, but they were intruded by later members of the dyke swarms. Within the complexes the magmatic sequences were rarely straightforward: in Mull early granite intrusions were followed by numerous basaltic, intermediate and acid cone-sheets, by gabbros and peridotites and by further granites; in Skye, the sequence was apparently simpler, the gabbros and peridotites of the Cuillins were cut by numerous basaltic cone-sheets and subsequently by granites of the Red Hills. Thus, the central complexes record varied intrusive sequences in which basaltic and granitic magmas have been intimately associated. Occasionally, it may be shown that contrasted magmas must have co-existed, forming, for example, the composite basalt/quartz–porphyry sheets and dykes of the Province and the complicated intrusion breccias found in Ardnamurchan, Skye and other central complexes.

Within the central complexes, individual centres of activity are defined by arcuate intrusions – cone-sheets, ring-dykes and stocks – which have a common focus. In most central complexes, and in particular Ardnamurchan, Skye and Mull, the intrusions of one centre may cut those of an earlier one, recording movements in the focus of magmatic activity with time. Ardnamurchan provides an exceptionally clear and often cited example of this phenomenon.

The central complexes are generally spatially

well-separated from one another. This has made correlation between centres difficult; furthermore, the present precision of radiometric age determinations is not sufficiently good to obtain relative ages, let alone distinguish between events within a central complex (cf. Table 1.1). Some progress has been made in correlating the centres in the north of the Province, where the activity on Skye appears to post-date that of Rum and most of the Small Isles (Table 1.1). The tentative sequence of these is as follows: basaltic lavas on Muck and Eigg are cut by a dyke swarm which extends to Rum where it is truncated by gabbros and peridotites formed late in the Rum central complex. In north-west Rum, lavas and fluviatile conglomerates rest on eroded granophyre of the Rum central complex and the sediments contain clasts from the granophyre and most other members of the Rum intrusive suite, including ultrabasic rocks. Granophyre and felsite, similar to rocks on Rum, also occur in conglomerates interbedded with lavas on Canna and Sanday and in the Skye Main Lava Series near Glen Brittle. The Skye lavas and sediments were, in turn, intruded by the Cuillin gabbros, the earliest members of the Skye central complex. The Skye central complex therefore post-dates the Rum central complex which had been deeply eroded and unroofed by the time the intervening lavas were erupted. This sequence also demonstrates that there were at least two periods when lava piles built up, one before and one after emplacement of the Rum central complex.

By combining the field evidence and the available radiometric age determinations with detailed palaeomagnetic measurements, Mussett *et al.* (1988) have been able to build up a stratigraphic framework for the BTVP, and at the same time have shown that intense igneous activity in any one lava field, or within a given centre, was usually of geologically short duration (cf. Table 1.1; Mussett *et al.*, 1988, Fig. 2); their pioneering approach is clearly of wide application.

A REVIEW OF RESEARCH

'The gradual development of opinion regarding the nature and history of volcanic rocks is thus in no small measure bound up with the progress of observation and inference in regard to the Tertiary volcanic series.' Sir Archibald Geikie's words, written nearly a hundred years ago about the fresh and extensive exposures of volcanic rocks in western Scotland and Northern Ireland (Geikie, 1897), remain true to the present day. Since the early days of geology in the late eighteenth century, investigations of these rocks have been to the fore in the advancement of geology and mineralogy, and, in the last few decades, of geochemistry and geophysics.

Major advances to *c.* 1945

The early history of investigation of the Province is admirably recounted in the second volume of Geikie's *Ancient Volcanoes of Great Britain* (Geikie, 1897), in which the important role of these rocks in the controversies of the Neptunist and Plutonist schools is recorded. Continental visitors, particularly from France, who were familiar with active and recently active volcanoes, unequivocally interpreted the basalts as the result of volcanic eruptions. Notwithstanding, others including Robert Jameson, the great Scottish mineralogist and Professor of Natural History at Edinburgh University, strongly advocated the Neptunists', or Wernerian, viewpoint and ridiculed any connection between these rocks and volcanic activity. The controversy continued well into the nineteenth century, with the gradual ascendency of the Plutonist viewpoint, partly through observations on the dykes of north-east England. Recognition of the Tertiary age of the rocks was due to the Duke of Argyll (1851) who discovered and described plant remains in sediments intercalcated with basalt flows on Mull, although arguments about the ages persisted and older dates were assigned to the rocks of Skye where difficulties were experienced in distinguishing sills in the Mesozoic rocks from true lava flows.

In the latter half of the nineteenth century very significant advances were achieved. The noteworthy investigations by A. Geikie (eg. 1888, 1894, 1897) and J.W. Judd (eg. 1885, 1886, 1889) were mainly concerned with the overall structure and mode of accumulation of the Hebridean lava plateau. Geikie drew on his experiences in the western USA and Iceland to assert that the lavas had built up by successive fissure eruptions, whereas Judd recognized that the central complexes were the eroded remains of major Tertiary volcanoes, and was strongly of the opinion that they were the sources of the lavas.

Partly in an attempt to resolve this controversy, A. Geikie, the then Director of the Geological Survey, arranged for Alfred Harker of Cambridge University to make a detailed survey of the Tertiary igneous rocks of Skye, thereby initiating Survey investigations which were to cover Skye (Harker, 1904), Rum and the other Small Isles of Inverness-shire (Harker, 1908), Mull (Bailey *et al.*, 1924), Arran (Tyrrell, 1928) and Ardnamurchan (Richey and Thomas, 1930). The impact of these memoirs and the accompanying maps was immense. The Skye Memoir detailed swarms of basaltic sheets inclined towards a common focus beneath the gabbros and granites, the coexistence and hybridization of magmas of contrasted compositions, and the formation of a central complex by multiple intrusions. The Mull Memoir confirmed the abundance of centrally inclined sheets (termed cone-sheets) in the central complex, described ring-dykes, including the type example at Loch Bà, and demonstrated the migration of igneous activity within a central complex with time. The relatively large number of chemical analyses made on the Mull igneous rocks allowed the Memoir authors to introduce the concepts of distinct chemical lineages of igneous rocks forming 'Magma Types' and 'Magma Series'; the importance of crystal differentiation in the development of these associations was also stressed. Richey and Thomas's work on Ardnamurchan (1930) built on ideas generated in the Mull Memoir, numerous ring-dykes and several suites of cone-sheets were recognized and the presence of three distinct centres was demonstrated.

Research in the BTVP was relatively limited between the early 1930s and the mid 1940s. The lavas and sills of northern Skye were mapped by the Geological Survey (Anderson and Dunham, 1966). A reassessment of the composite Glen More ring-dyke of Mull by Holmes (1936) and by Fenner (1937) concluded that the upward passage from olivine dolerite to granitic rocks resulted from mixing of compositionally contrasted magmas rather than by crystal fractionation, as originally proposed by Bailey *et al.* (1924) and subsequently supported by Koomans and Kuenen (1938). Valuable theoretical contributions based largely on data from the BTVP were made by Kennedy (1931a, 1933) and Kennedy and Anderson (1938) on the possible origins of the basaltic magmas, and by Anderson (1936) on the conditions of formation of ring-dykes and cone-sheets.

Research in the post-war years

The immediate post-war period saw the culmination of a controversy on the origins of granite, and other plutonic rocks, rivalling that between the early nineteenth century Plutonists and Neptunists. The question was, did plutonic granites form from the crystallization of magmas, or were pre-existing rocks transformed, or granitized, *in situ* to form granites? Evidence from the Province was regarded as crucial to the arguments of both sides, granites in the eastern Red Hills, Skye and western Rum were regarded as transformed Torridonian and Jurassic sediments respectively (e.g. Black, 1954, 1955), and it was also suggested that feldspar-phyric gabbros on Skye, were transformed amygdaloidal basalts (Reynolds, 1951; King, 1953). While these claims were strongly contested as they arose, a key discovery which decisively helped to swing the debate on granite origins in favour of the magmatists was made on the Beinn an Dubhaich granite on Skye, where Tuttle and Keith (1954) conclusively demonstrated that the mineralogy of this granite had characteristics of *both* the problematical plutonic granites and extrusive rhyolites of undoubted magmatic origin. It, and other BTVP granites, provided the essential 'missing link'.

The layered rocks

N. L. Bowen's classic work *The Evolution of the Igneous Rocks* (1928), included arguments strongly favouring crystal fractionation as a major mechanism in determining rock compositions. He drew heavily on examples from the BTVP, in particular citing the Mull feldspar-phyric Porphyritic Central Magma-type rocks and the peridotite dykes of Skye as products of the process. The efficacy of this mechanism was brought into sharp focus by Wager and Deer's (1939) investigations of the East Greenland Skaergaard Intrusion. The strikingly layered rocks of Skaergaard were considered to have formed by the gravitational accumulation of crops of crystals from fractionating basaltic magma and its derivatives. The literature of the BTVP was already full of examples of banded peridotites and gabbros (for example, Geikie and Teall, 1894; Harker, 1904, 1908; Bailey *et al.*, 1924; Richey and Thomas, 1930) and it was immediately recognized that there were at least superficial similarities to the

Skaergaard layering, especially in the Skye Cuillins and on Rum. Brown's (1956) study of the latter provided convincing evidence of crystal sorting and settling forming the layered rocks, but in contrast to Skaergaard, this was not accompanied by the systematic compositional ('cryptic') variations in rock and mineral chemistry. The explanation offered was that the Rum layered rocks built up from a succession of pulses (15) of basaltic magma, whereas there was only a single input of magma at Skaergaard. Thus Rum became, and remains, a classic example of an 'open' magma chamber, fed by successive batches of magma from each of which early, high-temperature crystals separated and accumulated to form the layered rocks of Hallival and Askival.

Minerals in layered igneous rocks frequently exhibit distinctive textural relationships thought to be due to the accumulation of some (cumulus) crystals and the precipitation of others from the liquid (intercumulus) trapped between sedimented crystals. The fresh, unaltered rocks of Skye and Rum provide type examples of the textures and structures found in these 'magmatic sediments' and the two centres are classic examples of layered igneous rocks and 'igneous cumulates' (Wager *et al.*, 1960; Wager and Brown, 1968). During the past decade, ideas on the origins of igneous rocks have undergone radical reappraisal (cf. McBirney and Noyes, 1979) and it has increasingly been recognized that *in situ* crystallization, double-diffusive convection and movement of trapped, high-temperature liquids through crystal mushes may contribute to the formation of layered igneous rocks or their modification, and that crystal fractionation aided by convection-driven magmatic currents may not always be applicable. The exceptionally well-exposed layered rocks of Rum have provided an excellent natural laboratory where the new hypotheses have been tested and developed (see for example, papers in the *Geological Magazine*, **122**, 1985).

The contribution from geochemical and isotopic studies

The 1950s and 1960s saw the extensive application of the analytical techniques of optical spectroscopy, spectrophotometry and X-ray fluorescence to geological and mineralogical problems. Geochemistry blossomed as a subject, as it became possible to analyse rocks rapidly, and to detect a whole range of hitherto geologically inaccessible elements in concentrations down to a few parts per million. Many of the pioneering applications of these techniques took place in the BTVP: trace element distributions in the Skye rocks were examined (Nockolds and Allen, 1954); the geochemistry of whole suites from the Province analysed and compared with other classic areas (Wager, 1956; Tilley and Muir, 1962, 1967); and detailed successions of lavas and intrusions were analysed (for example, the lavas of Skye, Thompson *et al.*, 1972), the Ardnamurchan cone-sheets (Holland and Brown, 1972) and the dykes of Skye (Mattey *et al.*, 1977). Several distinctive groups of basaltic and associated compositions were recognized in the Hebrides of which the Skye Main Lava Series and the Mull Plateau Group were the most abundant, including olivine basalts, hawaiites and mugearites which varied from transitional to mildly alkaline compositions (Fig. 1.2). Silica-rich, alkali-poor, tholeiitic rocks form late flows on Skye and are found as dykes on Skye and occur elsewhere in the Province; these were termed the Preshal More group (Table 2.2) after the type occurrence in south-west Skye. The BTVP lavas show significant differences in the compositional range when compared with lavas from another classic area, the Hawaiian Islands. The well-defined division into silica-poor, alkali-rich (alkali olivine basalts) and silica-rich, alkali-poor (tholeiitic) compositional fields found in Hawaii is not present in the Hebrides, signifying different conditions of both generation and transport to the surface of the magmas.

The geological application of mass spectrometry and the development of isotope geochemistry since the 1960s have provided crucial new information about the ages and origins of igneous rocks. The BTVP has figured prominently in the application of these techniques. The Skye granites, gabbros and lavas (Moorbath and Bell, 1965; Moorbath and Welke, 1969) yielded Palaeocene ages, together with strontium and lead isotopic data which showed that, whereas the basaltic rocks had distinct mantle signatures, the granites on the whole did not and were probably derived by partial melting from crustal sources. Lead derived from Lewisian sources was present and it was proposed that the granites came from partial melting of the gneisses. Subsequent oxygen isotope investigations (Taylor and Forester, 1971; Forester and Taylor, 1976, 1977) showed that the granites had been pervasively invaded by heated

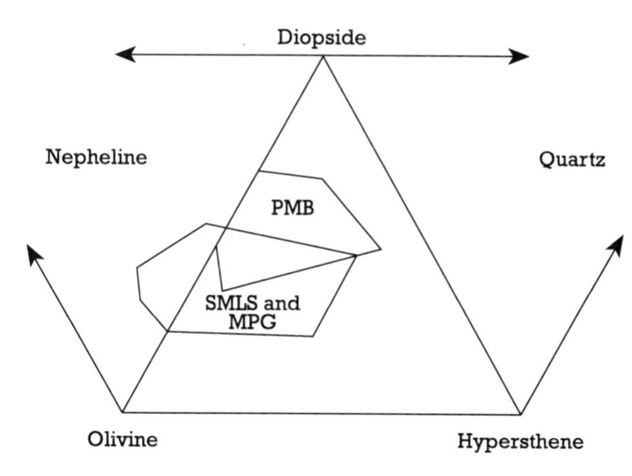

Figure 1.2 Diagram showing fields of the Preshal More Basalts (PMB), and the Skye Main Lava Series (SMLS) and Mull Plateau Group (MPG) when projected into normative nepheline–diopside–olivine–hypersthene–quartz (after Thompson, 1982, figure 2).

groundwaters as had the rocks around the central complexes. This raised the possibility of significant post-consolidation changes in the compositions of the rocks and consequently cast doubt on the validity of the earlier Sr- and Pb-isotope studies, with the added corollary that detailed elemental and isotopic studies of rocks from the BTVP (especially in and near central complexes) might be of very limited petrogenetic application. This crucial problem was specifically addressed by Pankhurst *et al.* (1978) who showed, by means of a careful study of the Loch Uisg Mull granite, that, provided that the samples were carefully selected, post-consolidation changes were in fact minimal.

Precision, accuracy and detection limits in elemental geochemistry have also been significantly improved in the past two decades, partly due to the introduction of induced neutron activation analysis (INAA) and inductively coupled plasma mass spectrometry (ICPMS). These, and other new and improved techniques, have been applied to the Province, notably to the problem of granite genesis. Both on Mull (Walsh *et al.*, 1979) and on Skye (Dickin, 1981) there is now a consensus that the granites contain variable, but always significant contributions from both mantle and crustal sources (for example, Dickin *et al.*, 1984). There is considerable compositional and isotopic variation within the basaltic rocks and their derivatives which has, for Mull, been attributed by Beckinsale *et al.* (1978) to variations in mantle sources. Essentially the same data are amenable to alternative interpretations: while agreeing with the ultimate mantle source, Morrison *et al.* (1985) and Thompson *et al.* (1986) suggest that a major factor contributing to the compositional variability is the extent to which a particular batch of magma has interacted with the lower and/or upper crust during its progress towards the surface. Using data obtained from flows low in the Skye and south-west Mull lava successions, they argued that each separate batch of magma may have had a distinctive and complicated history of intermittent ascent, ponding at different levels in the crust where rock types caused density traps, and reacting with the adjoining rocks to a greater or lesser extent. The Skye and Mull lavas have thus been used as geochemical probes to investigate the magmatic plumbing of part of the feeder system for the BTVP lavas.

A tholeiite dyke from the north of England was among the first intrusions to be dated radiometrically (Dubey and Holmes, 1929) and numerous age determinations have now been made on rocks from the BTVP (Table 1.1). As already mentioned, the time span for the Province is about eleven million years, between *c.* 63 Ma and 52 Ma (for example, Mussett *et al.*, 1988). Age-determination techniques have proved to be of somewhat limited application in the BTVP since individual igneous events, and groups of events, took place over very short times in geological terms. However, work is now well under way towards defining a reliable stratigraphy for the Province, using high-quality age data in combination with detailed determinations of remanent magnetization of rocks for which there are good geological data on relative ages. Table 1.1 provides a summary of some of the results to date (Mussett *et al.*, 1988).

Deep structure of the central complexes: the geological input

Studies of the gravitational field over the BTVP have yielded far-reaching and important results. In addition to regional surveys (by the British Geological Survey), much detailed local information has come from a series of investigations since the early 1950s (for example, Tuson, 1959). The central complexes are almost invariably the sites of positive Bouguer gravity anomalies which include some of the most pronounced in the British Isles (for example, the submarine Blackstones Bank complex; McQuillin *et al.*, 1975). The interpretations of the Skye, Mull and Ardnamurchan (Bott and Tuson, 1973), Arran and Rum (McQuillin and Tuson, 1963) anomalies show that each centre is underlain by dense rock (3.0–3.3 g cm^{-1}) of the composition of very olivine-rich gabbro or feldspathic peridotite. The masses of mafic rocks are broadly cylindrical or else steep-sided truncated cones, they extend for depths of 12 km or more through the crust, and they may be over 20 km in diameter (cf. Bott and Tuson, 1973, Figs 1 and 2). They are thus major features of the Hebridean crust and they underlie not only the mafic parts of centres but also the granites. The implications of the interpretations of the anomalies are far-reaching: mafic magmas must have dominated the BTVP igneous activity for, although areally extensive, the granites are only superficial bodies, perhaps only 2 km thick (Bott and Tantrigoda, 1987). The mafic rocks could have given rise to the granites by magmatic fractionation but equally they were heat sources capable of effecting melting in the surrounding crustal rocks, thus generating the granitic magmas, or a component thereof. The gabbroic and ultrabasic rocks of Rum are probably the upper tip of one of these mafic bodies.

The precise conditions governing the siting of the central complexes are not fully understood, but their location at or near the intersections of the basic dyke swarms with major faults has long attracted attention, and a causative connection has been suggested (Richey, 1937; Vann, 1978; Upton, 1988). If the major faults, such as the Great Glen and Highland Boundary Faults, have influenced siting of the central complexes, then that influence has extended to the base of the Palaeocene crust.

Modelling magmatic evolution

A major problem of the ultrabasic rocks of the BTVP, and elsewhere, has been to decide whether they formed from ultrabasic or basaltic magmas. Their frequent close association with rocks of basaltic compositions makes the latter an attractive proposition (Brown, 1956); however, the possibility that high-temperature ultrabasic magmas may have been present at upper crustal levels in the BTVP has become accepted in recent years. The idea was advanced by Drever and Johnston (1958) and Wyllie and Drever (1963) from examinations of minor ultrabasic intrusions around the Skye Cuillin site and on the nearby Isle of Soay. The close similarities between the minor ultrabasic intrusions of Skye and Rum and the layered ultrabasic rocks of those islands suggested to Gibb (1976) that the ultrabasic rocks had formed from parental magmas consisting of a suspension of olivine crystals in ultrabasic ('eucritic') liquid. Further research by Donaldson (1975) on the harrisites and ultrabasic breccias of south-west Rum led him to postulate the participation of feldspathic ultrabasic liquids (possibly hydrous) in the formation of these rocks. In addition, the subsequent discovery of quenched, aphyric ultrabasic dykes intruding the Rum layered rocks (McClurg, 1982), the recognition that some at least of the feldspathic peridotite layers in the Rum succession were intrusive (Renner and Palacz, 1987; Bédard *et al.*, 1988), and the presence of a quenched ultrabasic margin to this complex at Beinn nan Stac (Greenwood *et al.*, 1990), provided further strong supporting evidence for ultrabasic liquids. It thus appears inescapable from the evidence furnished by the BTVP that conditions in the central complexes and their immediate surroundings favoured the rise of hot, dense picritic liquids to within short distances (*c.* 1 km) of the Earth's surface. Although not unique, these examples are unusual and it appears that the special conditions involving rapid, strongly focused throughput of hot basaltic magmas, resulted in the formation of preheated pathways along which the ultrabasic liquids rose to high structural levels before crystallizing and congealing.

The availability of new, quantitative data on the physical and chemical properties of magmas and increasingly sophisticated computing techniques has made it possible to model magmatic processes. Noteworthy studies using examples from the BTVP have included examination of the emplacement of the Cleveland Dyke of north-east England and of the layering in the Rum ultrabasic rocks. MacDonald *et al.* (1988) demonstrated a close chemical similarity between the Cleveland Dyke tholeiite and rocks in the Mull central complex. They postulated that the dyke had been fed laterally from a source beneath Mull, flowing '. . . in a manner transitional between laminar and turbulent conditions'. It was calculated that the magma took between one and five days to reach North Yorkshire. Brown (1956) described contrasted olivine- and feldspar-rich layering in the Rum ultrabasic rocks, attributing the layering to successive gravitational settling of olivine and feldspar, the residual low temperature magma being extruded during concomitant surface volcanism. Recent attempts to model the layering (see Young *et al.*, 1988 for summary) accept Brown's concept of a repeatedly replenished open magma chamber but envisage pulses of olivine-phyric, Mg-rich basaltic (olivine-phyric picrite) magma with or without contemporaneous basaltic liquid. In one model, picrite crystallizes copious olivine during strong convective circulation, the crystals only settling when movement ceases. The residual, overlying liquid then crystallizes feldspar-rich cumulates by basalt fractionation (Huppert and Sparks, 1980; Sparks *et al.*, 1984; Tait, 1985).

An alternative view (Young *et al.*, 1988) is that simultaneous crystallization of olivine, and olivine plus plagioclase, occurs in a magma chamber zoned upwards from picritic to basaltic compositions. The model assumes a dipping floor to the magma chamber, corresponding more or less with the dipping layering observed on eastern Rum (cf. Brown, 1956; Volker and Upton, 1990). Simultaneous crystallization in picritic and basaltic liquids would result in the formation of peridotite at low levels on the magma chamber floor and of troctolitic (allivalitic) and gabbroic rocks at higher levels. The conspicuous major layering in eastern Rum would therefore result from variations in the rate of magma injection into the chamber, peridotite forming when this was high, and troctolite and gabbro when it waned and olivine precipitation depleted the picritic proportion of the resident magma. The model finds its support in the manner in which the proportion of peridotite in the layered rocks of eastern Rum increases towards the west where it has been suggested that the magma arose along the line of the present Long Loch Fault (McClurg, 1982; Volker and Upton, 1990).

Chapter 2

The Isle of Skye

INTRODUCTION

Skye is one of the classic areas of Great Britain for the study of igneous geology. The sea cliffs and hills in the north of the island magnificently expose a thick succession of mainly basaltic lavas which overlie Mesozoic sediments intruded by a suite of dolerite sills. Basaltic dykes of the north-west-trending swarm cut all these rocks. The Cuillins and Red Hills of central Skye have been eroded from the roots of a major volcano; they form a central igneous complex consisting of numerous intrusions of granite, gabbro and peridotite (Fig. 2.1). Suites of dolerite cone-sheets intrude the gabbros and peridotites and all members of the central igneous complex are also cut by at least some members of the dyke swarm. The island contains a long and complicated record of igneous extrusion and intrusion (Table 2.1) and is a particularly suitable area for the study of the field geology of extrusive and intrusive igneous rocks. Investigations throughout the nineteenth century elucidated and highlighted the diversity and abundance of igneous rocks present and firmly assigned the activity to the Tertiary (Geikie, 1888, 1894, 1897; Judd, 1874, 1878).

The geological importance of the island was acknowledged when, at the turn of the century, the Geological Survey of Great Britain commissioned Dr Alfred Harker to make a thorough examination of all aspects of the Tertiary igneous rocks. Harker's contributions on Skye include a set of extremely detailed (published) Six-Inch to One-Mile geological maps of the Cuillin and Red Hills centres together with surrounding rocks. These outstanding maps form the basis to the Survey's One-Inch scale geological maps (Minginish, Sheet 70 and Glenelg, Sheet 71) while Harker's detailed observations and deductions were published as the classic Memoir on *The Tertiary Igneous Rocks of Skye* (Harker, 1904).

The extensive literature on the Tertiary igneous geology of Skye has been summarized in several publications over the past decade or so (J.D. Bell, 1976; Emeleus, 1982, 1983; Bell and Harris, 1986). A considerable amount of this work has concentrated on the mineralogy, petrology and geochemistry of the igneous rocks, and it has included important contributions to the theory of the petrogenesis of both basaltic and granitic rocks (summarized by Thompson, 1982). Fourteen SSSIs have been selected to cover the Tertiary igneous geology of Skye (Fig. 2.2).

Tertiary volcanic activity commenced in the Palaeocene, when extensive NW-trending fissures acted as feeders for several phases of extrusion of basaltic and related magmas which built up the plateau lavas of the northern and south-western parts of the island (Table 2.2). The lavas were predominantly extruded subaerially, with periods of deep erosion and laterite formation between successive flows. Although the majority of the lava flows are mildly alkaline to transitional olivine basalts, more olivine-rich picritic flows occur and fractionated flows of hawaiite, mugearite, benmoreite and trachyte are present, particularly in the higher parts of the succession. The north Skye lavas were extensively subdivided by the Geological Survey (Anderson and Dunham, 1966; Table 2.2); they are now collectively grouped as the Skye Main Lava Series (SMLS), Thompson *et al.*, 1972). A few distinctly different tholeiitic flows fill valleys eroded in the SMLS of south-west Skye; these are the Preshal More type basalts (Thompson *et al.*, 1972; see Table 2.2).

Thompson *et al.* (1972) noted a tendency for progressive compositional changes from hypersthene-normative basalts, upwards to nepheline-normative basalts, alkali hawaiites and mugearites in the Beinn Edra, Ramascaig and Totaig groups (Table 2.2). However, both nepheline-normative and hypersthene-normative basalts and associated lavas occur locally throughout the SMLS. To explain the absence of any obvious strong evolutionary trend, the authors envisaged a complex plumbing system beneath Skye during the eruption of the Palaeocene lavas: there was probably no major magma chamber, but rather a whole series of small reservoirs fed from the mantle where the primary magmas were generated by partial melting of upper-mantle garnet lherzolite. Magma was generated at this source throughout the accumulation of the SMLS, some batches travelled quickly to the surface, others resided in reservoirs within the crust for varying times. Thus, the magmas which formed the lava flows followed a variety of paths to the surface, allowing for different amounts of fractionation, and of contamination by reaction with reservoir walls. In this way the somewhat random distribution of basalt and other effusive rock could be accounted for. Studies on the distribution of strontium, lead and neodymium isotopes in the lavas strongly indicate contamination by crustal rocks (Moorbath and Thompson, 1980; Thirlwall and Jones, 1983), supporting the model described above; furthermore, Thirlwall and Jones

Figure 2.1 Sgurr nan Gillean and the Cuillin Mountains viewed from Sligachan, Isle of Skye. (Photo: David Noton Photography.)

Table 2.1 Summary of the Palaeocene igneous geology of the Isle of Skye (based on Bell, J.D., 1976, table 1; Bell, B.R. and Harris, 1986)

Late dykes (dolerite, felsite and peridotite)

Eastern Red Hills Centre
- Composite acid/basic sheets
- Five granite intrusions
- Kilchrist hybrids (possibly post-date some of the granites)
- Broadford and Beinn nan Cro gabbros
- Acid lavas, ignimbrites, tuffs and agglomerates of Kilchrist vent (may pre-date this Centre by a considerable amount)

Dykes (dolerite, pitchstone)

Western Red Hills Centre
- Marsco and Meall Buidhe granites
- Marscoite suite of hybrids, etc.
- Nine granite and major felsite intrusions
- Marsco Summit Gabbro
- Belig vent

Dykes (dolerite)

Strath na Crèitheach Centre
- Three granite intrusions
- Loch na Crèitheach vent

Dykes (dolerite)

Cuillin Centre
- Cone-sheets (dolerite)
- Coire Uaigneich Granophyre (but see text)
- Intrusive tholeiites
- Druim na Ramh Eucrite
- Explosive vents (of several ages)
- Inner Layered Series: allivalite, eucrite, gabbro
- Outer Layered Series: allivalite, eucrite, gabbro
- Layered Peridotite Series
- Border Group: gabbro, allivalite
- Cone-sheets and dykes (overlap with many of the above)

Palaeocene lavas
- Preshal More tholeiitic flows
- Skye Main Lava Series (SMLS) flows (with sparse clastic sedimentary horizons, and basal sediments and tuffs)

N.B. Additional details through text.

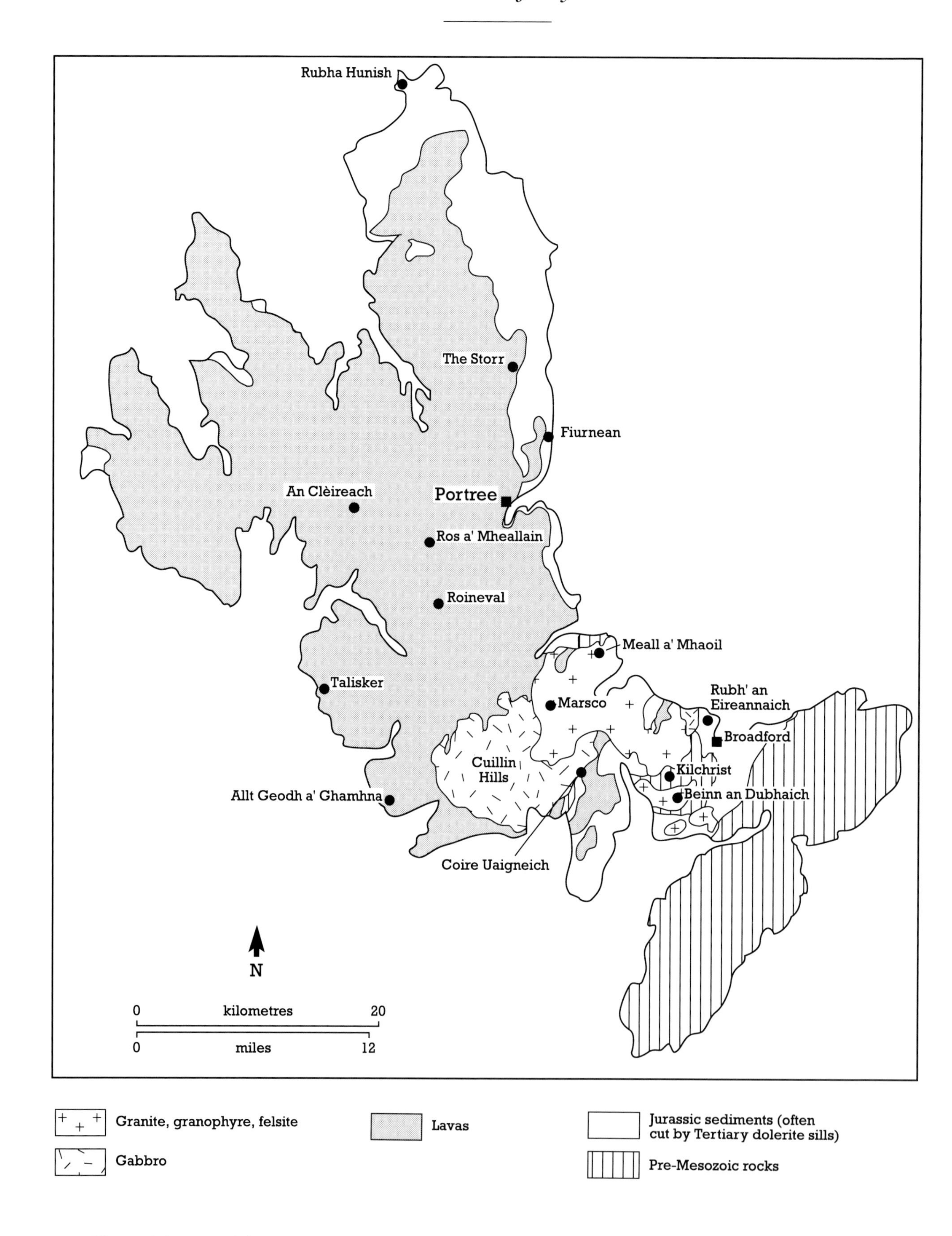

Figure 2.2 Map of the Isle of Skye, showing localities mentioned in the text.

Table 2.2 Correlation of the divisions of the Palaeocene lavas of the Isle of Skye (mainly after Williamson, 1979, table 1).

NORTHERN SKYE (1) Anderson and Dunham (1966)	WEST-CENTRAL SKYE (2) Williamson (1979)	Based mainly on NORTHERN SKYE (3) Thompson *et al.* (1972)
	7. Talisker Group	Preshal Mhor tholeiitic basalts
5. Osdale Group	6. Loch Dubh Group 5. Arnaval Group	
4. Bracadale Group	4. Tusdale Group	Skye Main Lava Series
3. Beinn Totaig Group	3. Cruachan Group*	Transitional and alkali-olivine basalts, hawaiites, mugearites, benmoreites and trachytes. More fractionated types are more common in the higher groups.
2. Ramascaig Group	2. Bualintur Group	
1. Beinn Edra Group	1. Meacnaish Group	

Individual groups are probably geographically restricted (see, for example, Anderson and Dunham, 1966, figure 13).

* The thick fluviatile conglomerates of the Allt Geodh a' Ghamhna site are at the base of this group.

found that contamination was most pronounced in the most primitive (hot) compositions, and least pronounced in the most evolved (cool) ones. Further investigations of lavas in Mull have shown similar distributions and the problem of magma plumbing has been examined using data from both islands (Figure 5.3; Morrison *et al.*, 1985). The Skye, Mull and other BTVP magmas appear to have been derived from mantle sources already depleted by partial melting and magma extraction during the Permo-Carboniferous (for example, Thompson and Morrison, 1988).

After extrusion of the SMLS flows, activity became focused in the area of the present-day Cuillin Hills where mafic magmas intruded to high crustal levels to form coarse-grained gabbros, eucrites, allivalites and peridotites, suites of cone-sheets and some of the dense swarms of dolerite dykes. At, or towards the end of this mafic magmatism, there was widespread explosive volcanicity in the Strath na Crèitheach area. Up to this stage, very little acid magma had been generated, except for the arcuate Coire Uaigneich granophyre.

Thereafter, the central igneous complex experienced changes in both the focus of activity and in magma composition. Three separate centres were established in succession:

1. the Strath na Crèitheach centre;
2. the Western Red Hills centre;
3. the Eastern Red Hills centre.

With time, activity moved progressively eastwards and granite became dominant at the present level of erosion, providing a particularly clear record of evolutionary changes taking place in the underlying magma chamber. However, despite the apparent dominance of granite in the later centres, basaltic magma was still clearly available and on occasions this mixed with acid magma to give distinctive hybrid bodies such as the marscoite suite, forming a ring-dyke in the Western Red Hills centre, and the Kilchrist Hybrids of the Eastern Red Hills centre. Further evidence of mixed magmas (acid–basic) comes from composite minor intrusions, such as the sills in the Broadford area. The final episode of igneous activity on Skye is represented by sparse NW-trending dolerite dykes.

Skye has been the site of notable geophysical experiments aimed at elucidating the deep structure of central intrusive complexes. Gravity surveys show that both the Cuillin Hills and the Red Hills are the site of a sharply defined, strong, positive Bouguer gravity anomaly. This is attributed to the presence of a large, steep-sided cylindrical or inverted cone-shaped mass of dense mafic rock extending to at least 15 km beneath all the central complexes (Bott and Tuson, 1973). An immediate implication of this is that the Red Hill granites are, despite their great areal extent, relatively superficial bodies probably not more than 2 km thick. Palaeomagnetic measurements made on the Cuillin gabbros and peridotites show that these rocks are reversely magnetized, as are the earlier SMLS lavas. However, some of the granites have normal polarities and magnetic investigations by Brown and Mussett (1976) indicate normal polarities over much of the area known, from the gravity studies, to be underlain by mafic rocks. Thus there must have been two periods when large quantities of mafic magmas were involved in the Skye centre; those producing the Cuillin centre were intruded when the Earth's magnetism was reversed, and those giving rise to the dense rocks under the granites were emplaced later when the polarity was normal. Radiometric studies indicate that the bulk of the igneous activity took place between about 60 Ma and 57 Ma ago, but that some intrusions were emplaced as recently as 53.5 Ma ago (Table 1.1).

Investigations of the oxygen isotope geochemistry of the rocks of the Skye central igneous complex and the surrounding lavas and sediments, show that there has been massive circulation of heated meteoric waters through and around the complex (Taylor and Forester, 1971; Forester and Taylor, 1977). This pervasive circulation, which was driven by heat from within the central complex, caused considerable metasomatic and hydrothermal alteration of both the igneous intrusions and adjoining basalts, frequently overprinting original igneous features in the intrusive rocks and the high-grade metamorphic effects in their surroundings. Thus doubts were cast on the igneous origins of some of the rock textures (for example, granophyric quartz–feldspar intergrowths in some of the granites) and on the validity of petrogenetic conclusions drawn from earlier geochemical and isotopic studies. Among the last were the isotopic investigations by Moorbath and Bell (1965) and Moorbath and Welke (1969) which had indicated that the granitic rocks of the Red Hills were largely derived from partial melting of Lewisian gneisses. As recounted in Chapter 1, these doubts and uncertainties were resolved by work on the Mull granites; subsequently isotopic data from Skye were refined and augmented by Dickin (1981) who demonstrated that both crustal and mantle sources have significantly contributed to the granite magmas of Skye.

FIURNEAN TO RUBHA NA H-AIRDE GLAISE

Highlights

The site provides excellent sections through the sediments and volcaniclastic rocks which form the base of the Skye Main Lava Series (SMLS, see Table 2.2). The presence of pillow lavas, glassy lava fragments and water-lain, plant-bearing volcanogenic sediments shows that the initial lava eruptions were into shallow water.

Introduction

The cliff exposures between Fiurnean and Rubha na h-Airde Glaise (Fig. 2.3), to the north of Portree Harbour, provide extensive sections through volcaniclastic rocks and associated lavas and sediments which mark the onset of Tertiary volcanism on Skye. Accordingly, the site represents the type locality for these rocks. Tuffs, including hyaloclastites produced by the eruption of magma into water, are associated with thin basaltic lavas and fossiliferous sediments; these lie between Jurassic sediments and the lavas of the Beinn Edra Group of northern Skye (Table 2.2; Anderson and Dunham, 1966).

Description

Good exposures of tuff crop out in the cliffs and scree between Rubha na h-Airde Glaise and Sithean Bhealaich Chumhaing (NG 509 467). The lower part of the succession is crystal-rich and contains well-formed and broken crystals of olivine, labradorite and augite in partly devitrified

Figure 2.3 Geological map of the Fiurnean to Rubha na h-Airde Glaise site (adapted from the British Geological Survey 'One-Inch' map, Northern Skye Sheet 80 and parts of 81, 90 and 91).

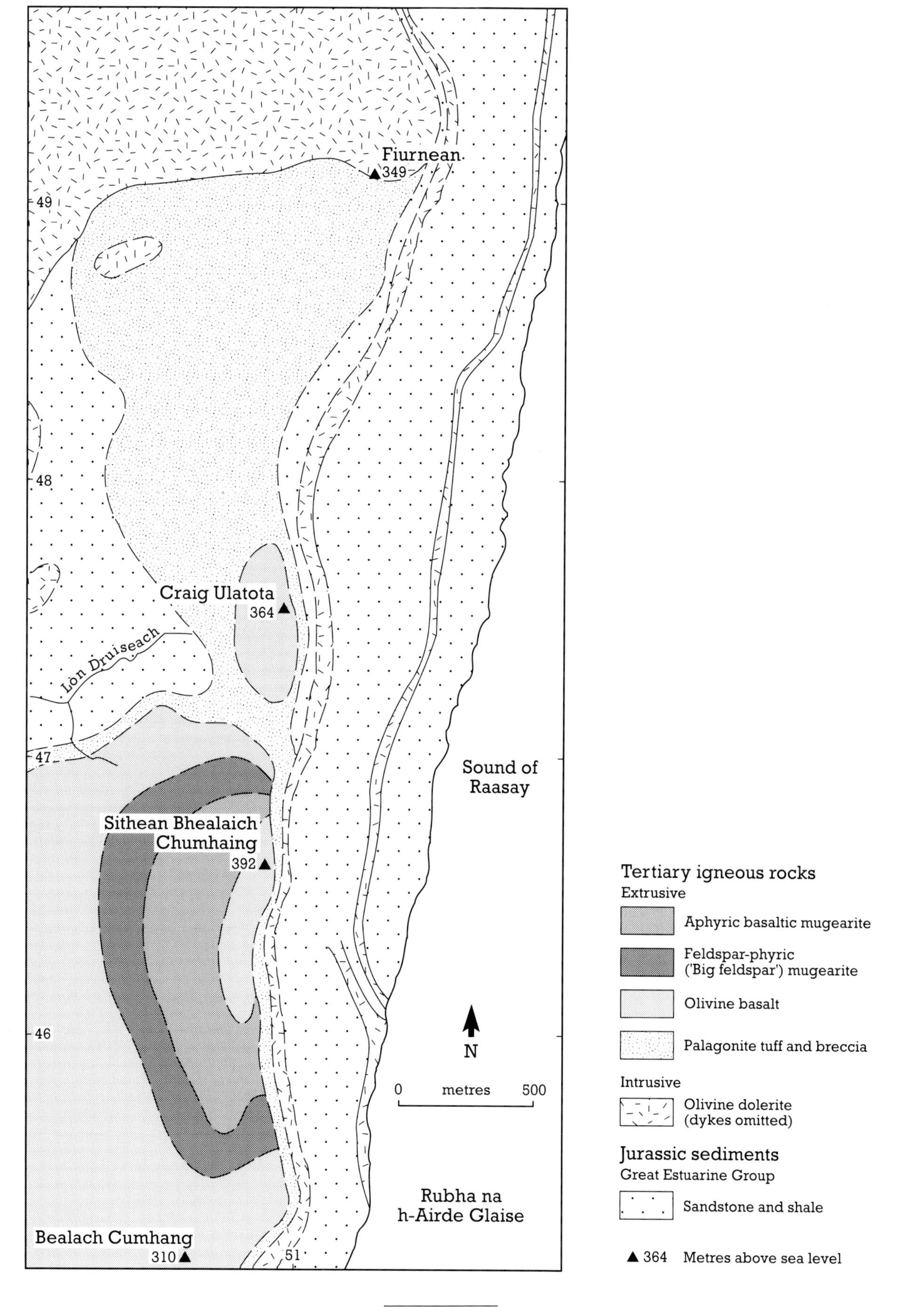

Fiurnean
349
49
48
Craig Ulatota
364
Lòn Druiseach
47
Sithean Bhealaich
Chumhaing
392
46
Bealach Cumhang
310
51
Sound of
Raasay
Rubha na
h-Airde Glaise
N
0 metres 500
Tertiary igneous rocks
Extrusive
Aphyric basaltic mugearite
Feldspar-phyric
('Big feldspar') mugearite
Olivine basalt
Palagonite tuff and breccia
Intrusive
Olivine dolerite
(dykes omitted)
Jurassic sediments
Great Estuarine Group
Sandstone and shale
▲ 364 Metres above sea level

basaltic glass. Above this are somewhat amygdaloidal tuffs, with thin sandy mudstone horizons and impersistent, minor lava flows. The lower portions of the flows show a limited development of pillow-like structures and, in places, well-formed pillows occur throughout the whole thickness of the flow. Where this happens, the individual pillows are separated by thin skins of red, lateritic material. The higher parts of the succession are largely scree-covered, but occasional exposures reveal tuffs containing basalt bombs several centimetres in diameter. These exposures resemble the upper tuffs of Camas Bàn (NG 493 423) on the south side of Portree Harbour.

Sections at, and up to 500 m to the north of, Craig Ulatota (NG 510 475) expose bedded tuffs, palagonite layers and thin basalt lavas, all of which may show reddening. The sequence, which attains a maximum thickness of about 60 m at Fiurnean, can be summarized as follows:

6. Thin shales and tuffaceous sandstones in which the bedding has been disturbed by massive basalt bombs. > 15 m
5. Brown, vitreous tuff with plant fragments. 10–15 m
4. Impersistent olivine basalt lava flow. 0–1.5 m
3. Coarse, fragmental tuff with large, pillowed masses of basalt with glassy selvedges. about 25 m
2. Pale greenish-brown sandstone containing wind-rounded quartz grains. up to 5 m
1. Jurassic strata.

Plant remains occur in thin, grey, ashy mudstone near the middle of the Fiurnean section.

A much condensed sequence of tuffs and lavas occurs about 700 m WSW of Craig Ulatola in a southern tributary of the Lon Druiseach (NG 502 471). The sequence is as follows (after Anderson and Dunham, 1966 and Brown, 1969):

6. Basalt lava with pipe amygdales at its base
5. *Palagonite tuff with pillow lava 6.7 m
4. Thin lava flow with pipe amygdales 0.23 m
3. *Thin-bedded, brown tuff 0.71 cm
2. *Hard, olive-green tuff 0.05 cm
1. Soft Jurassic sandstone with calcareous bands 4.3 m

* contain charred fragments of wood and other obscure plant remains (Wilson, 1937).

Interpretation

The site provides clear evidence that the first Tertiary deposits were water-lain, fossiliferous sediments derived from the weathering of contemporaneous volcanic debris, with some contributions from weathering of Jurassic rocks. There was an abundant flora, and plant remains are preserved in the finer-grained sediments which probably accumulated in shallow lakes. Initially, the basalt lavas erupted into the lakes to form hyaloclastite deposits, pillow lavas and thin, impersistent flows. In the hyaloclastites some of the original basaltic glass remains; this is crowded with microphenocrysts of olivine, feldspar and pyroxene. However, subsequent circulation of heated waters caused most of the basaltic glass to degrade to palagonite and the rocks to become heavily impregnated by zeolites. Sometimes these deposits appear to have been subjected to subaerial weathering, for individual pillows are occasionally coated by red lateritic material ('bole').

The overlying olivine-basalt lavas are probably similar to the transitional basalt flows found at the base of the Storr succession (q.v.). Individual flows often have reddened tops attesting to subaerial accumulation and vigorous weathering between successive eruptions. However, Anderson and Dunham (1966) quote an analysis of a (somewhat altered) pillow lava from Creag Mor, a few hundred metres south-west of the site, which they compare with tholeiitic basalts from Staffa and Antrim. If correct, this would imply that the first basalt magma formed under different conditions to the large number of overlying flows in the Ben Edra Group.

Conclusions

The site provides excellent opportunities for the study of sediments, volcaniclastic deposits and thin lava flows which mark the commencement of Tertiary igneous activity on Skye. In particular, it provides conclusive evidence that the initial activity was mildly explosive and that the deposits accumulated under shallow water, in contrast to most of the overlying lava flows which were erupted subaerially. The precise age of the deposits is not known. Late Oligocene or younger ages have been suggested from studies on these and similar plant remains elsewhere (for example,

Simpson, 1961) but palaeomagnetic studies and radiometric age determinations on the Tertiary lavas elsewhere in the BTVP indicate an age about 60 Ma, about the middle of the Palaeocene.

THE STORR

Highlights

The site contains excellent continuous exposures through lavas of the Beinn Edra Group, which is the oldest in the Skye Main Lava Series (SMLS). There is clear evidence that the lavas were erupted subaerially and weathered under warm, wet conditions. The abundant and varied suites of zeolite minerals formed under hydrothermal conditions after the lavas had solidified. The lavas show subtle variations in composition, and these have helped to elucidate the petrogenesis of the SMLS.

Introduction

North of Portree, the Trotternish escarpment provides classic exposures of transitional to mildly alkaline olivine basalt lava flows (Fig. 2.4). These lavas form the Beinn Edra Group of the Skye Main Lava Series (Table 2.2). Secondary hydrothermal mineralization within parts of the lava flows has produced an extensive suite of zeolite and associated minerals infilling vesicles.

The succession has been described in detail by Anderson and Dunham (1966) and Thompson *et al.* (1972); the late-stage mineralization has been investigated by King (1977).

Description

Between Beinn Dearg (NG 477 504) and Coire Scamadal (NG 495 547), the spectacular east-facing Trotternish escarpment is formed by massive lava flows belonging to the Beinn Edra Group (Anderson and Dunham, 1966; cover photograph and Fig. 2.5). The visible succession is about 250 m thick and it is estimated that a further 120 m is concealed beneath the scree and landslip which mantle the lower slopes. In a detailed section measured up the Storr Gully (NG 495 539) and in the cliff south of Coire Faoin (NG 497 535), Anderson and Dunham (op. cit.) identified at least 24 lava flows with an aggregate thickness of about 250 m. Individual flows vary from about a metre to over 30 m in thickness, are frequently separated by red bole horizons and are cut by a number of NW-trending basalt dykes. Almost all of the flows are olivine-phyric basalts; two near the base are feldspar-phyric and the succession is capped by a flow of columnar-jointed hawaiite and finally by a mugearite. Several of the flows contain coarse pegmatoid segregations and virtually all are conspicuously amygdaloidal. Geochemical investigations by Thompson *et al.* (1972) have shown that the lowermost flows tend to be transitional, hypersthene-normative basalts, whereas the higher ones are mildly alkaline, nepheline-normative olivine basalts. Several of the flows used in the geochemical study by Thompson *et al.* (1972) and subsequently by Moorbath and Thompson (1980) in a study of the strontium isotope geochemistry of the SMLS were obtained from A'Chorra-bheinn (NG 484 489) about 3 km SSW of the site. These lavas are the lateral equivalent of lavas in the Storr succession.

The basaltic lavas which form the landslipped masses and scree at the base of the Storr cliffs (see cover photograph and Fig. 2.5) (NG 495 540) and in Coire Faoin (NG 497 537) are highly amygdaloidal. The lavas are exceptionally rich in secondary, hydrothermal zeolites, allied silicates and other materials. The minerals include analcite, apophyllite, calcite, chabazite, chlorite, gyrolite, heulandite, levynite, mesolite, scolecite, stilbite and thomsonite.

The late-stage amygdale minerals are most abundant in the scoriaceous flow tops and flow bases and are relatively uncommon in the more massive (and unaltered) central parts of the flows.

Interpretation

The basalts of the Beinn Edra Group heralded the beginning of flood basalt volcanism in northern Skye following initial explosive activity (see Fiurnean to Rubha na h-Airde Glaise). The fissures from which the lavas were extruded were probably located in Trotternish (Anderson and Dunham, 1966). These may now be occupied by

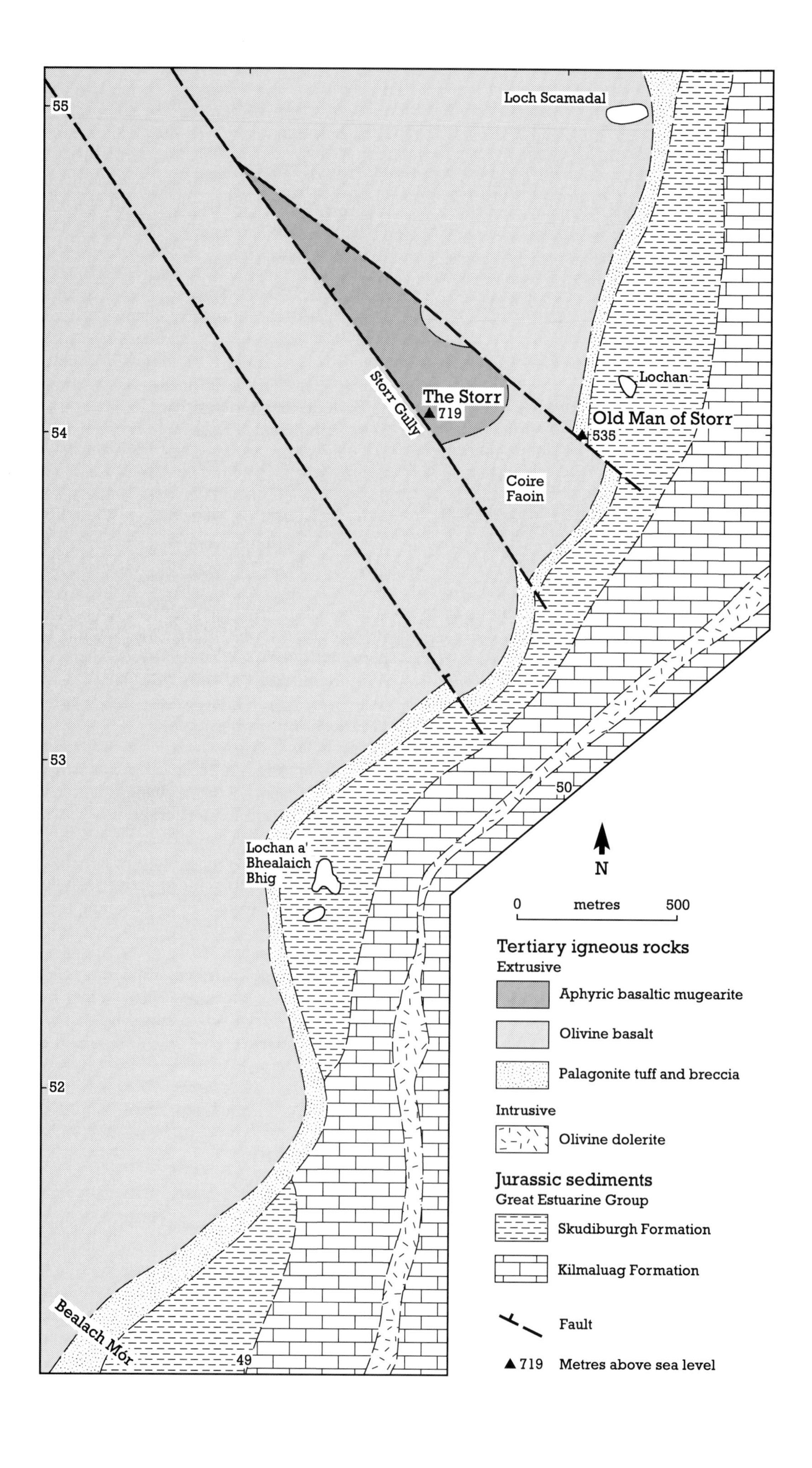

55
54
53
52
Loch Scamadal
Lochan
Storr Gully
The Storr
719
Old Man of Storr
535
Coire
Faoin
50
Lochan a'
Bhealaich
Bhig
Bealach Mór
49
N
0
metres
500
Tertiary igneous rocks
Extrusive
Aphyric basaltic mugearite
Olivine basalt
Palagonite tuff and breccia
Intrusive
Olivine dolerite
Jurassic sediments
Great Estuarine Group
Skudiburgh Formation
Kilmaluag Formation
Fault
▲719 Metres above sea level

Figure 2.5 Basalt lavas of the Skye Main Lava Series. Slipped masses of lava, including the Old Man of Storr pinnacle, occur in the foreground and to the right. Storr site, Skye. (Photo: C.H. Emeleus.)

NNW-trending picrite dykes such as those at Beinn Tuath (NG 435 530) and Glen-uachdarach (NG 430 585) to the west of the site. The lavas were erupted subaerially and subjected to intense weathering. This caused leaching and oxidation and formed the bright red bole horizons which now separate many of the flows. Similar present-day weathering occurs where there is a combination of a warm climate and abundant rainfall; thus the presence of bole horizons in the Palaeocene lavas of Skye and other areas in the BTVP implies tropical or subtropical conditions.

The geochemical data obtained by Thompson *et al.* (1972) and Moorbath and Thompson (1980) from flows in the Beinn Edra Group provided some of the evidence for variability in both bulk chemistry and isotopes which lead to the novel hypothesis of magma plumbing beneath northern Skye (see Introduction, above; also cf. Fig. 5.3).

Figure 2.4 Geological map of the Storr site (adapted from the British Geological Survey 'One-Inch' map, Northern Skye Sheet 80 and parts of 81, 90 and 91).

The distribution of the zeolites and associated minerals in lava piles is known to be controlled by a combination of temperature, pressure, circulation of heated aqueous fluids and the bulk composition of the rock (Walker, 1960). A study of the lavas of the Beinn Edra Group by King (1977) has shown that much of the group contains zeolite assemblages similar to the analcime–natrolite zone defined by Walker (1960). In the alkali olivine basalts of the upper part of the sequence at the Storr, analcite is lacking and instead, first mesolite and then thomsonite–chabazite assemblages are present. However, the reappearance of analcite in the hawaiite which caps the cliff adds support to the suggestion by Walker that the chemistry of the host rock plays an important role in determining the types of zeolites and associated minerals formed by the circulating fluids.

Conclusions

The importance of this site lies in the continuous excellent exposure through the Beinn Edra

Group which forms the base of the Skye Main Lava Series (Table 2.2). At least 24 flows are exposed (Anderson and Dunham, 1966, pp. 83–4); the interrelationships show that they accumulated subaerially and were subjected to deep weathering under wet, warm tropical or subtropical conditions. The site is particularly noted for the abundance and variety of zeolite minerals within the lavas which provide conclusive evidence of hydrothermal activity after solidification.

The site is a vital link in the chain of geochemical evidence obtained from the Skye lavas which suggests that they were derived from the upper mantle and rose towards the surface in small batches, each batch having its own history of crystal fractionation and contamination by crustal rocks as it passed through the Palaeocene crust.

ROINEVAL

Highlights

The site contains an excellent example of a composite mugearite flow with an aphyric lower member and a strongly feldspar-phyric upper part.

Introduction

A type example of a composite mugearite lava flow forms the summit area of Roineval. Composite lavas on Skye were first described by Harker (1904) who interpreted them as sills, but they were later shown (by Kennedy, 1931b) to be lava flows. More recent investigations include those of Muir and Tilley (1961) and Boyd (1974).

Description

The Roineval flow lies within the Beinn Totaig Group of lavas in northern Skye (Table 2.2; Anderson and Dunham, 1966) and rests upon the weathered vesicular top of a non-porphyritic mugearite. The composite flow consists of an upper member with phenocrysts of plagioclase, olivine and titaniferous magnetite overlying an essentially aphyric mugearite. The boundary between the units is gradational over several centimetres, in contrast to other composite flows such as those at Druim na Criche (NG 435 375), in which the two components are separated by a sharp, non-erosional boundary. The upper porphyritic unit at Roineval contains abundant plagioclase phenocrysts together with lesser amounts of altered olivine and magnetite. Boyd (1974) reports homogeneous (An_{55}) cores to the phenocrysts, which are euhedral but often broken, and thin, more calcic rims (*c.* An_{60}), and jackets zoned from An_{51} to An_{40}. A small amount (15 vol.%) of similar plagioclase phenocrysts (An_{52-50}) occurs in parts of the otherwise essentially non-porphyritic lower unit. The bulk composition of the upper unit is hawaiitic, that of the lower unit is mugearitic. The porphyritic mugearites show a marked degree of iron enrichment, especially when recalculated to a phenocryst-free base (cf. Anderson and Dunham, 1966, tables 8 and 9; Muir and Tilley, 1961, table 4).

Interpretation

Composite lava flows are a feature of special petrological interest within the Skye plateau lava succession; the Roineval flow is one of several particularly clear examples. The strongly feldspar-phyric top member of the composite flow, overlying rock with a similar matrix composition but lacking numerous phenocrysts, suggests that the source was a differentiated body of magma in which large, labradorite crystals had sunk, together with olivine and opaque oxides, leaving an upper, crystal-free layer of magma. An initial eruption evacuated the aphyric portion, forming the basal mugearite which was rapidly followed by the remaining, phenocryst-rich magma of the upper unit. Boyd (1974) suggests that the slight compositional differences between the upper and lower units may be attributed to the concentration of phenocrysts in the upper part. The lack of chilling between the units, and the manner in which they merge over 10–20 cm, provides proof that there was very little time between the eruption of each unit; the porphyritic unit may well have followed virtually instantaneously. The well-defined break between the porphyritic and aphyric members of the composite flow suggests that there were tranquil conditions in the magma chamber for some time prior to eruption; strong convective or other movement would have mixed the two parts.

The relationships seen in the Skye composite flows strongly suggest that plagioclase feldspar might sink and accumulate in an evolved magma of hawaiitic or, as in this instance, mugearitic composition. This has important petrological implications since it has been claimed, in connection with the formation of layered gabbro complexes, that plagioclase flotation would be more likely (cf. McBirney and Noyes, 1979), in contrast with the earlier view that settling of feldspar occurred in these bodies (for example, Wager and Brown, 1968).

Conclusions

The composite lava flow of Roineval provides evidence that labradorite feldspar phenocrysts might have sunk in the iron-enriched magmas represented by the Skye mugearites. The relationships found at this site therefore have an important bearing on the crystallization processes in magma chambers and on the origin of layered structures frequently present in gabbroic plutons. They support the original models for the origin of igneous layering proposed by Wager and Deer (1939) and elaborated by Wager and Brown (1968), rather than alternative explanations offered by McBirney and Noyes (1979) and others.

TALISKER

Highlights

The spectacular coastal cliffs here provide some of the best sections through the highest part of the Skye Main Lava Series (SMLS) which is notable for the variety of lava types within it. The exposures on Preshal More and Preshal Beg demonstrate a thick, ponded flow of low-alkali, high-calcium olivine tholeiite which was erupted after a long erosional interval, following the last flows of the SMLS. The site is the type locality for this distinctive tholeiite which is matched by only one other flow in Skye, although many of the north-west dykes are of similar composition.

Introduction

A coastal cliff section and three well-exposed inland areas have been selected to demonstrate the diversity and complexity of the plateau lava sequence of west-central Skye (Fig. 2.6). Collectively, they provide evidence of a thick lava succession (>400 m) comprising flows of picrite, olivine basalt, hawaiite, mugearite and olivine tholeiite compositions belonging to the Arnaval and Talisker groups (Williamson, 1979). The area was subject to major reassessment by Williamson (1979) following earlier studies by Harker (1904), Anderson and Dunham (1966) and Esson *et al.* (1975).

Description

The Talisker Bay (NG 300 315–317 284), Stockval (NG 351 296) and Ard an t-Sabhail (NG 318 333) localities (Fig. 2.6) incorporate lavas belonging primarily to the Arnaval Group of Williamson (1979) which is probably equivalent to the youngest Osdale Group (Table 2.2; Anderson and Dunham, 1966) of northern Skye. The overall thickness of the Arnaval Group is probably in excess of 400 m, but nowhere is the whole sequence exposed and, in general terms, the following divisions are recognized (Williamson, 1979):

C	Rare, thin, massive olivine basalts and mugearites. Hawaiites and mugearites become progressively more common towards the top.	175–200 m
B	Porphyritic and non-porphyritic basalts with rare hawaiites and picritic basalts.	100 m
A	Highly amygdaloidal olivine basalts and picritic basalts with mugearite.	150 m

The cliff sections around Talisker Bay afford excellent exposures of picritic basalts, well-developed brown/grey/red boles, and abundant amygdales filled with zeolites. A thick mugearite flow near to the base of the succession at MacFarlane's Rocks (NG 301 314) to the south of Talisker Bay, displays characteristic flow-jointing which has been intricately overfolded and also forms antiforms and synforms in the upper part of the flow. Major dislocations to the east have faulted the succession, rendering correlation with other areas difficult. Inland, at the Stockval ridge locality south of Gleann Oraid (NG 320 305), the

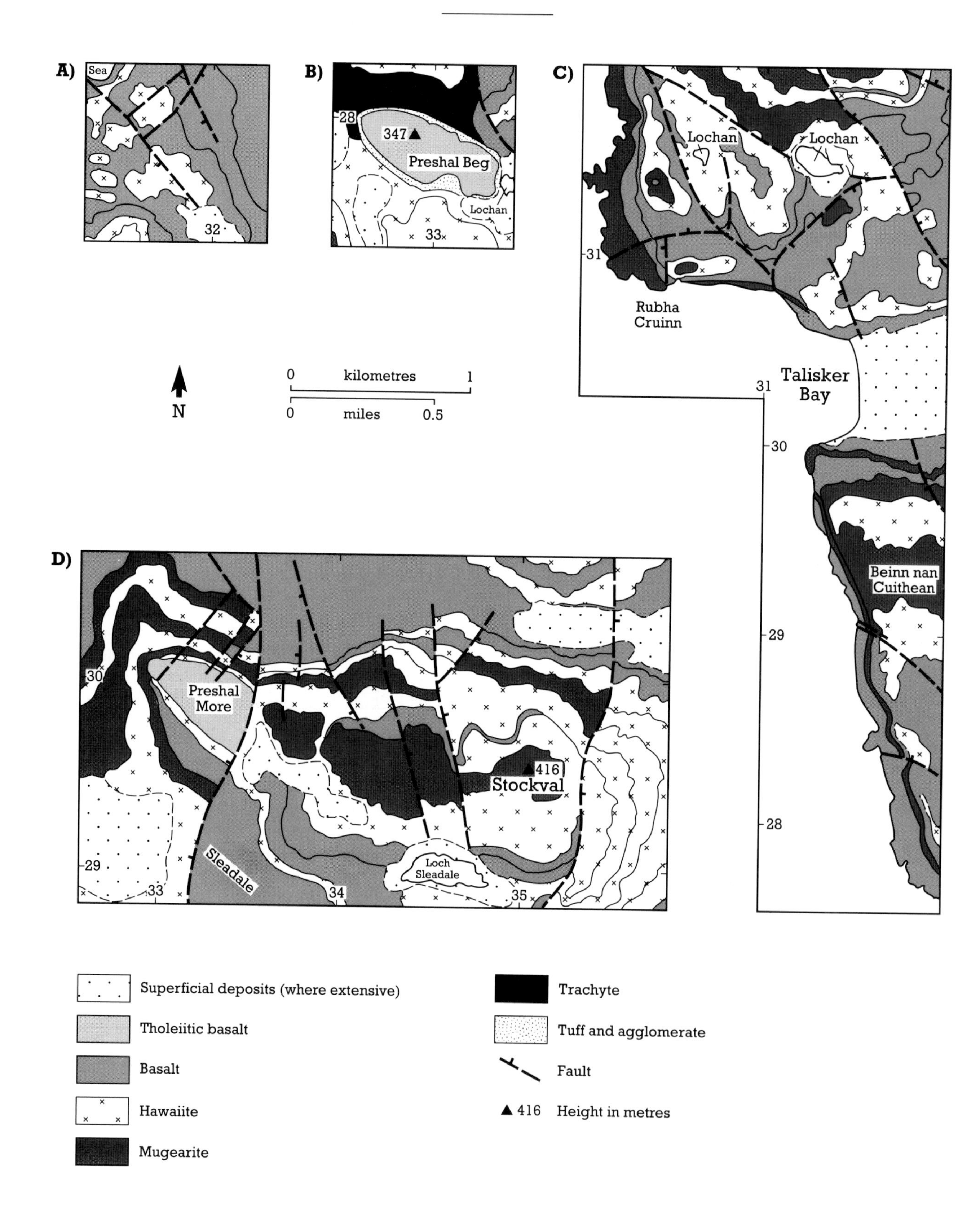

Figure 2.6 Geological map of the Talisker site (after Williamson, 1979). A) Area south-west of Fiskavaig; B) Preshal Beg area; C) coast north and south of Talisker Bay; D) area between Preshal More and Stockval.

upper part of the succession is exposed comprising mainly hawaiite and mugearite flows.

At Dun Ard an t-Sabhail, a composite lava flow overlies a sequence of alkali-olivine basalts. The lava is more basic than those at Roineval and at Druim na Criche (Harker, 1904; Kennedy, 1931b) but is similar in that the flow consists of a porphyritic member overlying a non-porphyritic member of a broadly similar rock type. The upper member carries between 14% and 27% volume plagioclase phenocrysts which are more numerous at the base and which appear to belong to two separate generations (Boyd, 1974). Labradorite predominates but rounded, resorbed andesine is also found. Williamson (1979) also recorded the presence of orthopyroxene phenocrysts (or xenocrysts) in the upper unit, thereby distinguishing this flow from the other composite flows of Roineval and Druim na Criche. Hercynite spinel microphenocrysts mantled by titaniferous magnetite are also present, these and the orthopyroxene may be relicts of a phase of high-pressure crystallization prior to eruption.

The twin hills of Preshal More (Fig. 2.7) and Preshal Beg consist mainly of a 100 m thick flow of olivine tholeiite belonging to the Talisker Group (Williamson, 1979). The two outliers exhibit spectacular columnar jointing. At Preshal More, a series of fine-grained, water-lain tuffs underlie the tholeiite flow along the north-eastern margins of its outcrop. In contrast, to the west the tuffs are absent and the flow rests directly upon a series of hawaiites, mugearites and basalts of the Arnaval Group. On Preshal Beg the tholeiite rests upon a coarse agglomerate associated with minor tuff horizons. The agglomerate overlies leucocratic, clinopyroxene-phyric trachytes or trachyandesites in the north, but at the southern margins of its outcrop it lies directly upon fine-grained hawaiite. The tholeiite has an unusual chemistry within the Skye lava succession and is only matched by a flow high in the Osdale Group (Table 2.2) near Edinbane (Esson *et al.*, 1975). It is an olivine tholeiite which is unusually depleted in alkalis and rich in calcium with a distinctive trace-element chem-

Figure 2.7 Late tholeiite lava forming Preshal More and infilling a former valley eroded in flows of the Skye Main Lava Series that form pale-coloured scarps below and to the right. Talisker site, Skye. (Photo: A.P. McKirdy.)

istry (Fig. 1.2; Esson *et al.*, 1975; Williamson, 1979). The thick flow is interpreted as a ponded lava lying on top of the Arnaval lavas (Esson *et al.*, 1975); ponding occurred either owing to a downstream obstruction, or when a thick flow outwith the area flooded back into the valley (cf. Fionchra, on Rum).

Interpretation

The site is largely occupied by the varied lavas of the Arnaval (Osdale) Group at the top of the SMLS. As with the lavas of the Storr (see above), these are transitional in geochemical character, but have a more varied composition (picrites, olivine basalts, hawaiites, mugearites and trachytes). Striking evidence for a major compositional change in the basaltic magma comes from the thick lava flow which forms Preshal More and Preshal Beg. This flow occupies a valley eroded in the upper flows of the transitional lavas of the Arnaval Group. The valley appears to have been partly floored by agglomerates and bedded water-lain tuffs, and it is most probable that the abnormally thick flow which overlies these deposits was ponded within the steep-sided valley. Thus, it is clear that the Preshal More–Preshal Beg flow was erupted after a significant erosional interval, a situation comparable with that of some basaltic andesites and icelandites on Rum (see Fionchra below). The analogy may be taken further since, as on Rum, there is major change in the lava composition; in this instance to a highly distinctive, low-alkali, high-calcium olivine tholeiite generally depleted in incompatible elements. Although the effusive representatives of this Preshal More magma type are limited on Skye to the two hills of this site, and one earlier flow, basic rocks of similar composition occur widely as intrusions and are particularly abundant in the Skye dyke swarm (Mattey *et al.*, 1977); the magma type is important in the British Tertiary Volcanic Province in general (Thompson, 1982) and has many similarities to mid-ocean ridge basalts (cf. Bell and Harris, 1986).

Conclusions

This is an important site for demonstrating both the structural and compositional complexity of the subaerially erupted lavas of the upper part of the SMLS. It also contains a particularly clear record of an abrupt change in lava composition from earlier flows after a major erosional interval. The site contains the type locality for the distinctive Preshal More type of olivine basalt, which is known to have a widespread occurrence within the BTVP and which is comparable in many respects with mid-ocean ridge basalts.

ROS A' MHEALLAIN

Highlights

The site contains excellent examples of trachyte, mugearite and benmoreite which are some of the most chemically evolved lavas in the Skye Main Lava Series. Chemical evidence shows that some (mugearite–benmoreite) formed under high-pressure conditions near the base of the crust; by contrast, others (iron-poor mugearites–trachytes) reflect lower-pressure conditions of formation high in the crust.

Introduction

Exposures between Portree and Bracadale on the hills of Ros a' Mheallain (NG 375 405), Ben Scudaig (NG 357 410) and Beinn na Cloiche (NG 366 418) display some of the most petrogenetically evolved lavas found on Skye (Fig. 2.8). The flows are members of the Bracadale Group (Table 2.2; Anderson and Dunham, 1966) and are principally porphyritic and non-porphyritic mugearites with subordinate trachytes and benmoreites. Rare basaltic/lavas and trachytic tuffs also occur.

Description

The summit of Ros a' Mheallain (Fig. 2.8) consists of a strongly feldspar-phyric mugearite which is petrologically allied to the composite flow of Roineval and Druim na Criche. The flow is underlain by scoriaceous mugearite and perhaps two trachyte flows, the lower of which is admirably exposed nearby in a small quarry and in cuttings along the Portree–Bracadale road. This trachyte is rather leucocratic and, although

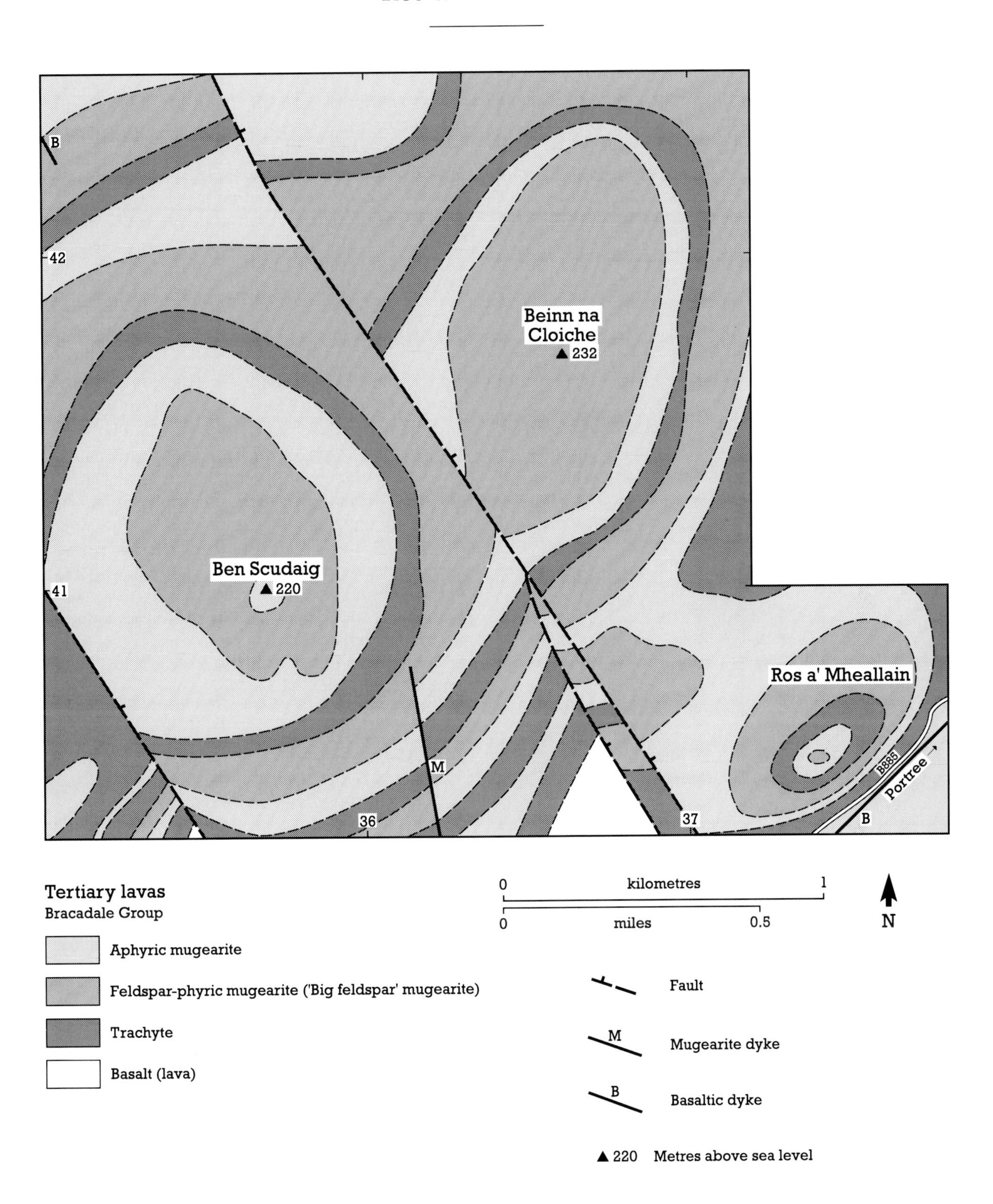

Figure 2.8 Geological map of the Ros a' Mheallain site (adapted from the British Geological Survey 'One-Inch' map, Northern Skye Sheet 80 and parts of 81, 90 and 91).

often deeply weathered, original flow banding, reasonably fresh alkali-feldspar and biotite phenocrysts and amygdales containing stilbite, mesolite, chabazite and heulandite are present. A short distance to the north-east, a melanocratic trachyte or benmoreite overlain by a thin red bole is exposed at the roadside. Large, often flattened, amygdales filled by quartz, agate and pyromorphite are common in this flow.

The conspicuous feldspar-phyric mugearite forms scarps on Ros a' Mheallain and Beinn na Cloiche where another, higher feldspar-phyric mugearite forms the hilltop. The rocks of Beinn na Cloiche are separated from those of Beinn Scudaig by a NW–SE-trending fault which downthrows to the north-east. On Beinn Scudaig, although exposure is poor, the trachytic flows here are seen to reach their maximum development and four or five individual flows can be distinguished. The lavas are intercalated with mugearites and they thin away from a NW-trending axis lying along a line from Beinn Scudaig to Beinn Aketil (NG 327 463).

Interpretation

The flows of trachyte, mugearite and benmoreite within the site are part of the Bracadale Group, high in the Skye Main Lava Series, and are some of the most chemically evolved members of this basalt-dominated series. According to Thompson *et al.* (1972), the hawaiite–mugearite–benmoreite suite evolved from transitional basalts by crystal fractionation under high-pressure conditions at the base of the crust. Conversely, the iron-poor mugearites and trachytes evolved from a similar magma, but this took place under low-pressure conditions in small magma chambers established at higher crustal levels. The diversity of lava compositions, the compositional evidence for both high-pressure and low-pressure controls on magmatic evolution and the apparently random distribution of the compositionally varied lavas through the pile, provide strong support for the complex model of magmatic plumbing proposed by Thompson *et al.* (1980) and Morrison *et al.* (1985).

Conclusions

The site is of particular value because it contains excellent examples of some of the most chemically evolved members of the Skye Main Lava Series. The chemical evidence is interpreted by Thompson *et al.* (1980) to show that the lavas evolved from parental basaltic magmas in reservoirs situated at both deep (high-pressure) and shallow (low-pressure) crustal levels.

ALLT GEODH A' GHAMHNA

Highlights

The site forms a vital link in a chain of localities extending from Muck and Eigg to Rum, Canna and Sanday and Skye which demonstrate that the Skye centre is younger than the Rum centre and that there were two periods during which substantial lava piles built up, one before and one after emplacement of the Rum centre. The site provides a superb section through water-lain sediments intercalated in the lava succession. The dominance of sandstone and arkose clasts of Torridonian facies indicates a Torridonian upland source; granophyre clasts were most probably derived from Rum, but the provenance of porphyritic rhyolite clasts is unknown.

Introduction

The Allt Geodh a' Ghamhna site contains the finest exposures on Skye of interstratified sediments and basic lavas. The sediments occur towards the local base of the plateau lava succession and consist of lenticular, channel-shaped bodies of conglomerates, sandstones and thin coal horizons. Harker (1904) recorded the succession at this locality but misinterpreted the sandstones as tuffs. Subsequently, a comprehensive field investigation of the Tertiary lava succession in west-central Skye, incorporating a study of the sedimentary deposits in this site, has been carried out by Williamson (1979) and the granophyre clasts have been chemically analysed (Meighan *et al.*, 1981).

Description

The inter-lava sedimentary deposits at Geodh a' Ghamhna are well exposed where the Allt Geodh a' Ghamhna reaches the sea cliffs immediately east of Rubha Thearna Sgurr (NG 365 197).

Table 2.3 The succession at Allt Geodh a' Ghamhna (after Williamson, 1979, table 2)

Bed	Description	Thickness
14	Thin, alkali olivine basalts with scoriaceous tops	7 m
13	Massive basaltic lava with pillow structures towards the base	5 m
12	Thin white ash	0.03 m
11	Coal	0.05 m
10	Sandstone with obscure plant remains occurring as diffuse carbonaceous streaks and rootlets, possibly seat earth	0.2 m
9	Coal	0.01–0.05 m
8	Conglomerate with well-packed, rounded pebbles and cobbles of granophyre, quartzite, porphyritic rhyolite and red arkose. Clasts have a maximum diameter of 0.10–0.15 m, and are set in a pale sandy matrix	3.2 m
7	Sandstone with micaceous partings	0.2 m
6	Coal	0.02 m
5	Sandstone with plant remains	1.8 m
4	Conglomerate with a more sandy matrix than Bed 2, and a smaller proportion of acid igneous to arenaceous sediments than Bed 8. Rare pebbles of amygdaloidal and feldspar macroporphyritic basalt. Clast size <0.30 m, averaging 0.10–0.15 m. Thin lenses of white sandstone in lower horizons	2.3 m
3	Fine-grained sandstone, laminated base	1.1 m
2	Massive conglomerate with densely packed, crudely imbricated clasts of red arkose up to 0.30 m in diameter. Contains green siltstones with a sandstone wedge thickening to the north	2.75 m
1	Highly amygdaloidal basaltic lavas forming the top of the cliff at about 125 m elevation	10 m

Williamson (1979) has recorded in detail the succession on the south side of the stream (Table 2.3). A few obscure, poorly preserved plant remains have been found in sandstones beneath the thin coal seams (Table 2.3). There is an abundance of arkosic sandstone clasts (Torridonian?) and Palaeocene granophyre and porphyritic rhyolite, but the scarcity of basalt pebbles in the conglomerates is notable.

At Geodh a' Ghamhna, the conglomerates are irregularly intruded by a sill with andesitic affinities, and the lavas forming the sea cliffs are traversed by several tholeiitic sheets dipping to the north. The lavas exposed on the north side of the stream, beneath the conglomerate, show many well-developed red and purple-red bole horizons.

Interpretation

Inter-lava sedimentary deposits are found at three main localities on Skye, namely Glen Osdale, Glen Brittle and Allt Geodh a' Ghamhna. All of these localities contain sediments of a similar nature, but those within the Allt Geodh a' Ghamhna are particularly well developed and exposed.

The channel-like nature of the sedimentary bodies is consistent with deposition in a fluviatile environment which was active during volcanic activity and the extrusion of the plateau lavas. The characteristics of the conglomerates, dominated by arkosic sandstone clasts, suggests that they are largely derived from the erosion of a high relief, Torridonian source area. The scarcity of basaltic clasts also suggests that the lavas were not deeply dissected and probably occupied a gently subsiding basin flanked by Torridonian highlands.

Williamson (1979) has suggested that the conglomerates were deposited during periods of flash flooding; probably the sandstones and certainly the coals belong to quieter periods of sedimentation. He also suggested (1979) that the source for the Torridonian clastic material may have lain to the south and that the sediments of

Skye, Rum (Fionchra) and Canna (see Sanday and Compass Hill) belong to a more-or-less penecontemporaneous fluviatile system.

The origin of the granophyre and rhyolite pebbles found in the conglomerates is problematic since they could suggest the existence of early Tertiary acid intrusions on Skye. Although the clasts are broadly similar to some of the western Red Hills granitic rocks they cannot have come from that source, for the Red Hills granites demonstrably post-date the Cuillin gabbros which, in turn, intrude the Skye Main Lava Series. Clearly, a pre-lava source is required and this could have been on Skye, in the general area of the central complex (cf. J.D. Bell, 1966, 1976; Walker, 1975). However, a study by Meighan *et al.* (1981) strongly suggests that the acid rocks may ultimately have been derived from the Rum central complex, where both granophyres and rhyolite (felsites) were deeply eroded during the early Tertiary along with Torridonian sandstones. Furthermore, detritus from Rum is known to have spread at least as far as Canna–Sanday (Emeleus, 1973). Thus, this site is an important link in a chain of sites which strongly indicates that the Skye Main Lava Series and the Skye central complex post-dated the Rum central complex (for example, Mussett *et al.*, 1988).

Conclusions

The dominance of sandstone and arkose clasts and the paucity of basalt pebbles in the conglomerates of this site indicate the former presence of a lava field traversed by streams and rivers which drained from a hilly, or possibly even mountainous, Torridonian hinterland. The additional presence of porphyritic rhyolite and granophyre clasts points to (presumably) Tertiary plutonic rocks intruding the Torridonian and also to the possibility of early acid lavas or ash flows of Tertiary age (cf. Cnapan Breaca, Rum). The provenance of the porphyritic rhyolite (felsite) cobbles and pebbles is not yet known; however, the petrography and geochemistry of the granophyre clasts singles out Rum as their most likely source. If this hypothesis is correct, this is one of the few instances where it is possible to obtain relative ages of the Tertiary central complexes. The site is thus a vital link in demonstrating that the Skye centre post-dates the central complex of Rum and that there must have been at least two periods of plateau lava eruption, one of which (the lavas of Eigg and Muck) preceded emplacement of the central complex of Rum and another (Canna–Sanday; north-west Rum; Skye Main Lava Series) which formed after the Rum volcano had been unroofed.

AN CLÈIREACH

Highlights

The site contains an excellent example of a coarse gabbroic-anorthosite (allivalite) sheet cutting the Skye Main Lava Series. The rock provides textural evidence that the sheet became choked with early-formed plagioclase megacrysts crystallized from a low-alkali, high-calcium tholeiitic basalt magma, and it demonstrates that this distinctive magma was available at a late stage in the igneous history of northern Skye.

Introduction

Substantial dyke-like bodies and occasional sills, containing coarse gabbroic anorthosite (allivalite) and gabbro (Fig. 2.9), intrude the plateau lavas of north-west Skye, following the NNW trend of the regional dyke-swarm. The characteristics of these intrusions are demonstrated by the gabbroic anorthosite intrusion at An Clèireach. The dykes are of significant petrogenetic importance in understanding the igneous activity of northern Skye.

Following their discovery by Harker (1904), they were described by Anderson and Dunham (1966). In a subsequent investigation, Martin (1969) coined the term 'Oseitic Group' for the intrusions. Donaldson (1977a) has given a detailed account of their petrology and suggested that they form volcanic plugs.

Description

At An Clèireach (NG 335 443; Fig. 2.9), a gabbroic anorthosite intrusion cuts the mugearitic and trachytic lava flows of the Bracadale Group (Anderson and Dunham, 1966). It belongs to the group of coarse-grained, basic and ultrabasic dyke and pod-like intrusions in the area between

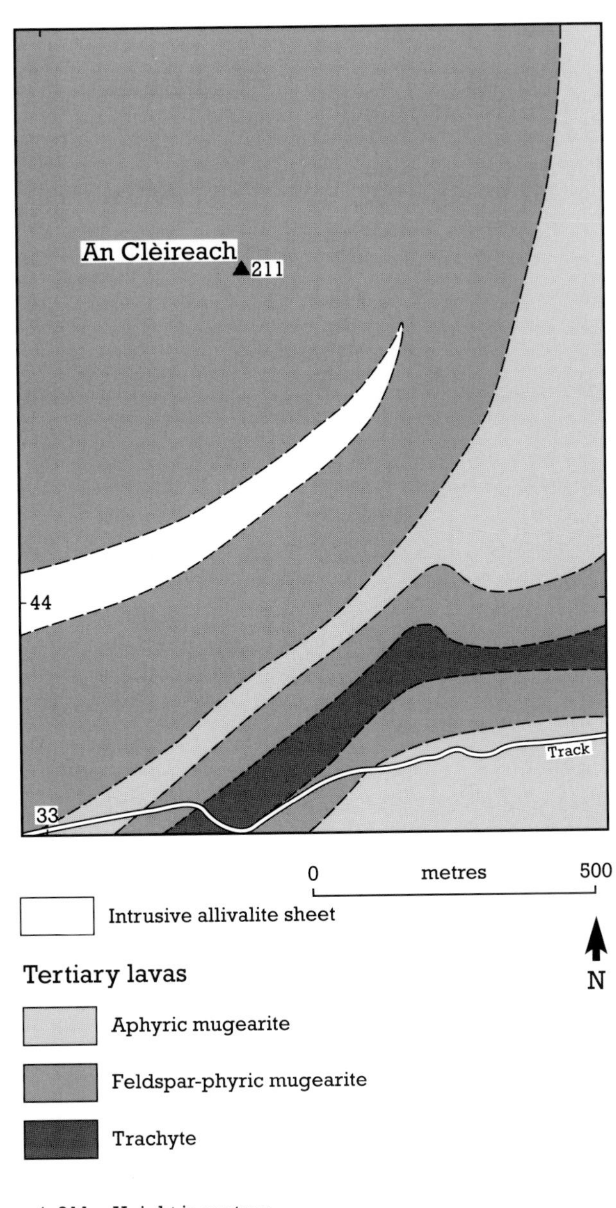

Figure 2.9 Geological map of the An Clèireach site (adapted from the British Geological Survey 'One-Inch' map, Northern Skye Sheet 80 and parts of 81, 90 and 91).

Bracadale and Beinn a Chlèirich (NG 332 451) mapped by Anderson and Dunham (1966) and collectively termed the Oseitic Group by Martin (1969) after the River Ose.

The intrusion at An Clèireach is poorly exposed, and Donaldson (1977a, Fig. 2) shows two disconnected outcrops. It is depicted on the British Geological Survey maps (Sheet 80) as a kilometre-long NNW-trending dyke, up to 150 m wide and, within the site, a sill-like apophysis skirting around the southern slopes of An Clèireach; this is one of the few sills shown to intrude the lavas (cf. Harker, 1904). The intrusion consists of coarse-grained, gabbroic anorthosite (allivalite on the BGS maps) with thin, basaltic marginal facies at the contact with the lavas. The anorthosite is dominated by calcic plagioclase megacrysts (bytownite/anorthite) subophitically enclosed by large augite crystals. Donaldson (1977a) provides detailed descriptions of the petrography and mineralogy of this distinctive group of intrusions.

Interpretation

The coarse-grained dykes lie on the axis of maximum dilation of the regional dyke swarm (Speight *et al.*, 1982). The spatially linked outcrops of the group suggest that they mark a major fissure from which the, now mostly eroded, youngest lavas on Skye (Upper Bracadale Group) were erupted.

Donaldson (1977a) has suggested that the dykes represent a concentration of solids, mainly plagioclase megacrysts, mobilized and subsequently sedimented by a basaltic liquid which erupted at the surface as lava. The plagioclase megacrysts, together with subordinate olivine, were envisaged to have crystallized in a small, shallow magma chamber from low-alkali, high-calcium olivine tholeiite liquids. The crystals were carried up by pulses of this magma into the feeder dykes, where they were concentrated and settled out, possibly by flow differentiation. The host magma subsequently erupted at the surface as low-alkali, high-calcium olivine tholeiites, the relicts of which are found in the vicinity of the dykes (Donaldson, 1977a). After the dyke had become 'choked' with crystals, the eruption site shifted to form new dykes/pods. Thus, according to Donaldson, this distinctive group of dykes represents a series of volcanic feeder plugs.

The geochemistry of these intrusions confirms the existence of low-alkali, high-calcium tholeiitic magmas, similar to those found in the Talisker site (Preshal More and Preshal Beg), towards the end of Tertiary volcanism in northern Skye. Their exact age and position in the igneous stratigraphy of Skye is not known but their petrography and chemistry suggests links with gabbros in the

Cuillin centre (see below). These rocks provide a further indication that this distinctive magma type developed widely with time in the British Tertiary Volcanic Province (see Talisker site).

Conclusions

The intrusion within this site is a representative of a group, intruding the Skye Main Lava Series of northern Skye, which demonstrates that low-alkali, high-calcium tholeiitic magmas became widely available in Skye, and elsewhere in the BTVP, after the eruption of substantial amounts of alkali-olivine and transitional basaltic magmas. The dykes are late in the igneous sequence in northern Skye, but may be coeval with intrusions in the Cuillin centre. The petrography of the dykes shows that they accumulated large amounts of early-formed crystals from the tholeiitic magmas.

RUBHA HUNISH

Highlights

The sills here show textbook examples of columnar jointing, chilled margins against sedimentary country rocks and transgressive relationships to the sediments. The thicker sills have a wide range of rock types, ranging from very olivine-enriched dolerites to pegmatitic dolerites rich in zeolites and poor in olivine. The marked variation in mineral proportions may be explained by the sinking of olivine during early crystallization of the sill magma and the subsequent redistribution of the mineral during flow of the magma.

Introduction

The Trotternish Sill Complex is one of the most remarkable features of the geology and scenery of northern Skye; excellent sill exposures lie within the Rubha Hunish site (Fig. 2.10). The complex consists of a great sheet of basic–ultrabasic rock, typically split into a number of leaves which transgress through folded Jurassic strata while retaining a constant level below the base of the Tertiary lavas. The total thickness of the sill does not depart substantially from 230 m. The sheets provide fine examples of crystal differentiation and many textbook illustrations of the relationships between sills and their host sediments, as well as columnar jointing.

Walker (1932) and Anderson and Dunham (1966) have produced detailed descriptions of the Trotternish Sill Complex and Simkin (1967) has considered the role of flow differentiation during its emplacement. Bell and Harris (1986) give a useful synthesis of the Trotternish intrusions, and the geochemistry, mineralogy, petrology and structure of the sill complex have been examined in detail by Gibson (1988, 1990) and Gibson and Jones (1990).

Description

Many of the petrographic and structural features of the Trotternish sill complex, including columnar jointing and transgressive relationships to the sediments, are demonstrated in the shore and cliff sections of this site. Inland, exposure is often indifferent. The site contains most of the facies found in these sills, including olivine dolerite, crinanitic dolerite, picro-dolerite, picrite, pegmatitic dolerite and finer-grained dolerite and basalt in the chilled margins against Jurassic strata.

A leaf of the sill at Duntulm Castle (NG 410 743) forms the headland beneath the castle. The rock is an olivine-rich, orange-brown weathering dolerite. It contains about 20 dark-coloured doleritic bands, each a metre or so in thickness, situated approximately midway up the steep cliff face. A similar feature is seen at Rubha Smellavaig and at Rubha Voreven (NG 406 758) to the north, and in the Meall Deas–Meall Tuath cliffs (NG 410 761). At the former locality, the banding is on a similar scale; individual dark bands averaging 0.3 m in thickness. Drever (*in* Brown, 1969) noted that the banding was conformable with wedging out of the sill leaf at Rubha Voreven. Gibson and Jones (1990) provide descriptions and illustrations of the layering from a number of localities and give the first detailed account of the petrography amd mineralogy of the layered rocks. The layers have well-defined mafic bases and grade upwards within centimetres (or rarely, metres) to felsic tops which are also more resistant to weathering. The layering reflects variation in the modal proportions of olivine, augite and plagioclase; at the base of layers at Rubha nam Brathairean (NG 444 758) olivine (14%) and plagioclase are poikilitically enclosed by large augite crystals (29%) and up to

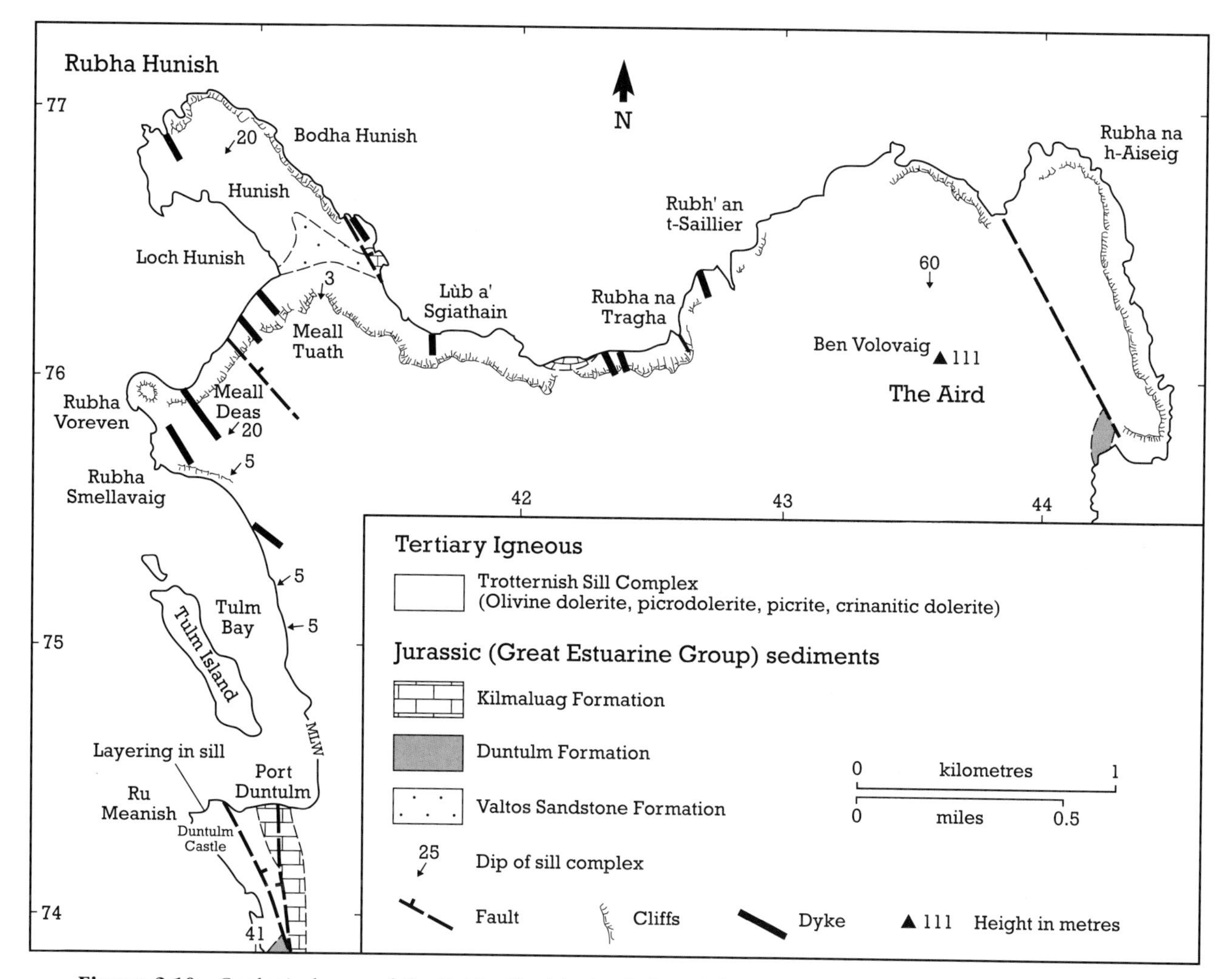

Figure 2.10 Geological map of the Rubha Hunish site (adapted from the British Geological Survey 'One-Inch' map, Northern Skye Sheet 80 and parts of 81, 90 and 91).

6% magnetite is present. Upwards, the proportion of plagioclase increases (59%), there is a decrease in olivine (7%) and augite (22%) and the habit of the pyroxene changes from large poikilitic (ophitic) crystals to small granules. No pronounced compositional variation ('cryptic layering') has been detected. Gibson and Jones consider that the layering developed *in situ*, influenced by strong thermal gradients across the contacts of the sills. It resembles layering found in other thick doleritic or fine-grained gabbroic intrusions, for example, the Camas Mòr dyke on Muck (Camas Mòr site). A thin dyke cutting the Duntulm sill exhibits banding on a centimetre scale parallel to its edges. The layering appears to be caused, in part, by alternations of bands rich in frond-like plagioclase and darker bands rich in augite; the exposure is reminiscent of the 'Mystery Dyke' exposed on the foreshore at Bornaskitaig (NG 372 715) which has been described by Drever (*in* Brown, 1969).

Immediately to the east of Rubha Voreven, the thickness of the sill complex increases substantially to over a hundred metres and is comprised of dolerite, picro-dolerite and picrite. Several distinct zones can be recognized in this exposure, banding being prominent in the lower third of the section, for which Anderson and Dunham (1966) give a comprehensive description. The basal contact of the sill is not seen here but is thought to occur just below the lowest exposures. Olivine dolerite at the base is at first succeeded upwards, through rapid gradation, by more olivine-rich dolerite (picro-dolerite) which forms the layered/banded portion. This is replaced at higher levels by olivine dolerite, the upper contact of which is missing. Simkin (1965, 1967) has described the modal mineral variation in the

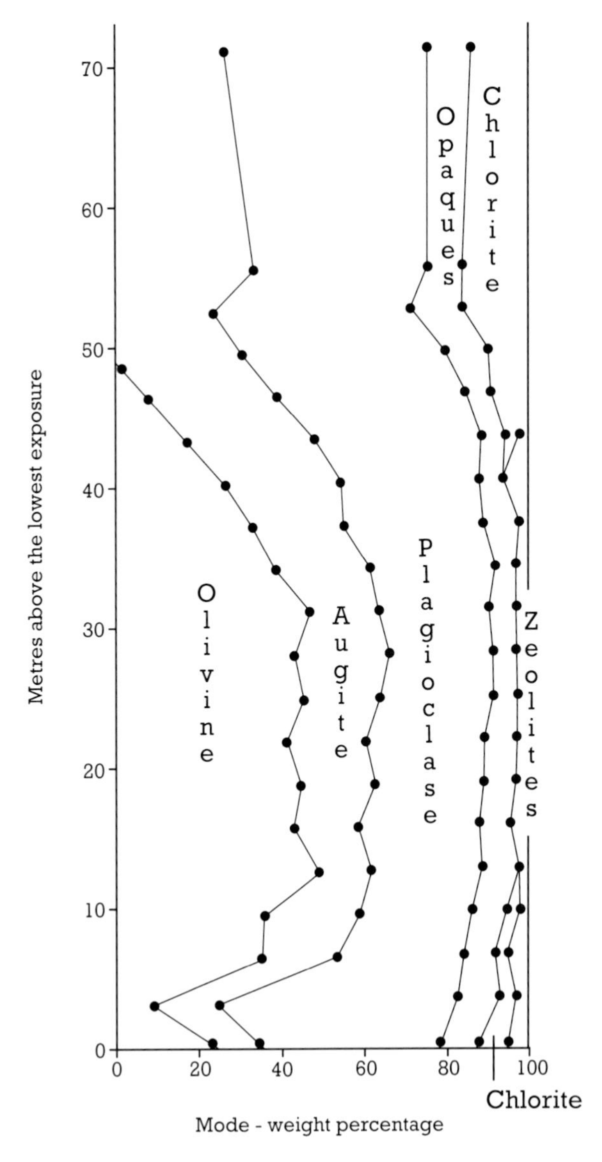

Figure 2.11 Modal variation in the Rubha Hunish sill (after Anderson and Dunham, 1966, fig. 15).

Trotternish sills, including the fine example at Meall Tuath (Anderson and Dunham, 1966, fig. 15; Fig. 2.11). Simkin attributed the distribution of olivine in the intrusions to the process of 'flowage differentiation' and on this hypothesis, some of the banding may therefore be a fluxion feature.

Much of the sill displays columnar, or prismatic jointing, fine examples occurring at Lub a' Sgiathain (NG 416 761). Here, spectacular fan jointing is observed which can be attributed to either intrusion into locally irregular bedding surfaces or unusual heat-loss conditions during cooling (Bell and Harris, 1986). At Rubha Hunish (NM 407 769), part of a sill leaf of crinanitic dolerite crops out beneath the olivine dolerite of Meall Tuath. In addition, patches of teschenitic dolerite occur within the sill at Ben Volovaig (NM 436 761) in association with veins and bands of pegmatitic dolerite. These carry rare olivine, a pale-brown idiomorphic clinopyroxene and large dendritic titanomagnetite. Zeolites are abundant, and large segregation veins, often sealing joint systems, are seen between Port Duntulm and Rubha Voreven.

Thin wedges of sediment separate the various sill leaves at Rubha Hunish and small pockets of highly baked, often fused, sediments are recorded from many localities within the complex throughout the Trotternish Peninsula (for example, at Ru Meanish, NM 408 743). The sediments are recognized as belonging to the Great Estuarine Group (Lower Ostrea Beds and Ostracod Limestones) by Anderson and Dunham (1966), now termed respectively the Duntulm and Kilmaluag formations (Harris and Hudson, 1980).

Interpretation

Thick, mildly alkaline, olivine dolerite sills are a prominent feature of the successions beneath the plateau lavas of the British Tertiary Volcanic Province. In northern Skye, as elsewhere, they seldom intrude the lavas and are virtually restricted to the underlying Mesozoic (or older) sediments. However, immediately south of the site, at Creag Sneosdal (NG 415 689), the uppermost leaf of the complex intrudes the basal tuffs and sediments of the lava succession. The sill complex is remarkable in that it remains at a fairly constant level at or near the base of the lavas and therefore behaves in a broadly transgressive manner towards the open folds in the Jurassic strata. This suggests that lithostatic loading by the lava pile was an important factor controlling sill emplacement (Anderson and Dunham, 1966) implying a fairly constant thickness of lavas over northern Skye when the sills were intruded.

The relative ages of the sill complex and the Skye Main Lava Series lavas are fairly closely defined: the sills cut tuffs at the base of the lavas and may intrude the lowermost lavas at Camas Ban on the south of Portree Harbour (NG 493 428) and at Oisgill Bay (NG 135 495) on the western edge of the lavas. Both the sill complex and the lavas are cut by numerous dykes of the NW-trending swarm. The broad geochemical similarity between the sills and lavas of the Skye Main Lava Series was noted by Anderson and Dunham (1966). This was confirmed in detail by Gibson (1988) who concluded that the sills and lavas were related, but not exactly contemporaneous; the detailed elemental and isotopic data indicated that the Trotternish sill magmas reached higher crustal levels than the Skye Main Lava Series magmas before being significantly contaminated by wall rocks or fractionating. Thus, the sills and Skye Main Lava Series lavas belong to the same general phase of magmatism but emplacement of the sills was somewhat after at least the basal Skye Main Lava Series flows.

The striking mineral variation, and hence rock types, within the sill complex can be largely explained in terms of modal variation in the amount of olivine (Fig. 2.11). Olivine enrichment in this sill complex and in many others is often explained in terms of gravitational settling of dense, early-formed olivine phenocrysts (for example, Walker, 1930). However, detailed investigations of the northern Skye sills by Simkin (1965, 1967) showed that such a process is an inadequate explanation, since the thicker parts of the sills do not display correspondingly thicker olivine-rich lower layers and the margins tend to be phenocryst poor. Using observations from the experimental studies of Bhattacharji and Smith (1964), Simkin applied their concept of flowage differentiation to explain the modal mineral variations in the Skye sills.

Conclusions

The north Skye sills result from the injection of mildly alkaline olivine basalt magmas into the Mesozoic sediments underlying the plateau lavas. The general level of sill intrusion conforms to the base of the lava pile but transgresses the folded sediments; the load imposed by the lava pile is thus seen as having exercised pressure control on the level to which the sill magmas rose.

The thick Trotternish sills show considerable internal variation in the proportions of their minerals, which is attributed to the combined effects of settling early-formed, dense minerals (olivine, possibly spinel) under gravity and their redistribution and concentration during flow of the intruding magma. The chemical composition of the less olivine-enriched facies of the sills is broadly comparable with alkali olivine basalts in the Skye Main Lava Series (SMLS). The sills must be of an age comparable with the SMLS, or slightly younger, since they appear to intrude only the lowermost part of the lavas and both sills and lavas are cut by many members of the north-west dyke swarm.

MARSCO AND MHEALL A' MHAOIL

Highlights

Segments of the narrow Marscoite Suite mixed-magma ring-dyke are excellently exposed on both sites. The Marsco site is the type site for the Marscoite Suite and provides key evidence from a variety of contacts for the relative ages of several of the superimposed granite intrusions forming the Western Red Hills centre.

Introduction

The Marsco and Mheall a' Mhaoil sites are dominated by various granite and felsite intrusions which belong to the Western Red Hills centre. The contact between the granitic intrusions and the Precambrian country rock is exposed in the Mheall a' Mhaoil site. However, both sites are of particular petrological importance because of the classic exposures of a composite, annular dyke-like intrusion within the granites containing ferrodiorite, felsite and hybrid rocks which are the product of magma mixing. These belong to the Marscoite Suite, the type locality for which is at Marsco. In addition, a gabbro forms the Marsco summit area.

Harker (1904) published the results of the first detailed survey of the central parts of Skye which included the Western Red Hills centre. He recognized the hybrid nature of some of the intrusions at Marsco, Glamaig and Mheall a' Mhaoil which he termed marscoites. Further detailed work in the area did not appear until that of Richey *et al.* (1946) followed by Wager and Vincent (1962), Wager *et al.* (1965) and Thompson (1969). These studies led to the subdivision of the Western Red Hills centre into at least 12 separate acid intrusions and provided a detailed knowledge of the petrology of the Marscoite Suite. Further work by Bell (1983) and Vogel *et al.* (1984) confirmed the hybrid character of the suite. A recent synthesis covering the sites is contained in the Skye field guide of Bell and Harris (1986) and the 1:50 000 compilation

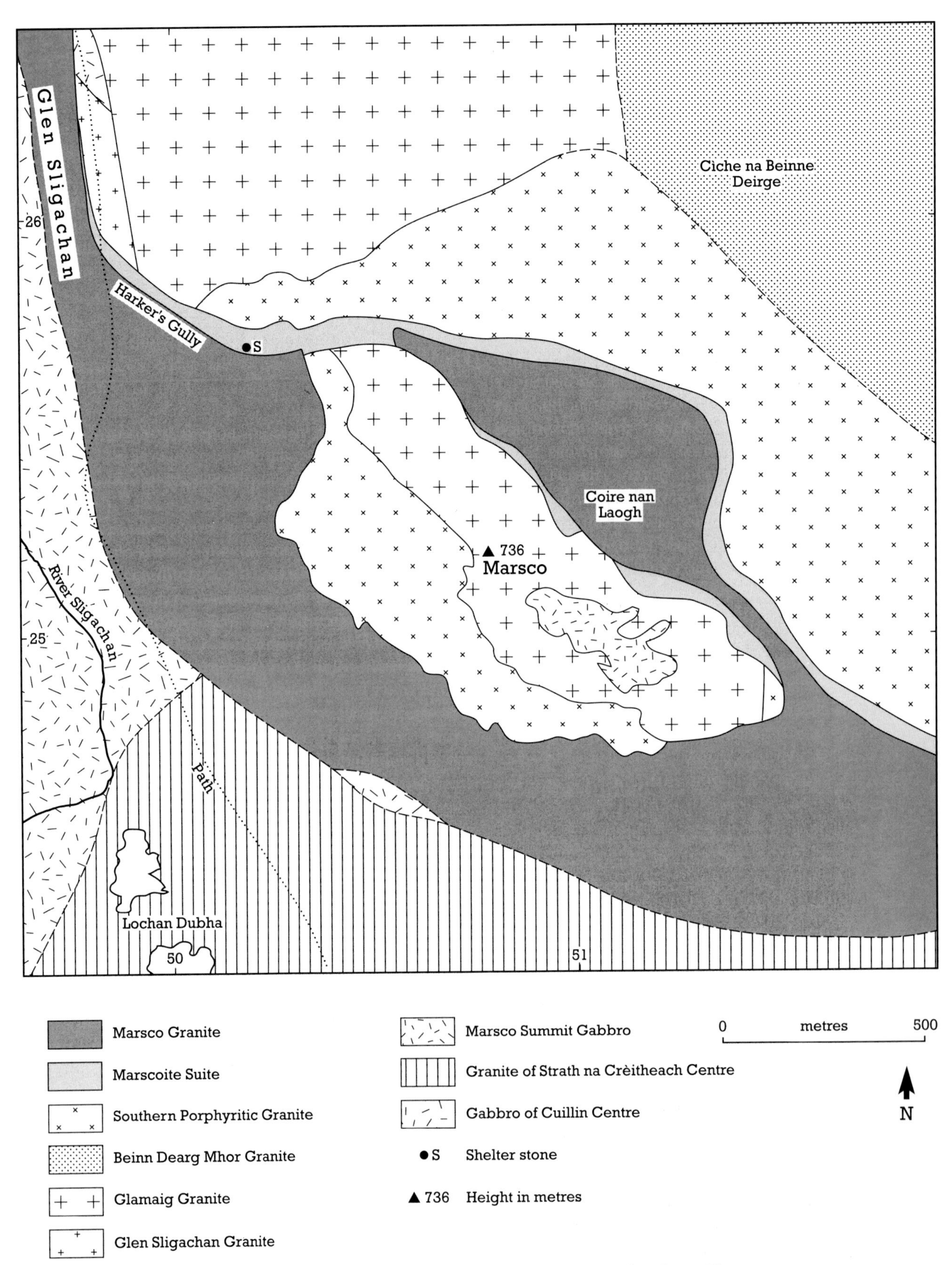

Figure 2.12 Geological map of the Marsco site (after Thompson, 1969, plate 18).

map of the Skye intrusive centres (published by the Open University) is particularly useful.

The term epigranite was introduced by Wager *et al.* (1965) to distinguish the high-level granites of the British Tertiary Volcanic Province from other, deeper-level granites such as those of Caledonian age in Scotland. The granites are characteristically found as stocks and ring-dykes emplaced into brittle crust and their chemical compositions indicate equilibration at low pressures, equivalent to one or two kilometres of cover. They are frequently drusy, or miarolitic, and the cavities indicate that a gas phase developed; they typically did not develop pegmatites. The rocks are, strictly speaking, alkali-feldspar granites, granites or alkali granites on Streckeisen's classification (1978). Although epigranite has been used widely in connection with the Skye granites, use of the term has not spread to the other centres in the Province and it is not used by Bell and Harris (1986). The term will not be used in this account.

Figure 2.13 Sketch of the west side of Marsco, showing the relationships of the Marscoite Suite rocks to the granite intrusions (reproduced from Sutherland, 1982, fig. 29.5; after Brown, 1969, figure 13).

Description – Marsco

Granites and felsites form several distinct intrusions within the site (Fig. 2.12; Wager *et al.*, 1965; Thompson, 1969). These either have steep-sided, wall-like contacts or else their margins dip at low angles, suggesting roof-like relationships. The contacts are sometimes marked by zones of crushing and are frequently fine-grained and chilled (cf. Thompson, 1969, plate 17B). This, together with veining of one acid intrusion by another has enabled their relative ages to be established; the intrusive sequence within the Western Red Hills centre is summarized by J.D. Bell (1976) and by Bell and Harris (1986).

The sequence of intrusions within the site is as follows:

(Youngest) Marsco Granite
Marscoite Suite
Ferrodiorite core
Marscoite
Porphyritic Felsite (Southern Porphyritic Felsite)
Southern Porphyritic Granite
Glamaig Granite
Marsco Summit Gabbro

The intrusive sequence may be determined at various points on Marsco; the age relationships, and the varied types of igneous contacts present, are shown in cartoon form in Fig. 2.13.

The Glamaig Granite, which is exposed around the summit and northern slopes of Marsco, is the earliest of the acid intrusions in the Western Red Hills centre (Wager *et al.*, 1965). The rock is characteristically medium-grained, dull-grey in colour and contains both biotite and amphibole. A persistent feature is the presence of small (0.5–5 cm diameter), partly digested mafic inclusions which constitute about 5% of the rock. Less commonly, somewhat larger (up to 0.3 m diameter) rounded and lobate inclusions of felsitic rock also occur. About 1 km to the north-west of the site, this granite is in contact with crushed gabbros of the Cuillin centre. On Marsco it is capped by the Marsco Summit Gabbro, which is fine-grained where it comes into contact with the granite. The granite, however, also 'net veins' the gabbro and is contaminated for some distance away from the contact by partly digested xenoliths of gabbro. On the basis of these relationships, the Glamaig Granite is considered to have been intruded before the Summit Gabbro was completely solidified. The Summit Gabbro is not cut by cone-sheets or dykes and is considered to be significantly younger than the Cuillin centre gabbros (Thompson, 1969).

The Southern Porphyritic Granite is preserved as a steep intrusion on the northern slopes of Marsco whereas, in the Marsco summit area, it forms a flat-lying, wedge-shaped body below and chilled against the Glamaig Granite; it overlies the younger Marsco Granite which intrudes it. Both the upper and lower contacts of this body dip at low angles and are considered to be roof-like. The intrusion is a leucocratic granophyre carrying quartz and alkali-feldspar phenocrysts.

The rocks of the Marscoite Suite intrude the Glamaig and Southern Porphyritic granites as a discontinuous ring-dyke (Figs 2.12 and 2.13). In the site, the sheet is up to 50 m thick and is steeply inclined to the south. It normally consists of outer zones of porphyritic felsite (the Southern Porphyritic Felsite) and a core of ferrodiorite which are separated by zones of hybrid rock (marscoite). The Marscoite Suite rocks are distinctly younger than the Glamaig Granite with which they are in sharp, chilled contact in Glen Sligachan just north-west of the site. The contact with the Southern Porphyritic Granite, seen just north of Harker's Gully, is again sharp but it is unchilled, suggesting that the marscoite is only slightly younger than the Southern Porphyritic Granite. The Marsco Granite intrudes the Marscoite Suite and breaks the symmetry of the latter in Harker's Gully, where it is in gradational contact with the ferrodiorite of the centre of the Marscoite Suite ring-dyke. The relationships of the several intrusions in and around Harker's Gully are shown diagrammatically in Fig. 2.13. Within the Marscoite Suite intrusion, near and above the Shelter Stone ('S' in Fig. 2.12), exposures on the east of the gully show marscoite in sharp, chilled contact with porphyritic felsite but the bulbous, crenulated contact indicates that the felsite was not consolidated when intruded by the marscoite as such a contact indicates liquid–liquid relationships. Inwards, over about 10 m, there is a complete gradation between the marscoite and the central ferrodiorite which was intruded and extensively mixed with marscoite before the former had consolidated.

The porphyritic felsite contains quartz and alkali-feldspar phenocrysts set in a fine-grained matrix; the phenocrysts are similar to those in the Southern Porphyritic Granite and the two intrusions are therefore considered to be related.

The marscoite is a fine-grained, grey rock containing xenocrysts of rounded andesine, orthoclase with embayed margins ('fingerprint' textures) and quartz rimmed by augite or amphibole. These minerals can be identified as phenocrysts in both the basic and acid rocks in the suite, suggesting that marscoite is a hybrid produced by the mingling of these contrasted types.

The ferrodiorite is variably porphyritic and non-porphyritic, having a fairly complex mineralogy consisting of andesine (as phenocrysts in the porphyritic variety), alkali-feldspar, quartz, hornblende, clinopyroxene (including inverted pigeonite), opaques, biotite and olivine (Wager and Vincent, 1962). Thompson (1969) noted the presence of quartz xenocrysts and concluded that the ferrodiorite, like marscoite, contains a component of felsite magma and that xenocrysts of alkali-feldspar have been dissolved during the slow cooling of the coarse ferrodiorite.

In the lower part of the Harker's Gully section, the later Marsco Granite has intruded along the southern margin of the ring-dyke and cut out the porphyritic felsite and marscoite members. However, the full symmetrical sequence is preserved above 500 m elevation. Within the ferrodiorites of the Marscoite Suite in Harker's Gully, below the level of the Shelter Stone, there are blocks of banded quartz–oligoclase–biotite gneiss of possible Laxfordian age. Such occurrences are also found elsewhere, for example in Coire nam Bruadaran (NM 523 254), and are important in that they provide evidence as to the nature of the deep basement beneath central Skye.

The Marsco Granite is the youngest intrusion in the site. As mentioned, it has an unchilled, completely gradational contact with the ferrodiorite of the Marsoite Suite, but develops sharp, chilled roof-like contacts against the Southern Porphyritic Granite and Glamaig Granite (Thompson, 1969). The detailed mineralogy has been described by Thompson; of particular interest is the occurrence of fayalitic olivine and hedenbergite clinopyroxene as well as calciferous amphibole, indicating crystallization in an

Figure 2.14 Geological map of the Mheall a' Mhaoil site (after Gass and Thorpe, 1976, fig. 6).

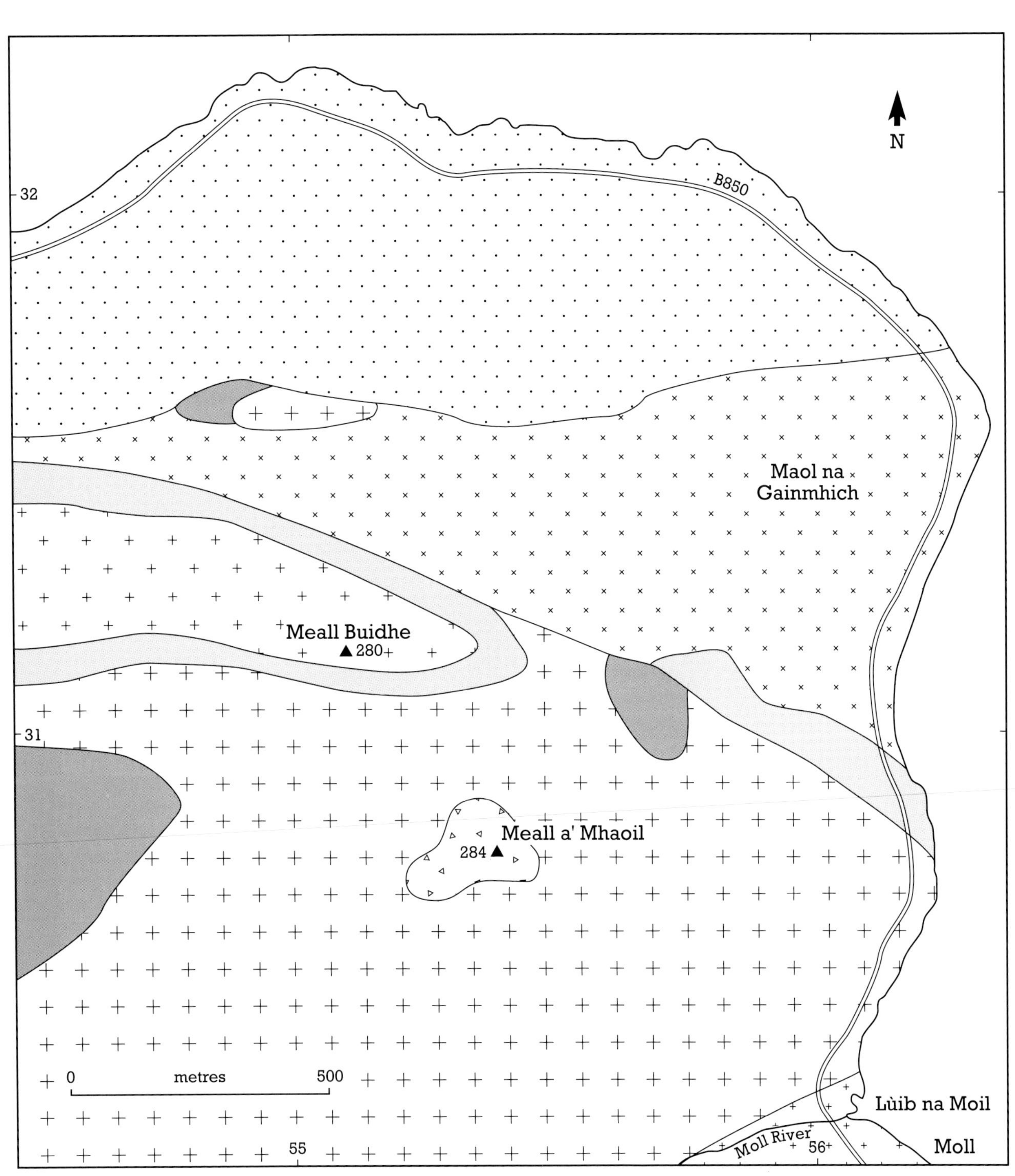

N
B850
32
31
Maol na Gainmhich
Meall Buidhe
▲280
Meall a' Mhaoil
284 ▲
0
metres
500
55
56
Lùib na Moil
Moll River
Moll
Granites - Western Red Hills
Meall Buidhe Granite
Marscoite/Glamaigite Suite
Northern Porphyritic Felsite
Loch Ainort Granite
Maol na Gainmhich Granite
Volcanic agglomerates
Tertiary basalts
Torridonian sandstone
▲ 280 Height in metres

anhydrous magma with a high Fe/Mg ratio, at initially high-temperature.

Description – Mheall a' Mhaoil

Within this site both early and late intrusions belonging to the Western Red Hills centre are exposed (Fig. 2.14). These granophyres and felsites have arcuate outcrops with steep, outward dipping contacts and may be presumed to be ring-dyke in form. The principal intrusions within the site are:

(Late) Meall Buidhe Granite
Northern Porphyritic Felsite
Marscoite – Glamaigite Hybrid Suite
Loch Ainort Granite
(Early) Maol na Gainmhich Granite

These intrusions are closely related to those of the Marsco site in the southern part of the Western Red Hills centre but differ in petrographic detail. In addition, subsidiary masses of vent agglomerate, crushed gabbro and basaltic lavas are preserved as screens within and between members of the complex.

The earliest Maol na Gainmhich Granite intrusion is in contact with Torridonian sandstones at Maol na Gainmhich and farther to the west. The contact is sharp and irregular in detail; a number of xenolithic masses of sandstone can be seen in granite exposures on the shore below the road. This is a coarse-grained, alkali granite containing arfvedsonite and a potassium-rich feldspar. It was formed early on in the evolution of the Western Red Hills centre although there is no direct evidence for its age relative to the early Glamaig Granite in the Marsco site. To the south, the granite is intruded by the later Northern Porphyritic Felsite, Meall Buidhe Granite and the Marsco–Glamaigite Hybrid Suite.

The intrusion of the large Loch Ainort Granite succeeded the Maol na Gainmhich Granite. In hand specimen, the Loch Ainort Granite is coarse grained with a blue-green to pink weathered surface. Its conspicuous, zoned, sodic feldspar phenocrysts are fringed by granophyric intergrowths; the principal mafic phases, fayalite and ferro-hedenbergite, serve to distinguish it from the similar Beinn Dearg Mhor Granite exposed to the west. A crushed contact zone between the granites is exposed near Moll, and along the valley of the Moll River between Druim na Cleochd and Leathad Chrithinn to the south of the site.

The Northern Porphyritic Felsite crops out to the south of Meall Buidhe. It contains phenocrysts of quartz and potash feldspar with rarer pyroxene and iron oxide. Apart from the significantly greater volume of phenocrysts, this felsite is mineralogically very similar to the Southern Porphyritic Felsite in the Marsco site. Inclusions of fine-grained basalt lavas and vent agglomerate are found in the felsite to the west and north-east of Mheall a' Mhaoil. The basalts are undoubtedly derived from the earlier lavas, while the agglomerates may be remnants of early vents between and around which the granite was emplaced.

Following these granitic intrusions and before the emplacement of the Meall Buidhe Granite, the Marscoite–Glamaigite Suite of hybrid rocks was emplaced as a discontinuous, inclined sheet in the form of an incomplete ring-dyke. These composite hybrid intrusions are essentially the same as the Marscoite Suite rocks described from the Marsco type locality, although there are some differences: instead of the ferrodiorite central member present at Marsco, the marscoite passes inwards into a rock with streaky, light-coloured areas in a darker matrix, or else shows acid net-veining of the darker component. These variants give way to a rock described as xenolithic granophyre by Harker (1904) but which Wager *et al.* (1965) termed glamaigite. Glamaigite consists of rounded, sometimes globular dark dioritic areas up to one centimetre in diameter set in a medium-grained microgranitic or granophyric matrix. The centre of the intrusion is a more homogeneous rock termed dioritic glamaigite (Wager *et al.*, 1965). Three intrusions of the Marscoite–Glamaigite Suite were described by Wager *et al.* (1965); two of these areas of hybrid suite rocks are within the site, on the coast at Moll and on Meall Buidhe (Fig. 2.14).

The Meall Buidhe hybrid rocks are exposed on Meall Buidhe (NM 551 312) and extend to Abhainn Torra-mhichaig west of the A850 Portree–Broadford road to the west of the site. The Moll Shore intrusion is a small, steeply inclined dyke-like mass on the coast and is well exposed in the road cuttings to the south of Maol na Gainmich, where the contact with the Northern Porphyritic Felsite shows signs of crushing.

Some of the hybrid rocks in the Mheall a' Mhaoil site are very similar to the marscoites at the Marsco site and they contain xenocrysts of quartz, sodic plagioclase and alkali feldspar. However, in this site they are lighter coloured and coarser grained, and grade into a net-veined, heterogeneous mottled rock termed glamaigite. Glamaigite is chemically similar to, and contains xenoliths of, marscoite. The characteristic mottled appearance is enhanced by weathering. Both the typically rounded dark patches and the lighter-coloured matrix in which they lie contain identical xenocrysts of andesine, quartz and orthoclase and both components are marscoitic hybrid rocks with similar chemical characteristics. Rounded xenoliths and inclusions of hawaiitic affinity are also found within the glamaigites.

Detailed traverses across the hybrid bodies at Meall Buidhe and Moll are described by Wager *et al.* (1965). The Meall Buidhe traverse, however, is interrupted centrally by the intrusion of a later granite called the Meall Buidhe Granite. Its north-eastern and south-eastern contacts with glamaigite are gradational where contamination of the granite with basic material is observed. Contamination has resulted in the presence of plagioclase phenocrysts mantled by alkali feldspar. The matrix contains pyroxene and interstitial hornblende in a micrograpnitic, rather than a granophyric groundmass. The granite, in fact, bears a striking resemblance to the Marsco Granite, but analyses by Wager *et al.* show it to be more basic than either the Marsco Granite or any of the other granites in the Western Red Hills centre.

Wager *et al.* (1965) have recorded the following traverse across the summit of Meall Buidhe through a steep composite dyke containing the Marscoite–Glamaigite Suite.

North
1. Maol na Gainmhich Granite
2. Chilled marscoite
3. Glamaigite (streaky and net-veined in the north but more patchy in the south)
4. Transitional zone between glamaigite and Meall Buidhe Granite
5. Meall Buidhe Granite – basic in places
6. Gradational Meall Buidhe Granite – glamaigite boundary
7. Glamaigite (on Meall Buidhe summit)
8. Marscoite
9 Chilled marscoite

South
10. Northern Porphyritic Felsite

The contacts between the intrusive members appear to be very steep or vertical, in contrast to the composite hybrid intrusion on Marsco where the contacts are inclined outwards relative to the centre of the ring-shaped intrusion. This may signify a deeper level of erosion of the marscoite ring-dyke at this site when compared with Marsco.

Interpretation

Richey's classic model for the emplacement of high-level granites involved subsidence of a central block of country rock bounded by steep, outward-dipping, arcuate fractures, terminated upwards by a cross-fracture (Richey, 1928, Fig. 7). The model proposed that magma subsequently moved both up the arcuate fractures to form ring-dyke intrusions with steep, wall-like contacts, and across the bounding cross-fracture to produce a roof-like contact with the country rock. The Marsco site demonstrates both wall- and roof-like contacts (Fig. 2.13). The former are well developed on either side of the Marscoite Suite which is a classic, if thin, example of a ring-dyke. Roof-like contacts are extremely well developed on the south-west face of Marsco above and below the Southern Porphyritic Granite. Richey's model involved the foundering of a detached block several kilometres in diameter, rather than piecemeal stoping. In this respect, it is significant that the contacts show remarkably few xenoliths either of country rock sediments (Mheall a' Mhaoill site) or earlier granite (Marsco site). Thus, the sites provide evidence strongly supportive of Richey's model, although this evidence does not prove that individual intrusions, with the exception of the Marscoite Suite, are true ring-dykes rather than a suite of nested granite stocks which become younger towards the centre of the complex.

Throughout the British Tertiary Volcanic Province there is compelling evidence that acid and basic magmas:

1. coexisted during much of the life of the central complexes and
2. that they were frequently intruded more or less contemporaneously.

The Marscoite Suite of Marsco and the Marscoite–Glamaigite Suite of Maol na Gainmhich furnish excellent examples of both magma mixing and hybridization, and of the nearly contemporaneous intrusion of contrasting acid and more basic magmas. At Marsco, the Southern Porphyritic Felsite, which contains distinctive quartz and alkali-feldspar phenocrysts, was intruded first and the ferrodiorite, with its equally distinctive plagioclase phenocrysts, came last to form the core of the Marscoite Suite ring-dyke. Between these two contrasted rocks there is the marscoite which contains as xenocrysts the same minerals that form the phenocrysts in its neighbours and, furthermore, chemical analysis shows that the bulk composition of the marscoite is intermediate between the compositions of the adjoining rocks. Thus, the petrographic and chemical evidence shows that the marscoite is a mixed, hybrid rock (Wager *et al.*, 1965; Vogel *et al.*, 1984) and Bell (1983) has demonstrated that it consists of ferrodiorite and porphyritic felsite mixed in the proportions 74:26. Further evidence for mixing is provided in hand specimen by the emulsion-textured glamaigite of Mheall a' Mhaoil.

The contact relationships seen within the Marscoite Suite, and between these rocks and both the Southern Porphyritic Granite and the Marsco Granite, suggest that all were intruded in close succession: the Southern Porphyritic Felsite is in sharp but unchilled contact with the Southern Porphyritic Granite; within the Marscoite Suite the bulbous, crenulated contact between the Southern Porphyritic Felsite and the marscoite is interpreted to show that both were liquid at the same time; and the marscoite is completely gradational towards the ferrodiorite. On the ring-dyke inner contact, a similar gradation from ferrodiorite towards the Marsco Granite means that the granite intruded shortly after the Marscoite Suite. It is also evident (Fig. 2.13) that the granite exploited the ring-dyke structure during emplacement.

The Marscoite Suite thus abounds with igneous intrusive contacts showing a spectrum from those which are sharp and chilled to others which are completely gradational. From careful consideration of the contacts, it appears that the Marsco Summit Gabbro and the Glamaig Granite form an early group of intrusions and that the Southern Porphyritic Granite, the Marscoite Suite and the Marsco Granite are a later suite. Sufficient time must have elapsed between the emplacement of the Southern Porphyritic Granite and the Marsco Granite for the latter to develop a chilled contact.

Conclusions

The field relationships of the granites, Marscoite Suite intrusions and country rocks provide evidence which supports the classic ring-fracture, block subsidence model for granite emplacement proposed by Richey (1928). The sites also contain classic examples of rocks formed by magma mixing; the marscoite member of the Marscoite Suite resulted from an approximate 3:1 mixing between porphyritic magmas of ferrodioritic and granitic compositions. In the ferrodiorite end-member there is subtle evidence for the incorporation of a small amount of porphyritic acid material, and the common occurrence of small, basic inclusions in the early Glamaig Granite suggests that some basic magma was mixed in with acid magma prior to emplacement of this intrusion.

COIRE UAIGNEICH

Highlights

The distinctive mineralogy of the Coire Uaigneich granophyre indicates that it crystallized under shallow, near-surface conditions and its bulk composition shows that it received a substantial contribution from the Torridonian sediments. It is the only sizeable granitic intrusion belonging to the Cuillin centre and it is older than the other Skye granites. Basalt lavas and calcareous country rocks adjoining the Cuillin gabbros have developed distinctive high-temperature mineralogies.

Introduction

This site partly demonstrates the nature of the south-east margin of the Cuillin basic/ultrabasic centre and its effects upon the country rock and, as such, should be considered in conjunction with the Cuillin Hills site. The site contains Lower Jurassic sediments and overlying Palaeocene lavas which show the effects of thermal metamorphism and structural disturbance from the emplacement of the adjacent Cuillin layered eucrites (Fig. 2.15). In addition, the site also encompasses the Coire Uaigneich Granophyre, the only granite associated with the Cuillin centre (Fig. 2.16). This intrusion has been of significant petrological importance in the study of the origin and emplacement of the Skye granites.

The gabbros of the northern part of the Blaven Range belong to the Cuillin centre and were

Figure 2.15 Gabbros of the Cuillin centre form the Bla Bheinn ridge in the background, Tertiary lavas on prominent ridge in top left, and Mesozoic sediments reinforced by Tertiary sill(s) in left foreground. Loch Slapin in foreground. Coire Uaigneich site, Skye. (Photo: C.H. Emeleus.)

briefly considered in a study of granites and associated rocks in the eastern part of the Western Red Hills centre by Bell (1966). The effects of the Cuillin centre on the surrounding country rocks around Strathaird were the subject of a detailed investigation by Almond (1960, 1964). Wager *et al.* (1953) and Brown (1963) have studied the origins of the Coire Uaigneich Granophyre. The contentious issue of granite petrogenesis on Skye is discussed in detail by Gass and Thorpe (1976) in the Open University's case study on the Tertiary igneous rocks of Skye and these problems were further considered by Dickin and Exley (1981). The oldest rocks within the site are Jurassic sediments. These are overlain by Palaeocene basalt lavas and both were intruded successively by layered eucrites, the Coire Uaigneich Granophyre, basaltic cone-sheets and numerous NW-trending dykes.

Description

The site comprises three principal units (in order of decreasing age):

1. The country rocks
2. Layered gabbros and other mafic intrusions of the Cuillins
3. The Coire Uaigneich Granophyre

i. Country rock

In Coire Uaigneich and along the course of Abhainn nan Leach to the south-west, Lower Jurassic strata overlain by Palaeocene lavas are in contact with the gabbros of the Cuillin centre. The emplacement of the latter has severely hornfelsed and faulted Jurassic shales, sandstones and thin limestones. Minor folding has also occurred immediately adjacent to the basic intrusion and a small anticlinal structure is present. Calc-silicate assemblages are developed in some of the dolomitic limestones and blocks of high-grade, thermally metamorphosed yellow, serpentinous marbles containing unusual calc-silicate minerals (for instance, spurrite and rankinite; Wyatt, 1952) are locally developed. Rheomorphism of the more leucocratic sediments has occurred along the contact; however, the most spectacular examples of this phenomenon

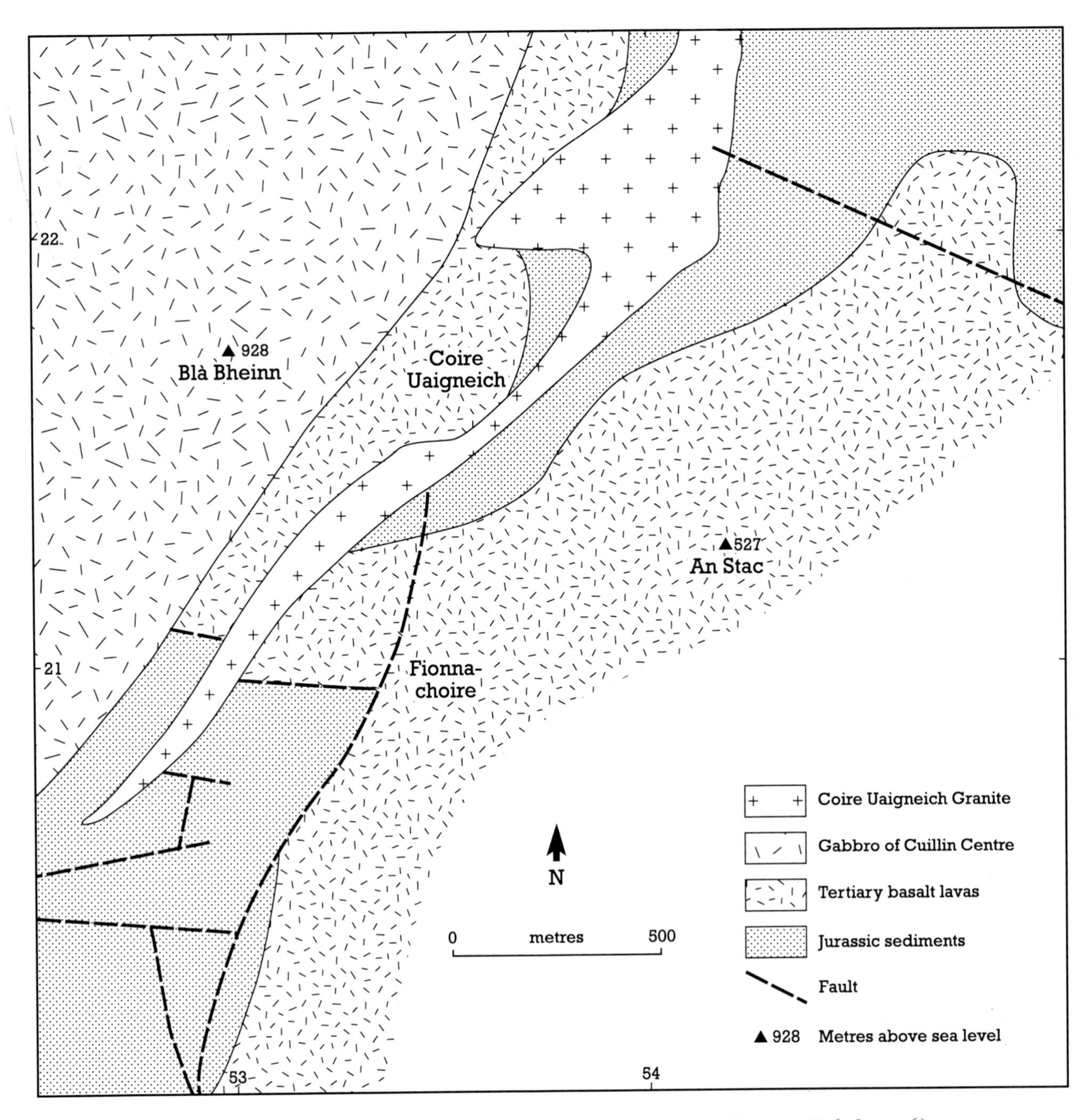

Figure 2.16 Geological map of the Coire Uaigneich site (after Gass and Thorpe, 1976, figure 6).

are found 2 km to the south-west of the site at Camasunary (NG 512 192).

The Palaeocene alkali olivine basaltic lavas are also hornfelsed, and Almond (1960, 1964) has recognized three zones of alteration.

a. Inner Zone: Granular hornfelses consisting of plagioclase, clino- and ortho-pyroxene with or without olivine – formed by complete, or partial, recrystallization. Amygdaloidal structures commonly retained as aggregates or calcic plagioclase.

b. Middle Zone: Actinolite–albite–epidote–chlorite assemblages.

c. Outer Zone: Hydrothermal alterations only.

Figure 2.17 East face of Bla Bheinn formed by gabbros cut by later dykes (weathering to give notches) and cone-sheets (forming terraces on faces). Pale rocks at lower levels are the Coire Uaigneich Granite and Mesozoic sediments against the gabbros. Coire Uaigneich site, Skye. (Photo: A.P. McKirdy.)

The lavas of the Coire Uaigneich site lie in zones a and b; the lava pile forming An Stac, though, lies totally within zone b.

ii. Cuillin layered gabbros

Layered gabbros and eucritic gabbros of the Garbh Bheinn–Blaven (Fig. 2.17) range dip towards the centre of the Cuillin intrusion. They form the high serrated ridge east of Strath na Creitheach and the Cuillin Hills. These rocks can be correlated to the layered gabbros and eucrites belonging to the Outer Layered Series of the Cuillin Hills centre (Bell, 1966) and, as such, are discussed within the Cuillin Hills site report below. Essentially, the gabbro is a layered, dark-coloured rock of variable grain size containing bytownite, clinopyroxene and opaque oxides with small amounts of olivine and orthopyroxene. In addition, the gabbros also contain lensoid inclusions of hornfels and ultrabasic rock which parallel the layered structures.

iii. Coire Uaigneich Granophyre

The site has been chosen to include part of the only granitic intrusion associated with the Cuillin centre. The granophyre is exposed in Coire Uaigneich and on the col at Fionna-choire as part of a narrow, discontinuous, ribbon-like intrusion dipping steeply to the south-east at 70–80°. The granophyre is light weathering and contains fine- and coarse-grained facies bearing needles of hypersthene, together with quartz paramorphs after tridymite indicative of crystallization under low-pressure conditions (Wager *et al.*, 1953; Brown, 1963). Outside the limits of the site, on a wave-cut platform at Abhainn Camas Fhionnairigh (NG 509 187), the clean exposures of Coire Uaigneich Granophyre reveal partly digested xenoliths of sandstone.

The margins of the granophyre are chilled against Jurassic country rocks which have been deformed by the intrusion. Eucritic and basaltic xenoliths also occur along the margins of the granophyre showing that it is younger than the emplacement of the Cuillin centre. A close association with the Cuillin centre and a pre-Western Red Hills age are proved by the presence of numerous dykes and cone-sheets, related to the Cuillin centre, cutting the granophyre. In Coire Uaigneich, gabbro xenoliths are enclosed by the granophyre, but on the shore exposures there is a suggestion that the granophyre may have been remelted by the adjoining gabbros since acid net-veining is common.

Interpretation

The very high-grade thermal metamorphism of the adjoining lavas, the presence of unusual calc-silicate hornfelses in the altered Jurassic strata and the occurrence of extensive net-veining and other rheomorphic phenomena close to the site at Camasunary, all provide unequivocal evidence that the Cuillin gabbros were a major heat source on emplacement. This view is compatible with *in situ* crystallization of mafic magmas during the formation of the layered rocks.

Both the Cuillin gabbros and the Coire Uaigneich Granophyre appear to have caused varying degrees of folding and faulting in the country rocks. In the case of the granophyre, the evidence for forcible emplacement contrasts strongly with the prevailing view that the majority of the Skye granites and granophyres were essentially passively emplaced by ring-dyke and block subsidence mechanisms. However, even a cursory examination of the existing maps suggests that this view may be an oversimplification, since disturbance of the sedimentary envelopes to the granites of both the Western Red Hills and the Eastern Red Hills centres may be verified at a number of localities (for example, on the north side of Glamaig, north-east of Mheall a' Mhaoil; on the south and west of Beinn na Caillich, west of Kilchrist). The emplacement of the Skye granites should provide a fruitful field for further research.

Much of the importance of the site lies in the fact that the Coire Uaigneich Granophyre has played a significant role in the understanding of the origin of the Skye granites. Attention was focused on this granite because of the occurrence of partly digested sandstone xenoliths (just outside the site), pointing to a possible origin from the anatexis of Torridonian basement rocks. Wager *et al.* (1953) studied the intrusion in detail, concluding that the granophyre represents partially melted Torridonian sandstone. This concept was supported by Brown (1963) on the evidence of experimental melting studies. Bell (1966) calculated that the heat output from the dense basic/ultrabasic magma body beneath much of southern Skye (see Cuillin Hills) would have been adequate to produce all of the granite magmas of Skye by melting of Lewisian/Torridonian basement rocks. Since these studies, ideas on the origin of the Skye granites have advanced significantly and a detailed discussion on the continuing controversy surrounding the origin of the Skye granites is contained in the Open University's igneous case study on Skye (Gass and Thorpe, 1976) and in Thompson (1982). More recently, in a detailed study on the Coire Uaigneich Granophyre, using trace-element and isotope data, Dickin and Exley (1981) suggested that the granophyre originated from the mixing of two contrasting magmas. These magmas were envisaged to be:

1. partially melted Torridonian sediments and
2. an acid differentiate of basic (Cuillin) magma, which mixed in the ratio 2:1.

The Coire Uagneich Granophyre extends around the south-east side of the Cuillin layered mafic rocks and is obviously structurally related to the Cuillin centre. Since the majority of Skye granites are significantly younger than this centre, it is of some importance to determine its relative age as closely as possible. Probably the most conclusive evidence is provided by the cone-sheets which cut the granophyre and belong to the same suite which cuts the Cuillin layered gabbros, focusing on the Cuillin centre. None of these cone-sheets cuts the other Skye granites and rather, the reverse is true. Thus, the Coire Uaigneich Granophyre is significantly earlier than the Strath na Creitheach and Western Red Hills granites. There is also clear evidence that the granophyre post-dates the Cuillin gabbros of the Blaven area (Fig. 2.17) as xenoliths of the gabbro have been recorded in the granophyre. More importantly however, in Coire Uaigneich the granophyre lies well within the zone of high-grade thermal metamorphism developed in basalts around the gabbros but shows no sign of metamorphism itself. Had the acid intrusion pre-dated the gabbros it would have exhibited a thermal overprint and the fine-grained chilled contact, with its distinctive quartz paramorphs after tridymite, would in all probability have been obliterated. The presence of the quartz paramorphs after tridymite indicate a depth of cover of about 1 km (Brown, 1963). Nevertheless, exposures on the shore west of Abhainn Camas Fhionnairigh suggest that the xenolithic granophyre at this locality has been rheomorphically melted by the nearby gabbros, which may exhibit acid net-veining; thus, the unit mapped as Coire Uaigneich Granophyre may in fact be two separate acid bodies, one pre-gabbro, the other post-dating it.

Conclusions

There is abundant evidence in the site that the Cuillin gabbros were intruded at high temperatures. The Tertiary olivine basalt lavas have been altered to high-temperature olivine–pyroxene–plagioclase granulitic hornfelses adjacent to the contact and to rocks showing wholesale hydrous alteration at a greater distance from the contact. In the impure (sometimes dolomitic) Jurassic limestones there is alteration to high-grade calc-silicate mineral assemblages. The arcuate outcrop of the Coire Uaigneich Granophyre around the south-east of the Cuillin gabbros and intrusion of the granophyre by basic cone-sheets which focus on, and also cut, the Cuillin gabbros, shows that this granitic intrusion is structurally related to the Cuillin centre and of a significantly earlier date than the majority of the Skye/Tertiary granites. Geochemical studies show that it contains major components from melted Torridonian sandstones and from fractionated basaltic magma. The absence of any evidence of thermal overprint in the granophyre, which intrudes the severely thermally metamorphosed basalts, shows that it post-dates the Cuillin gabbros. The presence of quartz paramorphs after tridymite provides mineralogical evidence that it was intruded at 1 km or less beneath the Tertiary land surface.

BEINN AN DUBHAICH

Highlights

Spectacular calc-silicate and skarn mineral assemblages are developed in carbonate sediments at the granite contacts, along with a limited development of alkali granite. The mineralogy of the granite provides a link between the minerals of rhyolite lavas and plutonic granites. Striking deformation of the carbonate sediments and pre-granite Tertiary dykes has occurred over a limited area close to the western end of the granite intrusion.

Introduction

The site contains the major Beinn an Dubhaich granite intrusion which has been emplaced into Cambro-Ordovician limestones and dolomites overlain by Mesozoic sediments and cut by pre-granite Tertiary basaltic dykes (Fig. 2.18). The granite has caused thermal metamorphic and metasomatic alteration of the carbonate sediments and there has been a small, locally spectacular amount of structural disturbance of the sediments and dykes. The Beinn an Dubhaich Granite, with its aureole, is one of the most intensively studied small intrusions in Britain. Following the regional investigations of Harker (1904) and Richey (1932), Tilley (1951) and Hoersch (1981) investigated the contact phenomena, King (1960), Whitten (1961) and Hoersch (1979) worked on the structure of the granite, and Raybould (1973) showed the granite to be a multiple intrusion. Stewart (1965) summarized much of the earlier work and a review of current ideas is contained in the field guide by Bell and Harris (1986). The granite mineralogy was also intensively studied (Tuttle and Keith, 1954; Tuttle and Bowen, 1958).

Description

The Beinn an Dubhaich Granite is the dominant feature of the site and is one of a group of intrusions which comprise an outer granite mass in the Eastern Red Hills centre. This granite mass extends from Beinn na Cro in the west to Beinn an Dubhaich and the Allt Fearna Granite to the south and east of Beinn na Caillich respectively (Bell and Harris, 1986). The contacts of the Beinn an Dubhaich Granite with the Cambro-Ordovician sedimentary country rock are well-exposed; the sediments have been folded into a broad anticline (the Broadford anticline; Bailey, 1954) and the granite intrusion is approximately coincident with the anticlinal axis. The Cambro-Ordovician strata have also suffered earlier Caledonian tectonism and to the north-east, on the Beinn Suardal summit area on the edge of the site, Torridonian sediments have been thrust over them. To the south, the Lower Palaeozoic and Precambrian sediments are unconformably over-

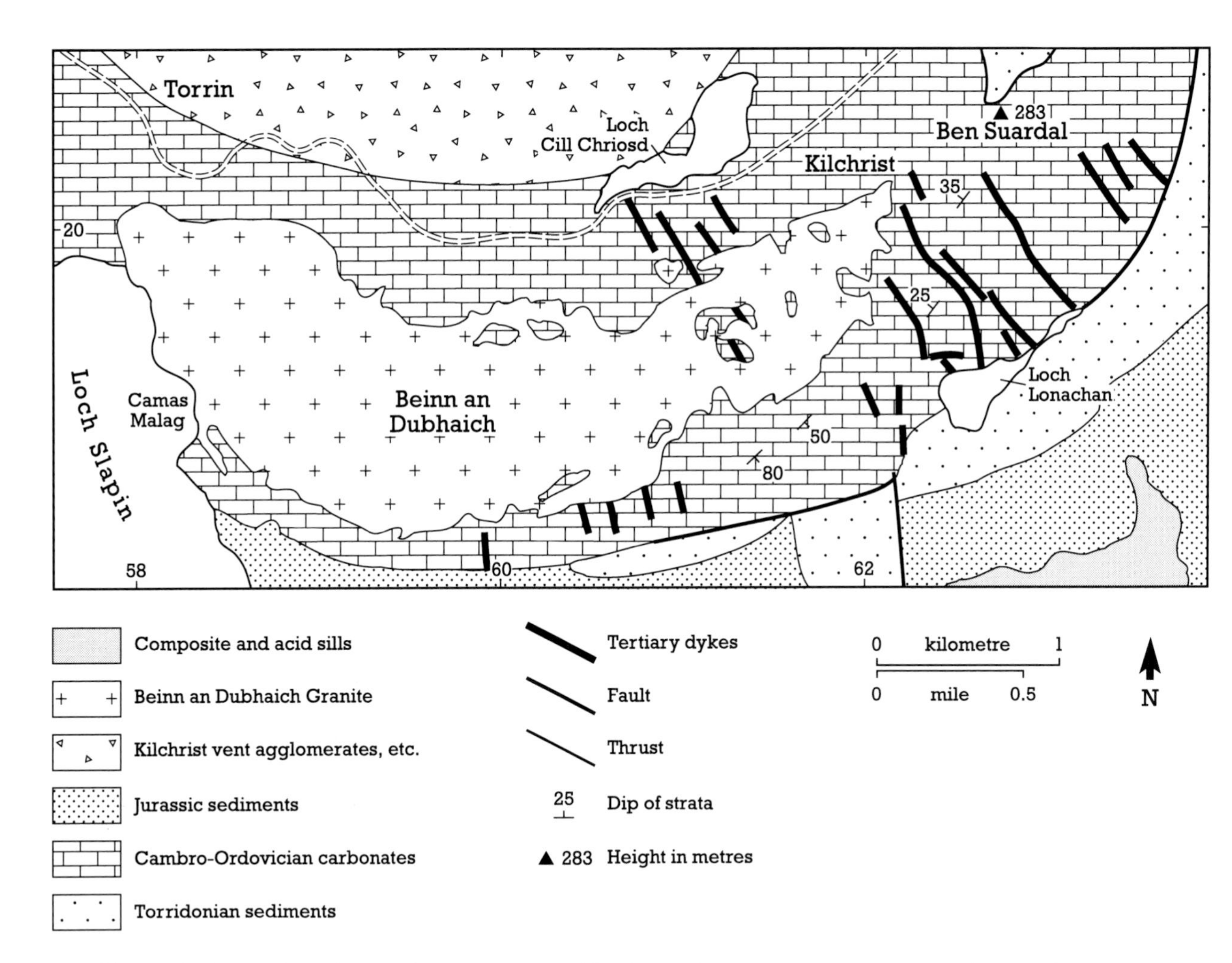

Figure 2.18 Geological map of the Beinn an Dubhaich site (after Gass and Thorpe, 1976, figure 6).

lain by Mesozoic (Triassic–Lower Jurassic) strata.

At its western end, and on the headland at the southern end of Camas Malag (NG 583 189), the granite is in steep contact with the Cambro-Ordovician sediments. However, to the east of the Beinn an Dubhaich summit area, and particularly in the vicinity of Kilchrist (NG 616 201), the relationships between the granite and metamorphosed carbonate sediments are complicated; individual contacts exposed in old quarries and pits often appear vertical, suggesting that many steep-sided rafts of limestone lie within the granite and that the granite–limestone contact is extremely irregular. The complex pattern of outcrops has resulted in numerous differing interpretations as to the form of the granite. These are discussed below.

The granite is a typical high-level Tertiary granite and has been the subject of extensive petrological investigation by Tuttle and Keith (1954) and Tuttle and Bowen (1958). It consists essentially of coarsely-perthitic alkali feldspar (including orthoclase cryptoperthite and sanidine cryptoperthite varieties), quartz and some oligoclase. Whitten (1961) examined modal variation in the intrusion and suggested that there was a definite stratification within the granite, with more basic lower parts merging upwards with more quartz-rich lithologies.

Raybould (1973) has distinguished at least four granite types forming a crude ring-structure within a small area near to the summit of Beinn an Dubhaich. In this area, the main type is a grey-coloured, porphyritic microgranite with a very fine-grained groundmass. The other types are: yellowish, sparsely quartz-phyric granophyre, greenish medium-grained pyroxene-bearing, granite and medium-grained, hornblende granite.

The intrusion of the granite into the Cambro-Ordovician country rock has resulted in sig-

Table 2.4 Minerals present in skarn zones (after Tilley, 1951, Table 1)

	Skarn zones		
Aureole beyond the skarn zones	Group 1 Primary skarns	Group 2 Boron–fluorine ore skarns	
Talc	Grossular–andradite*	Magnetite*	Grossular-andradite
Tremolite	Wollastonite solid solutions*	Tremolite	Hydro grossular
Forsterite	Diopside–hedenbergite	Forsterite*	Idocrase
Diopside	Spinel	Diopside*	Bornite
Periclase	Plagioclase	Monticellite*	Chalcosite
Wollastonite	Idocrase	Cuspidine*	Covellite
Spinel	Xanthophyllite	Fluorite	Chalcopyrite
Idocrase	Phlogopite	Chondrodite*	Pyrite
Grossular	Orthite	Humite	Blende
Phlogopite	Clinozoisite–epidote	Clinohumite	Galena
Brucite	Prehnite	Ludwigite	Chessylite
Serpentine	Apophyllite	Fluoborite	Malachite
Chlorite	Pectolite	Szailbelyite	
Hydromagnesite	Xonotlite	Datolite	
		Harkerite	

* most abundant minerals

nificant thermal metamorphic and metasomatic alteration of the cherty, dolomitic limestones. Within the carbonate country rocks immediately adjacent to the granite there are skarn zones up to 3 m wide which are characterized by assemblages of pneumatolytic minerals. Beyond this zone lies an aureole of marble. Tilley (1951) recognized two groups of skarns: Group 1, at the granite–country rock contact not containing boron–fluorine-bearing minerals, and Group 2 on the country rock side of Group 1 associated with boron–fluorine-bearing minerals. Table 2.4 shows the mineral assemblages recognized by Tilley within the skarn zones and in the outer aureole.

The discovery of magnetite led to the search for workable deposits, which included a ground magnetometer survey by Whetton and Myers (1949), which recorded numerous bodies of lenticular shape lying along the limestone–granite contact. Samples from these bodies proved to be relatively pure magnetite, accompanied by trace amounts of copper carbonate and sulphide.

Hoersch (1981) made a systematic study of cherty nodules in the carbonates which have reacted with the surrounding rock to produce reaction rims of calc-silicate minerals. Four mineralogically distinct zones of increasing temperature were recognized:

1. Talc-bearing, 350–425°C
2. Tremolite-bearing, 425–440°C
3. Diopside-bearing, 440–520°C
4. Forsterite-bearing, 520–600°C

At Kilchrist and Camas Malag, Tilley (1949) identified an alkali facies of the granite in contact with the skarn deposits. This rock is characterized by the presence of green alkali pyroxene (aegirine–hedenbergite) instead of the normal mafic minerals (biotite and hornblende) of the granite. At Kilchrist, the granite additionally contains metasomatized areas near the contact with veins of pyroxene (diopside–hedenbergite), plagioclase (oligoclase–andesine), fluorite, idocrase and garnet (grossular–andradite) with accessory epidote and orthite. Occasionally, phenocrysts of both microperthite alkali feldspar and plagioclase occur in the marginal facies.

At Camas Malag (NG 583 190), numerous basic and rare acidic sheets intrude near-vertical Cambro-Ordovician sediments which strike approximately parallel to the granite contact (Nicholson, 1985). Where concordant with the bedding, the sheets are clearly boudinaged and, where oblique, they are strongly folded. Individual boudins have chilled margins, even in the 'necked' portions, indicating that boudinage occurred during cooling (Longman and Coward, 1979). Nicholson (1970) considered these sheets to be of Tertiary age from their petrography; their deformation and that of the country rock is very localized and probably occurred during the initial stages of granite emplacement (Longman and Coward, 1979).

Interpretation

The form of the Beinn an Dubhaich Granite has been the subject of much controversy. Harker (1904) mapped numerous vertical contacts between the granite and limestones and concluded that the intrusion is boss shaped. Broadly similar conclusions were reached by Stewart (1965), Raybould (1973) and Bell (1982). However, radically different interpretations were advanced by King (1960) and by Whitten (1961) who considered that the granite had a sheet-like form and that the complicated contacts at the east end of the intrusion were in part caused by erosion through the granite sheet into underlying sediments. These authors suggested that the granite had been emplaced along the Kishorn Thrust Plane which separates the Torridonian and Cambro-Ordovician rocks hereabouts. However, limited borehole evidence (Raybould, 1973) and geophysical studies (Hoersch, 1979) support the interpretation that the granite is a steep-sided boss which extends to depth.

The coincidence of the granite with the Broadford anticlinal structure in the Cambro-Ordovician sediments invites the view that the fold formed in response to the forcible intrusion of the granite. This is not, however, the case as it can be shown that Tertiary dykes, intruded after the sediments had been folded but before the granite was intruded, show no signs of granite-induced deformation. The dykes intruding sediments in the large 'rafts' within the eastern end of the granite show little strike deviation from those in the country rock and, in some instances, dykes in the country rocks are seen to have once been continuous with dykes within the marble rafts. This suggests that these 'rafts' are more likely to have been attached to the roof of the granite rather than to have been 'free-floating' enclaves within the intrusion. Despite the lack of evidence for large-scale deformation around the granite, limited, but locally intense deformation of sediments and dykes occurs at the western end of the intrusion, where both Cambro-Ordovician and Mesozoic sediments have been affected by granite emplacement (Nicholson, 1970; Longman and Coward, 1979).

The calc-silicate mineral assemblages and skarns produced in the cherty dolomitic limestones along the granite contacts and within the rafts are very spectacular and internationally renowned. However, they are strictly limited to within a few metres of the contacts although marmorization is more extensive. Similarly, the chemical and related mineralogical modifications of the granite caused by reaction with the carbonate rocks or carbonate-derived metasomatizing fluids are limited; an alkali-granite facies is present but there is no suggestion from these occurrences that any large bodies of, for example, syenitic compositions, could have formed. Thus, the exposures at this site support the general view that limestone syntexis is probably of limited petrogenetic importance in the genesis of the alkali rock.

The detailed investigations of the mineralogy of the granite by Tuttle and Keith (1954) and Tuttle and Bowen (1958) demonstrated that this high-level, unmetamorphosed intrusion has mineralogical characteristics which are intermediate between those of deep-seated batholithic granites and rhyolites. In the 1950s, there was considerable controversy about the origin of granitic rocks and one school of thought held the view that granites were largely the result of metasomatic transformation of pre-existing rocks. These views were based essentially on observations made in high-grade metamorphic terranes and on large, deep-seated bodies of granite. Magmatists, who opposed this concept, maintained that granites resulted from the intrusion of granite magma and cited the evidence from small, high-level granite bodies and from rhyolite flows. Granitic rocks from these contrasted settings have different and distinct mineralogical features, which, it was argued, arose because of fundamental differences in their origins; the work by Tuttle and others on the Beinn an Dubhaich Granite, as well as other granites in the BTVP, demonstrated that some granites have mineralogical features supposedly characteristic of both settings which were, therefore, not separate and distinct but at the extremes of a series of continually varying mineralogies. Their work thus helped to pave the way for the present-day acceptance of the real existence of granite magma in the crust.

Conclusions

The granite of Beinn an Dubhaich intrudes Lower Palaeozoic carbonate sediments with limited deformation of the sediments and a group of pre-granite, Tertiary basaltic dykes. Reaction between the granites and the country rocks resulted in the production of a distinctive calc-silicate mineral assemblage in the baked carbonate rocks at the contacts, and more widespread recrystallization to form extensive tracts of marble. In addition, there are thin magnetite-rich skarns at some of the contacts, and reaction between the granite and the carbonate rocks has produced small amounts of a marginal alkali granite.

The mineralogy of the granite shows features intermediate between those of extrusive rhyolites and plutonic granites. It thus provides a vital link between these compositionally similar rocks which were formerly considered, by some, to be of fundamentally different origins.

Although it has been proposed that the granite is a sheet-like intrusion injected along a Caledonian thrust plane, the general consensus argues that it is boss shaped. The complicated contact relationships at the eastern end are due to the present level of erosion being nearly coincident with a flat, roof contact which is highly irregular in detail. Evidence for several separate granite intrusions found towards the western end of the body may indicate the presence of small, nested ring-dykes.

KILCHRIST

Highlights

At Kilchrist excellent examples of ignimbrites, tuffs and rhyolite flows are interbedded with agglomerates. Granite and gabbro fragments in the agglomerates demonstrate that a period of plutonic activity pre-dated the Eastern Red Hills centre. The lava flows and volcaniclastic rocks are preserved in a downfaulted block bounded by a ring-dyke of hybrid (mixed-magma) rocks and the Beinn na Caillich Granite.

Introduction

The site encompasses the internationally famous Kilchrist Vent which lies within the eastern part of the Eastern Red Hills centre and thus also includes part of the last major phase of igneous activity on Skye. The vent contains a highly varied association of agglomerates, ignimbrites, acidic rocks, hybrids and gabbros. The site also contains part of the large Inner, or Beinn na Caillich, Granite of the Eastern Red Hills centre (Fig. 2.19).

The Eastern Red Hills centre was, until recently, comparatively neglected. The results of the earlier investigations and some work in the immediate post-war years have been summarized by Stewart (1965). At about the same time, aspects of the Kilchrist Vent were investigated by Ray (1960, 1962, 1964, 1966 and 1972). An intensive study of the centre has since been carried out by B.R. Bell (1982, 1983, 1984a, 1984b and 1985) and these investigations are summarized in Bell and Harris (1986).

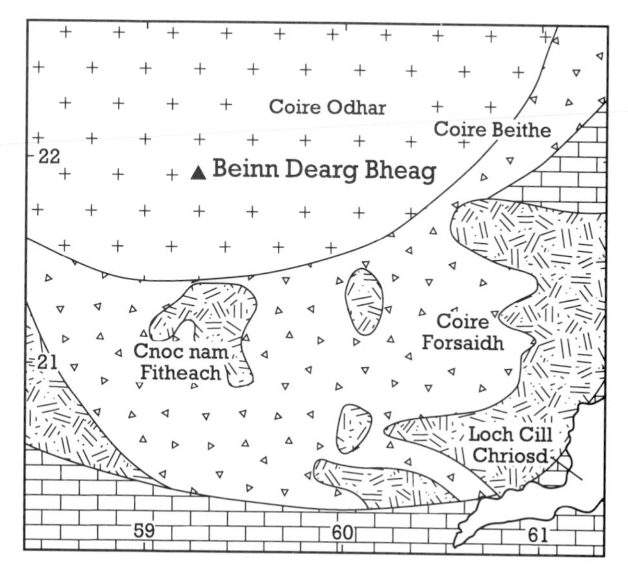

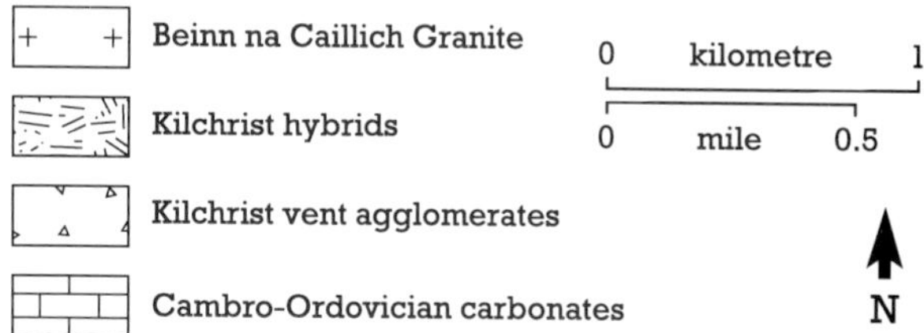

Figure 2.19 Geological map of the Kilchrist site (after Gass and Thorpe, 1976, figure 6).

Description

The volcaniclastic rocks and hybrids in the area between the middle slopes of Beinn Dearg Bheag (NG 593 219) and Loch Kilchrist are the principal features of interest in this site (Fig. 2.19). These rocks occupy the area defined in the earlier literature as the Kilchrist Vent but subsequently shown by Bell (1982) to be a down-faulted bedded volcaniclastic sequence later intruded by hybrids, rather than a chaotic vent infill. Predominant coarse agglomerates contain blocks and smaller clasts of Torridonian sandstones and shales, Cambrian carbonates and quartzites, Jurassic limestones, sandstones and siltstones and a variety of igneous rocks of probable Palaeocene age such as basalt, rhyolite, ignimbrite, granite, pitchstone and gabbro. Evidence for bedding comes from rare lateritic horizons, thin basic tuffs with lapilli, acid tuffs with wispy rhyolitic fragments, and thin rhyolites which probably represent flows (better evidence for rhyolite flows comes from exposures east of Beinn Dearg Mhor; Bell, 1985). In addition, in stream sections east of Cnoc nam Fitheach (NG 593 211), extremely well-developed ignimbrites with beautifully formed *fiamme* texture crop out. Ray (1960) initially recorded the ignimbrites and considered them to be intrusive but they have since been shown by Bell (1985) to be extrusive. The common occurrence of bedding argues against these rocks representing a chaotic vent infilling.

Five distinct masses of hybridized rock within the vent area together form a discontinuous ring-dyke known as the Kilchrist Hybrid Ring Dyke (Bell and Harris, 1986). The margins of this intrusion are steep against the Cambro-Ordovician and Jurassic country rock, but relatively flat-lying against the vent deposits. The hybrids are leucocratic to mesocratic and are medium grained, containing irregularly shaped crystals of quartz rimmed by clusters of amphibole and/or pyroxene, together with rounded white crystals of alkali feldspar. In addition, xenoliths of fine-grained basic material with irregular, diffuse margins also occur. An unusual feature of these hybrids is the presence of thin, flow-banded margins against the fragmental rocks. A good example of this feature occurs in the Allt Coire Forsaid, where small xenoliths of volcaniclastic material are also enclosed within the marginal hybrid.

The steep-sided Beinn na Caillich Granite intrusion occupies the northern part of the site. The rock is a granophyre or microgranite and contains amphibole and biotite as the main ferromagnesian phases. A fine-grained felsitic or spherulitic marginal facies, rarely more than a few metres wide, is developed at some localities, most notably in the gorge of the Allt Slapin in the extreme western tip of the site (NG 584 218). Here, fresh samples contain the iron-rich minerals ferrohedenbergite and fayalite which have been frequently replaced by hydrous alteration products. In this excellent section a succession of near-vertical basaltic and rhyolitic tuffaceous breccias strikes parallel to the granite margin and passes downstream into brecciated Jurassic sediments, which also dip steeply off the granite hereabouts. The volcaniclastic rocks are part of a narrow zone which separates the Beinn na Caillich, or Inner, Granite from Jurassic sediments and Palaeocene lavas to the west and north-west (Bell, 1985, fig. 2). Emplacement of the Inner Granite has caused severe deformation of the adjoining country rocks.

Interpretation

In the literature prior to 1980, the volcaniclastic rocks of this site were interpreted as being vent infill deposits intruded by hybrid rocks. It was termed the Kilchrist Vent (cf. J.D. Bell, 1976). B.R. Bell's detailed reinvestigation has clearly demonstrated that the fragmental rocks are not a chaotic vent infill and there is much evidence for bedded volcaniclastics, highly compacted ignimbrites and rhyolite flows (Bell, 1982). The hybrid status accorded by Harker (1904) to the highly variable intrusive rocks has been confirmed by the later studies and Bell (for example, in Bell and Harris, 1986) noted that the mixed magma nature of the rocks is very apparent in their heterogeneous field appearance. He suggested (Bell, 1982) that the Kilchrist hybrids could be produced by mixing variable proportions of basic and acid magmas similar to those which gave rise to the composite basic and acid sills of the Broadford area.

The form of the Kilchrist hybrid intrusions is important in the interpretation of the volcaniclastic rocks. The hybrids have steep outer margins against Cambro-Ordovician sediments but intrude the volcaniclastic rocks at low angles as sheet-like masses. Bell has interpreted the hybrids as a ring-dyke, the Kilchrist Hybrid Ring Dyke (Bell and

Harris, 1986), and considered that the volcaniclastic rocks form a central, downfaulted block preserved within this structure. Thus, the coarse-bedded volcaniclastic deposits, together with intercalated lava flows, may once have been more extensive, as is suggested by their occurrence on the western and north-western sides of the Beinn na Caillich (or Inner) Granite. Clearly, from the evidence within this site and that provided elsewhere around Beinn na Caillich, the emplacement of the Inner Granite involved, or was at least associated with, complex tectonic events which have not as yet been fully explained.

The varied assemblage of clasts in the fragmental rocks is of considerable interest, not least since it provides evidence for early Tertiary igneous rocks which pre-date the bedded volcaniclastics, hybrid intrusions and the Inner and Outer granites. These have been listed (see above) but attention is directed especially to the presence of blocks of granite which indicate, once again, that there may have been a distinctly early phase of plutonic acid intrusions in Skye (cf. Bell, 1976 and the evidence from Allt Geodh a' Ghamhna).

Conclusions

The site provides clear evidence that, during the Palaeocene, the Eastern Red Hills were an area of fairly extensive bedded volcaniclastic accumulations with interbedded acid and basic lava flows and excellent examples of ignimbrites. The clast content of the volcanic breccias shows that coarse-grained plutonic rocks (both gabbros and granites) were present, possibly representing a phase of central complex development on Skye which was obliterated by the emplacement of the presently exposed centres. These early volcaniclastic rocks are preserved as a downfaulted block within the Kilchrist Hybrid Ring Dyke; they may represent vent-infill material but their present margins are determined by the ring-dyke intrusion. It is therefore possible that they once formed part of an early, much more widespread cover of acid and basic lavas and volcaniclastic deposits.

The hybrid rocks provide yet another example of coexistence and mixing of acid and basic magmas found throughout the period of Tertiary magmatism on Skye. They augment the evidence from the Marscoite Suite of Marsco and elsewhere (Marsco and Mheall a' Mhaoil) and from the composite sills (Rubha' an Eireannaich). The intrusive behaviour of these mixed magmas is convincingly demonstrated in this site.

The Beinn na Caillich, or Inner Granite, is the youngest major intrusion in the site and also in the Eastern Red Hills centre. It is in sharp, chilled contact with the country rocks but does not appear to have produced very marked thermal effects on them. However, the very steeply dipping sequence of volcaniclastic rocks and Jurassic sediments, found at the granite contact in the extreme west of the site, provides convincing evidence that emplacement of this granite caused considerable tectonic disturbance of the adjoining country rocks.

CUILLIN HILLS

Highlights

Excellent exposure of a wide variety of arcuate ultrabasic and gabbroic intrusions allows the sequence of intrusion and the shape of individual bodies to be established in detail (Figs 2.20 and 2.21). Igneous layering, xenolith suites and other internal features of the Cuillin Hills provide evidence for consolidation mechanisms in the intrusions. There are superb examples of cone-sheet swarms. The site is the best area in the British Tertiary Volcanic Province where the roots of a major volcanic complex may be examined in detail. The Strath na Crèitheach vent provides evidence of subaerial activity after emplacement of the gabbros.

Introduction

The Cuillin Hills site is dominated by a major basic/ultrabasic central complex forming the Cuillin mountain range (Fig. 2.1). The intrusion, which is one of the largest of all the British Tertiary plutonic/volcanic centres, contains a series of arcuate bodies of coarse-grained gabbros, eucrites, dunites, peridotites and allivalites, some of which display strong layering. These are cut by numerous minor intrusions including basic dykes, cone-sheets and agglomerate pipes. The site also contains part of the later acidic Strath na Crèitheach centre comprising volcaniclastic deposits and several granite intrusions. The igneous succession is outlined in Table 2.5.

Figure 2.20 Gabbros of the Cuillin centre form rough ground around Loch na Crèitheach in foreground, gabbro peak of Sgurr nan Gillean on left, and smooth-weathering mass of Marsco (Western Red Hills granites) on right. Cuillin Hills and Marsco sites, Skye. (Photo: C.H. Emeleus.)

The site has attracted attention since the earliest days of geology but it was not until the latter part of the nineteenth century that the broad outlines of the geological structure were elucidated through the studies of Judd (1874), Geikie and Teall (1894) and Geikie (1897). Harker completed the first comprehensive field survey of central Skye between 1895 and 1901 which was published in his classic memoir, *Tertiary Igneous Rocks of Skye* (Harker, 1904), with accompanying maps (on both the One-Inch and Six-Inch scales). Further detailed studies were not carried out on Skye until the mid-1940s, many of which were reviewed by Wager and Brown (1968) and by Bell (1976). Further assessment of the Cuillin centre is made by Gass and Thorpe (1976), Sutherland (1982) and Bell and Harris (1986).

The term eucrite, as used in this site description, applies to a gabbro rich in bytownite plagioclase. Although the term has been criticized (Le Bas, 1959) and is not now recommended as a rock name (Le Maitre, 1989), it has been retained here as it is still commonly employed in the literature relating to the site, as well as in Rum and Ardnamurchan.

Description

The site has been selected to include much of the Cuillin basic and ultrabasic centre as well as representative parts of the later Strath na Crèitheach centre (Fig. 2.22). Exposure is generally excellent on the great ridges of the Cuillin range which rise to almost 1000 m elevation and extend in a horseshoe rampart about Loch Coruisk (Fig. 2.22).

The geology of the site is considered under three principal headings:

1. The Cuillin basic/ultrabasic centre.
   (a) The Outer Marginal Gabbros and Eucrites and their relationship with the country rock.
   (b) The layered basic and ultrabasic members of the central part of the intrusion, which are divided into the Outer Layered

Figure 2.21 Rough-weathering gabbro on Bla Bheinn (right) contrasting with smooth-weathering granites of the Strath na Crèitheach centre (left). Jurassic sediments occupy the right foreground. Cuillin Hills and Marsco sites, Skye. (Photo: C.H. Emeleus.)

Series and the Inner Layered Series.

2. The Strath na Crèitheach centre.
3. Minor intrusions.

The country rocks surrounding the Cuillin centre are almost entirely lavas belonging to the Skye Main Lava Series (Thompson *et al.*, 1972). However, on the southern edge of Sgurr na Stri (NG 501 193) Torridonian strata occur and, to the east of the site at Camasunary (NG 517 188) and to the south of Blaven (Bla Bheinn, NG 530 217), both Torridonian sediments and Jurassic limestones have been metamorphosed to high grade (Coire Uaigneich; Almond, 1960). Good examples of metamorphosed basalts occur close to a lochan to the north-east of An Sguman (NG 438 189; Harker, 1904, p. 52 and plate 16).

Outer Marginal Gabbros and Eucrites

Largely massive gabbros and eucrites extend in a broad arc from Glen Sligachan, south-west to the vicinity of Glen Brittle, and south-east to Gars Bheinn (NG 468 188) and Loch Scavaig. These intrusions are cut off to the east by the Glamaig Granite of the Western Red Hills centre in Glen Sligachan, and by younger mafic rocks to the south.

The published One-Inch geological map (Sheet 70, Minginish) and the Six-Inch sheets depict extremely irregular contacts between the Outer Marginal Gabbros and the lavas north-west of Bruach na Frithe (NG 461 253), NNE of Glen Brittle, and on the southern slopes of the Cuillin ridge near Gars Bheinn. These relationships have been interpreted as showing the gabbros fingering into the lava succession, with fairly flat-lying screens of metamorphosed basalt (and volcaniclastic rocks) caught up between (frequently fine-grained) gabbro. Harker (1904) considered that these features were a consequence of the laccolithic form of the Cuillin intrusions, but in the vicinity of Coire na Banadich (NG 435 219), at least, some of these fine-grained rocks are intrusive sheets of amygdaloidal tholeiite (Hutchison, 1966b).

Table 2.5 Succession in the Cuillin Hills site (after Bell and Harris, 1986, pp. 45–6)

Granites of the Strath na Crèitheach Centre
Volcaniclastic deposits of Strath na Crèitheach dolerite cone-sheets
Coire Uaigneich Granite
Intrusive tholeiites of the Outer and Main Ridge Complexes
Inner Layered Series
Inner Layered Gabbros
(?vent agglomerates in Harta Corrie)
Inner Layered Eucrites
Inner Layered Allivalites
Druim nan Ramh Eucrite
Agglomerates and explosion breccias of diatremes
Dykes
(Gars Bheinn ultrabasic sill?)
Outer Layered Series
Outer Layered Gabbros
Outer Layered Eucrites
Outer Layered Allivalites
Layered Peridotites
Border Group (including White Allivalite)
Cone-sheets
Dykes
Outer Marginal Gabbros and Eucrites
?Early Granites (may pre-date Palaeocene basalts of south-west Skye)
Basalt lavas
Torridonian sediments

In the Gars Bheinn area, Weedon (1961) showed that the southern margin of the Outer Unlayered Gabbros is formed by a 200 m wide, steep-sided intrusion termed the Ring Eucrite which intervenes between the wide Gars Bheinn Gabbro and the basalt lavas. To the north, the Gars Bheinn Gabbro assumes a distinctive dull matt-black appearance caused by intense clouding of the plagioclase feldspars. This clouding, caused by thermal metamorphism, was attributed by Weedon to alteration by the younger ultrabasic rocks which are in contact with the Gars Bheinn Gabbro on the south side of An Garbhchoire (NG 470 200). In this same area, south of Loch Coire a' Ghrunda (NG 452 203), an east–west striking eucrite (termed the Ghrunda Eucrite) intrudes darkened Gars Bheinn Gabbro. The eucrite has been correlated with the Border Group of allivalitic ultrabasic rocks which extend to the north and north-west (Hutchison and Bevan, 1977; Bell, 1976).

The layered basic and ultrabasic rocks of the central part of the intrusion

These intrusives may be divided into an outer, earlier sequence of layered peridotites, eucrites and allivalites known as the Outer Layered Series, and an inner later sequence of allivalites, eucrites and gabbros called the Inner Layered Series. The two are separated by the Druim nan Ramh eucrite ring-dyke. These layered rocks have been

Figure 2.22 Geological map of the Cuillin Hills site (after Gass and Thorpe, 1976, figure 6).

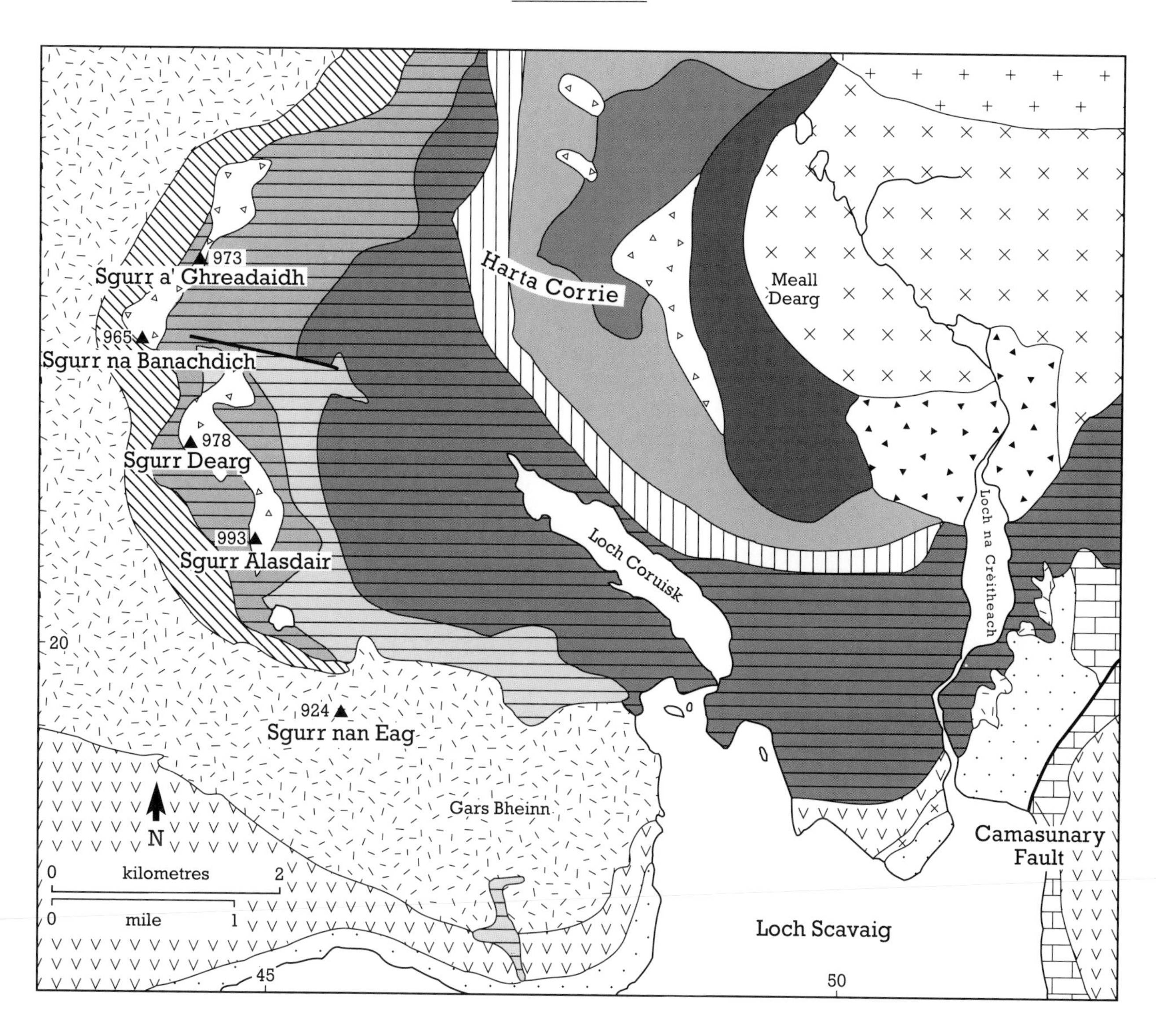
973
Sgurr a' Ghreadaidh
965
Sgurr na Banachdich
978
Sgurr Dearg
993
Sgurr Alasdair
20
924
Sgurr nan Eag
N
0 kilometres 2
0 mile 1
45
Gars Bheinn
Harta Corrie
Meall Dearg
Loch Coruisk
Loch na Crèitheach
Camasunary Fault
Loch Scavaig
50

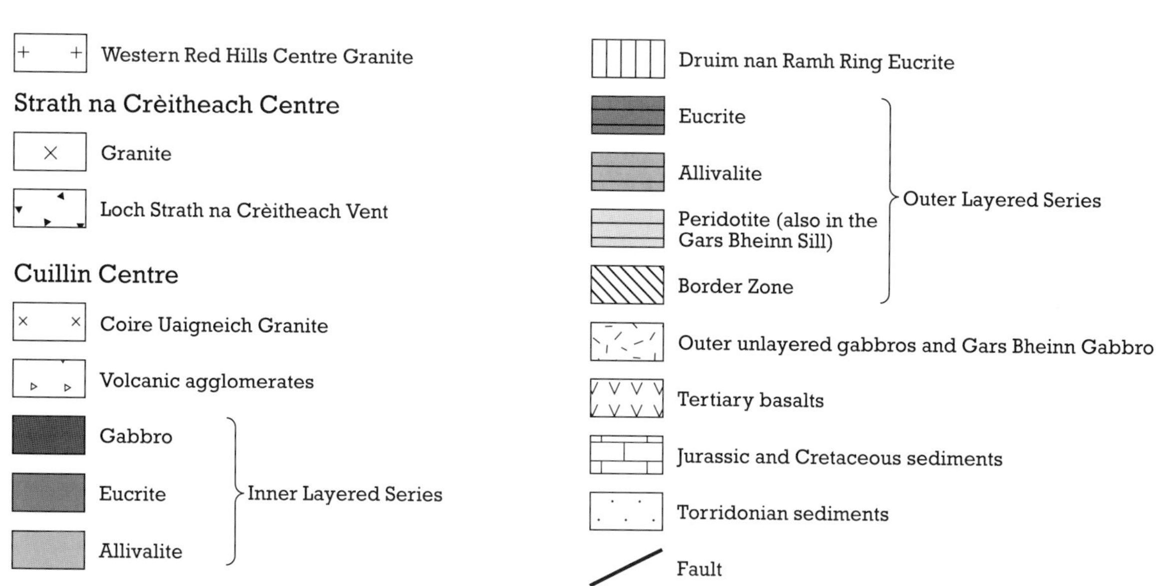
Western Red Hills Centre Granite
Strath na Crèitheach Centre
Granite
Loch Strath na Crèitheach Vent
Cuillin Centre
Coire Uaigneich Granite
Volcanic agglomerates
Gabbro
Eucrite
Allivalite
Inner Layered Series
Druim nan Ramh Ring Eucrite
Eucrite
Allivalite
Peridotite (also in the Gars Bheinn Sill)
Border Zone
Outer Layered Series
Outer unlayered gabbros and Gars Bheinn Gabbro
Tertiary basalts
Jurassic and Cretaceous sediments
Torridonian sediments
Fault

Figure 2.23 Gabbros of the Cuillin Hills site around Loch Scavaig, Skye. From Elgol. (Photo: C.H. Emeleus.)

investigated by Carr (1952), Weedon (1961, 1965), Zinovieff (1958), Wager and Brown (1968), Hutchison (1966a, 1968), and Hutchison and Bevan (1977).

(a) The Outer Layered Series

Zinovieff (1958) and Hutchison (1968) recognized a Border Zone to the Outer Layered Series at the contact with the outer gabbros which is well exposed in Coire Lagan. The Border Zone consists mainly of unlayered allivalites known collectively as the White Allivalite which displays wispy banding near to the contact (Hutchison, 1968). Intrusive tholeiites containing eucrite xenoliths occur at the contact. The White Allivalite is between 200 m and 600 m thick, thinning upwards and outwards, with the incoming of cumulate rocks marking the start of the Outer Layered Series (Bell and Harris, 1986). The Outer Layered Series is banked up against the White Allivalite along a steep contact.

The outermost part of the Outer Layered Series consists of a suite of dunites, peridotites and allivalites with layered structural and textural features typical of igneous cumulates (Wager and Brown, 1968). Weedon, who worked between Loch Coruisk and Loch Coire a' Ghrunda, showed that the dunites give way upwards to the allivalites found in the lower part of An Garbhchoire (NG 468 200). The lower, more mafic part of this succession of layered rocks was termed the Sgurr Dubh Peridotite (Weedon, 1965). To the east, the Outer Layered Series rocks are succeeded by a broad, cross-cutting, well-layered eucrite intrusion containing inclusions of earlier mafic rocks from the Outer Layered Series. These eucrites extend north-west from Sgurr na Stri and Loch Scavaig to the upper part of Harta Corrie (NG 470 235). The inner margin of the layered eucrites is marked by the highly transgressive Druim nan Raimh ring-dyke which marks the boundary between the Outer Layered Series and the Inner Layered Series.

(b) The Inner Layered Series

The outer perimeter of this group is formed by the 100–300-m-wide unlayered Druim nan Raimh eucrite ring-dyke. The ring-dyke is highly transgressive towards the older rocks, allivalites immediately next to the outer edge are crushed and there is some thermal metamorphism of pre-ring-dyke rocks.

Zinovieff (1958) mapped three zones of layered allivalitic rocks to the east of the ring-dyke, comprising the Inner Layered Series which is best exposed on the high ground to the east of Sgurr nan Gillean and on the summit of Sgurr Hain. The layering in these rocks, as in Outer Layered Series rocks, dips at between 20° and 40° towards a focus beneath Meall Dearg and the total thickness of allivalite has been estimated to be *c.* 500 m (Wadsworth, 1982). Different units were distinguished by Zinovieff on the basis of modal mineralogy and within each unit, basal plagioclase–clinopyroxene cumulates give way upwards to feldspar–olivine cumulates. Rhythmic layering also becomes increasingly pronounced up through the sequence. These allivalites were so similar to those in the Outer Layered Series that Zinovieff considered them all to be part of the same layered sequence and numbered two zones mapped on the main Cuillin ridge at Bidean Druim nan Raimh (NG 456 239) as units 1 and 2; those within the ring-dyke were numbered 3, 4 and 5. The allivalite units 3–5 appear to have been succeeded by a layered eucrite which transgresses the allivalites in the lower part of Harta Corrie. An elongate vent cuts these units on the south-east side of the layered eucrite which is, in turn, transgressed by a north–south elongated mass of strongly layered gabbros, termed the Inner Layered Gabbro. This appears to have been the last major mafic intrusion in the Cuillin centre. It crops out around Druim Hain (NG 495 224). The western and southern margins of the Inner Layered Gabbro are unusual since there is a marked development of mylonite and brecciation, suggesting that the gabbros were emplaced along an arcuate, steeply inclined fracture when in a solid state. They are truncated to the east by two granite intrusions.

(c) Agglomerates and explosion breccias

Zinovieff (1958) recorded at least forty small vents and diatremes in various stages of development cutting the rocks of the intrusive centre. The vents are particularly common in the northern Cuillins. These range in size from a few metres in diameter to major bodies over a kilometre across. It is clear that the agglomerates within these vents represent one of the youngest events in the Cuillin centre, post-dating the basic and ultrabasic intrusives which contribute the majority of clasts found in the volcaniclastic rocks. Small agglomerate masses occur immediately west and south-west of Sgurr nan Gillean (NG 472 253) and Harker (1904) noted that the highest point of the Cuillins, Sgurr Alasdair (NG 450 208) was formed by volcanic breccia. A large agglomerate body also crosses lower Harta Corrie cutting members of the Inner Layered Series and the Inner Layered Eucrite; it is in turn cut by the Inner Layered Gabbros and basaltic cone-sheets.

The Strath na Crèitheach Centre

The members of this small centre cut some of the youngest intrusions in the Cuillin basic and ultrabasic centre and in turn are cut by members of the Western Red Hills centre (Figs 2.20 and 2.21). The centre marks a major change in the composition of the intrusions; in contrast to the earlier basic and ultrabasic rocks of the Cuillin centre, the Strath na Crèitheach centre is acidic and the major intrusions are granophyres. The centre also post-dates the majority of minor intrusions which are so abundant in the Cuillin centre (see section on 'Minor intrusions', below).

The centre can be divided into four principal units:

Loch na Crèitheach vent
Rudha Stac Granite
Meall Dearg Granite
Blaven Granite

(a) The Loch na Crèitheach vent

The vent, which is about 2 km in diameter, forms the hillsides north and north-west of Loch na Crèitheach (NG 514 205) and extends to Druim Hain, crossing the path between Loch Coruisk and Glen Sligachan. It also extends to the east side of Strath na Crèitheach on the lower slopes of Blaven. The vent is filled with a mixture of gabbroic, doleritic and basaltic agglomerates, coarse tuffs and finer-grained bedded tuffs, the latter being evidence for contemporaneous subaerial activity. The bedded tuffs are best exposed about 0.5 km north-west of the head of Loch na Crèitheach consisting of alternating bands, on a millimetre scale, of coarse- and fine-grained fragments. Large sheets, or rafts, of gabbro up to 250 m long and 2–10 m thick are also found within the vent concordant with the bedding in the volcaniclastic rocks. It is highly likely that many of the basaltic clasts in the Loch na Crèitheach vent originated from minor basic intrusions rather than from the plateau lavas.

The vent deposits are up to 450 m thick and have contacts with older rocks which show signs of crushing and injection of basaltic magma into fractures and crush lines. It is possible that the margins mark the site of a small ring-fault and the volcaniclastic rocks may be lying within a caldera whose centre has subsided between 750 and 1000 m (Jassim and Gass, 1970).

(b) The granites

Three granite intrusions form the northern part of this centre. All are alkali granites; the Rudha Stac and Meall Dearg granites are feldspar-phyric and contain fayalite and hedenbergite or alkali amphibole; the Blaven Granite is amphibole-bearing. The Rhudha Stac Granite cuts the Loch na Crèitheach vent in the crags south of Loch an Athain (NG 512 227) and the Blaven Granite intrudes the vent on the lower north-west side of Blaven. The two granites are separated by a thin screen of gabbro along the course of the Allt Teanga Bradan, south-east of Rudha Stac (NG 515 333). Exposures on the eastern edge of this site, on the lower north-west side of Blaven, give a clear view of the pale-weathering Blaven Granite in contact with the darker, overlying (and older) gabbros and eucrites which form most of Blaven; this is one of the most visually striking igneous contacts in the BTVP.

The Meall Dearg Granite is probably the youngest of the three acid intrusions (Thompson, 1969). Its chilled contact with earlier gabbros is superbly exposed along the western edge of its outcrop. Fine-grained, flow-banded, spherulitic apophyses at the granite contacts were described by Harker (1904, p. 285) cutting the gabbros at Druim an Eidhne (Druim Hain).

There is a notable lack of minor intrusions in the Strath na Crèitheach centre, apart from a few NW-trending dolerite dykes which intrude the granites.

Minor intrusions

There are two principal types of minor intrusion associated with the Cuillin centre within this site:

1. abundant suites of centrally-focused inclined sheets, or cone-sheets, of basalt and dolerite; and
2. numerous radial dykes, generally of basaltic compositions but some ultrabasic, trachytic, felsitic examples also occur.

The minor intrusions frequently develop close-set joints which render them more susceptible to weathering than the surrounding massive intrusions, thus, many of the gullies and the serrated outlines of the Cuillin ridges owe their origins to the presence of easily-weathered dykes and cone-sheets (Figs 2.1 and 2.17).

A major NW–SE swarm of dominantly basaltic dykes crosses Skye from the Sleat Peninsula in the south-east, to the area around Waternish Point (NG 233 670). A subsidiary NE–SW swarm has been identified at Glen Brittle and on Scalpay (see Speight *et al.*, *in* Sutherland, 1982; especially figs 33.4 and 33.5) which appear to radiate about the Cuillin centre. From the field relationships of the dykes with the other intrusions in the Cuillin centre, it is clear that the injection of the NW–SE swarm occurred throughout the life of the centre and their inception probably pre-dated that of the centre. However, the majority of dykes in this swarm were intruded prior to the Strath na Crèitheach centre.

Several suites of exceptionally well-developed cone-sheets have also been recognized within the Cuillin centre, the latest of these intruding vent rocks in Harta Corrie but not the agglomerates or granites of the Strath na Crèitheach centre. The sheets focus on a centre beneath Meall Dearg and therefore have the same focus as the layering within the basic/ultrabasic cumulates of the Cuillin centre. Cone-sheets cutting the Outer Layered Series eucrites are exceptionally well exposed in the bare, glaciated slabs around the outlet to Loch Coruisk and on the southern face of Sgurr na Stri. The cone-sheets are frequently fine-grained, olivine-free dolerites and some, particularly those in the vicinity of Gars Bheinn, are strongly feldspar-phyric. The sheets dip at angles between 25° and 50° or more, with a tendency for steeper dips closer to the focus area of Meall Dearg. Bell (1976, table 1) has distinguished several generations of cone-sheets.

Ultrabasic minor intrusions represent the latest activity in the Cuillin centre and probably all belong to the same intrusive episode (Wyllie and Drever, 1963). Such intrusions include the Gars Bheinn layered peridotite sill (Weedon, 1960), an irregular body forming An Sguman (NG 436 188), and sparse ultrabasic dykes and sheets occurring in an arc from north of Sgurr nan Gillean to Sgurr nan Gobhar (NG 427 224) and

Gars Bheinn, and on Soay, to the south of the site. The dykes seem to focus on a centre towards the eastern side of the main body of the Cuillin intrusion. The Gars Bheinn sill is of significant interest as it demonstrates the presence of mineral and textural layering in feldspathic peridotite within a fairly thin intrusion, where it is difficult to envisage an origin by crystal settling. It seems likely that an explanation for layering within this sill may be one of double-diffusive convection (cf. McBirney and Noyes, 1979).

Interpretation

The Cuillin intrusion is one of the largest and most elaborately developed Tertiary central complexes in the British Isles, representing a major event in the evolution of the British Tertiary Volcanic Province. The great variety of igneous rock types gives this site its scientific importance; of particular note are the frequent, clear age relationships shown by the different intrusive members and the great variety of layered and other internal structures discussed above. The site therefore provides one of the best opportunities to establish the intrusive history of a central complex in the Province. Like many of the central complexes, the Cuillin centre has been strongly affected by structural activity during and after magmatism; the tectonics of the intrusion, however, remain poorly understood and require comprehensive reappraisal. The intrusive history of the site is summarized in Table 2.5 (based on Bell and Harris, 1986).

Reassessment of the Cuillin centre following the unrivalled field investigations of Harker (1904) has been aided by the work of Zinovieff (1958), Weedon (1960, 1961, 1965), Hutchison (1964, 1966a, 1966b, 1968), Wager and Brown (1968), J.D. Bell (1976), and Hutchison and Bevan (1977). Present interpretations of the Cuillin centre stem mainly from the syntheses of Wager and Brown (1968), Gass and Thorpe (1976) and Bell and Harris (1986). The high-level shape of the Cuillin centre is interpreted to be funnel-like, centred on Meall Dearg and intruding earlier unlayered gabbros and Tertiary lavas. The upward pressure caused by the emplacement of the centre to high structural levels probably caused inflation and doming of the overlying country rock, with the formation of cone-sheets and dykes radiating from the centre of the intrusion. Caldera-like subsidence within the intrusion probably occurred after its emplacement (Gass and Thorpe, 1976). Indications of surface or subsurface explosive volcanic activity are suggested by the prolific presence of vent agglomerates in pipes produced by gas fluidization cutting the layered rocks (Zinovieff, 1958). Some of the agglomeratic pipes may not have broken through to the surface but the presence of bedded tuffs in other pipes indicates that some were probably associated with subaerial volcanism; the bedded tuffs would represent air-fall or water-lain deposits. Variation between agglomerate bodies in respect to size, mixing and rounding of blocks and the nature and proportion of the matrix led Zinovieff to suggest that different stages in the evolution of a typical vent can be recognized:

1. country rock is brecciated and veined by the injection of basaltic material involving no movement of adjacent blocks;
2. a rise in vapour pressure results in a process of fluidization causing blocks and magma to expand and rise as one fluid-like phase;
3. blocks become rounded by attrition and mixed as the fluidized system develops with a tuffaceous matrix.

Wager and Brown (1968) proposed a relatively simple interpretation for the evolution of the centre, relating all of the layered rocks to a single episode of crystal accumulation in a basaltic magma chamber resulting in modal and cryptic layering from dunites/peridotites to gabbros. The mechanism of layering within igneous rocks is discussed in detail for the Askival–Hallival site on Rum. It was envisaged by Wager and Brown that an initial intrusion of the outer unlayered gabbro was succeeded by the intrusion of basic magmas from which the layered rocks formed at a high structural level. Complex tectonic activity during and after the emplacement and crystallization of the intrusion was suggested by Wager and Brown to be responsible for the disruption of the sequence. This explained the present configuration of the units, many of which are transgressive and have steep-sided contacts. The Druim nan Ramh Ring Eucrite was recognized by Wager and Brown (1968) to have important structural implications, separating the Outer Layered Series, towards which it transgresses, from the Inner Layered Series. It was postulated that the Inner Layered Series may either be a fault-controlled repetition of the Outer Layered Series, or the result of a later, separate intrusion of basic magma, bordered by the unlayered ring eucrite. Since a massive central uplift of *c.* 1 km is required to repeat the succession, the latter explanation is preferred (Wager and Brown, 1968; Bell, 1976; Wadsworth, 1982).

Hutchison and Bevan (1977) however, have suggested a multiple intrusion model in preference to the model of tectonic emplacement and crystallization from a single basaltic magma proposed by Wager and Brown. Several episodes of magma injection of ultrabasic and basic composition (Hutchison, 1968; Gibb, 1976) and subsequent accumulation processes have been proposed to account for the relationships between the different components of the layered intrusion. The high structural level of the formation of layered ultrabasic rocks in the intrusion precludes their accumulation from basic tholeiitic magmas as suggested by Wager and Brown, because this would require a magma chamber 9000 m thick, together with massive uplift for which there is no evidence. Hutchison (1968) has therefore suggested that the ultrabasic layered rocks accumulated from ultrabasic magmas. In addition, the dykes and ultrabasic minor intrusions of Skye and Soay have been considered by Drever and Johnston (1966) to provide unequivocal evidence for the presence of liquids appreciably more mafic than basalt, a view also advanced by Gibb (1976).

From a study of the Outer Layered Series on the western side of the Cuillin centre, Hutchison and Bevan (1977) have suggested the following sequence of events:

1. Injection of tholeiitic cone-sheets into earlier, unlayered gabbro during inflation caused by rising magma.
2. Major intrusion of ultrabasic magma producing a funnel-shaped magma chamber.
3. Outer and upper margins of the magma chamber chilled to form the allivalitic, unlayered chilled border zone. In the central part of the intrusion crystal accumulation occurred, producing dunites and peridotites followed by allivalites which banked up against the border series.
4. A second pulse of ultrabasic magma was injected into the chamber to produce an essentially allivalitic unit.
5. The cumulates were later cut by eucritic magmas injected into the chamber (Outer Eucrite Series).

Such a model of multiple magma injection is also supported by J.D. Bell (1976) who concluded that the Cuillin centre is likely to represent a series of partly confluent intrusions, each of which fractionated to varying degrees (cf. Walker, 1975). Wadsworth (1982) postulated a series of 'nested' layered intrusions, each bounded by ring fractures controlled by central subsidence of the funnel.

Gravity surveys of the Skye intrusive complex by Bott and Tuson (1973) indicated that there is a substantial volume of basic–ultrabasic rocks beneath the Cuillin centre. This study modelled an intrusive body widening downwards, from 6 km across at the top to 9 km wide at the base some 14 km below the surface. The volume of this intrusion was estimated to be in the region of 3500 km (assuming the rocks to be of gabbroic composition) with granites occupying, at most, 5% of this volume. Bott and Tuson suggested that the magma rose from the lower crust by piecemeal stoping; this is consistent with the downwards widening of the intrusion but conflicts somewhat with the funnel-shaped form postulated from surface exposure (see above).

The later intrusion of the acidic Strath na Crèitheach centre involved the emplacement of several granite bodies after the formation of a subaerial vent. The dilemma of the origin of the Skye granites is discussed in the Marsco site account.

Conclusions

The Cuillin Hills site is of special geological significance for the following reasons:

1. The Cuillins contain one of the largest and most elaborately developed of the Scottish Tertiary basic/ultrabasic intrusive complexes – the Cuillin centre. The relationships between this centre and a younger granitic centre are well demonstrated.
2. The site contains the first of the Scottish centres to be examined in detail and the Memoir (Harker, 1904) and the One-Inch and Six-Inch geological maps arising from these investigations are geological classics.
3. The intrusive relationships between the different rock types are very clearly displayed allowing unequivocal establishment of the history of intrusive activity. The Cuillin centre is thought to be composed of several superimposed phases of intrusion in associa-

tion with complex tectonic activity.

4. The systems of inclined sheets, or cone-sheets, are among the best known anywhere.
5. Geochemical investigations on the rocks from this site show, when combined with detailed field studies, that there are grounds for seriously considering the existence, at high crustal levels, of magmas significantly more mafic than basalt.
6. Geophysical studies show that this site is the surface expression of a very large, steep-sided body of dense, mafic rocks underlying central Skye and extending deep into the Earth's crust (to at least 14 km depth).

RUBHA' AN EIREANNAICH

Highlights

The site contains a fine, continuous section through a composite (felsite–basalt) sill intruded into Jurassic strata. Mixing of acid and basic magmas is demonstrated by the complete gradation from basalt at the margins of the intrusion to felsite in the core with thin, intervening hybrid zones.

Introduction

This site contains an exceptionally well-exposed example of a composite intrusion which demonstrates a virtually continuous variation from chilled upper and lower margins of basic rock, through hybrid rocks to a central felsite member. It is an excellent example of mixing between contrasting magma types. The sill lies at the northern end of a series of arcuate composite intrusions which focus on the Inner (Beinn na Caillich) Granite of the Eastern Red Hills centre. Buist (1959) has described the intrusion in some detail, and Bell (1983) worked on the geochemistry of the different components and produced a model for the formation of the intrusion. The main features of the intrusion have been summarized by Bell and Harris (1986).

Description

A sill of about 5 m in thickness intrudes sandstones and siltstones belonging to the Lower Jurassic Broadford Beds at Rubha' an Eireannaich, Broadford. In addition, two thin basic sills intrude the overlying sandstone and both sediments and the sill are intruded by basic dykes. The section through the sill can be summarized as follows (after Buist, 1959):

Upper basalt	up to 0.75 m
Hybrid zone	between 0.23 and 0.3 m
Felsite	up to 2.4 m
Hybrid zone	between 0.23 and 0.3 m
Lower basalt	up to 0.75 m

The lower and upper basic margins contain xenocrysts and phenocrysts of feldspar but the lower basic member also contains felsic stringers and small areas of fine-grained basic material, together with rare, partly resorbed gabbroic inclusions. The felsite core carries altered phenocrysts of sodic plagioclase and shows an increase in the proportion of groundmass quartz towards the centre of the sheet. The hybrid zones contain sodic plagioclase xenocrysts and phenocrysts of altered andesine and groundmass pyroxene is pigeonite, in contrast to the augite found in the basic margins. There is a complete gradation from one rock type into the other with no suggestion of chilling; this contrasts strongly with the external margins of the basic member which were chilled to (now devitrified) glass against the sedimentary rocks of the Broadford Beds.

Interpretation

The field evidence provided by the sill shows that basic and acid magmas were essentially available simultaneously. The initial injection of basic magma was followed by injection of the acid magma before the former had cooled and consolidated. The absence of a well-defined boundary between these contrasting magma types, a feature also observed in other composite intrusions (for example, in the Marscoite Suite in Harker's Gully; the composite sills of Arran, see below), led Bell (1983) to conclude that high-temperature diffusion occurred between the basic and acid members at their present level in the crust. In addition, the presence of feldspar xenocrysts in the basic margins indicates that some mechanical mixing occurred prior to intrusion.

Geochemical work by Bell (1983) has shown that for all of a range of elements determined, there is complete compositional continuity between basic and acid members of the sill, and that the chondrite-normalized, rare-earth element patterns for the basic and acid members are parallel.

From these data, Bell concluded that the basic and acid components were cogenetic. He envisaged two periods of hybridization of the acid and basic magmas. An early event involved limited addition of porphyritic acid magma to basic magma forming a basic hybrid with xenocrysts, the basic member was then intruded to form the present marginal rocks in the sill. This was rapidly followed by further porphyritic acid magma which formed the centre of the sill. At this stage, *in situ* hybridization occurred by diffusion of elements between the two contrasting magmas while they were still both close to their liquidus temperatures (Bell and Harris, 1986). This process formed the *c.* 0.30-m-thick hybrid zones which now separate the basic margins from the acid core.

Conclusions

The site provides a very clear example of a common phenomenon in the British Tertiary Volcanic Province, namely, the coexistence of basic and acid magmas. In this instance, there was limited mechanical mixing between the different magmas prior to intrusion; further limited high-temperature diffusion within the intrusion occurred during the emplacment of the basic magma, which was followed very rapidly by the central injection of acid magma. The exposures at Rubha' an Eireannaich provide a continuous section through all of the rock types which can be readily distinguished in the field.

Chapter 3

The Small Isles – Rum, Eigg, Muck, Canna–Sanday

INTRODUCTION

The Isle of Rum (Fig. 3.2) was the site of a major volcano which was active for a period of one or two million years in the Palaeocene, about 59 million years before the present. The volcano developed on a ridge of Precambrian rocks (Torridonian sandstones unconformably overlying Lewisian gneiss) which was covered by a veneer of Mesozoic sediments and Palaeocene lavas. The ridge was flanked by basin structures in the Minch to the west and the Sound of Rum to the east in which thick sequences of Mesozoic sediments accumulated. At the present time, the deeply eroded roots of the volcano are exposed, together with mainly Torridonian country rocks (Fig. 3.2). The north-westerly dipping Torridonian rocks crop out over most of northern and eastern Rum and form the scarp and dip-slope topography north of Kinloch Glen; the Tertiary igneous rocks form the high, rugged mountains occupying most of the southern part of the island.

Rum abounds in features of geological interest but owes its special significance to the spectacular geology of the eroded remains of the Tertiary volcano. Excellent examples of layered ultrabasic rocks, unrivalled in the British Isles, are an outstanding feature of the complex (Fig. 3.1). The ultrabasic Layered Series is of international importance in the development of theories relating to the origin of layering in igneous rocks (for example, Wager and Brown, 1968; Bédard *et al.*, 1988; Young *et al.*, 1988). The centre is also of significance because of its demonstration of the tectonics associated with its emplacement (for example, Bailey, 1945, 1956; Brown, 1956; Emeleus *et al.*, 1985). Spectacular examples of felsites, explosion breccias, and tuffisites related to the emplacement of the complex are well exposed (for example, Hughes, 1960a; Dunham, 1968). Good examples of contacts between mafic and acid rocks, with evidence for local generation of rheomorphic acid magmas, are also of special interest in the study of the Rum centre (Hughes, 1960a; Dunham and Emeleus, 1967; Greenwood, 1987).

Almost everywhere on Rum there is some significant feature of Tertiary igneous activity exposed. The sites selected cover most aspects, but the island is, in effect, a single site in its own right and it is fortunate that it is a National Nature Reserve. The geology of Rum has been sum-

Figure 3.1 Layered allivalite (light) and peridotite (dark) high in the Eastern Layered Series ultrabasic rocks, Hallival. Askival–Hallival site, Rum. (Photo: C.H. Emeleus.)

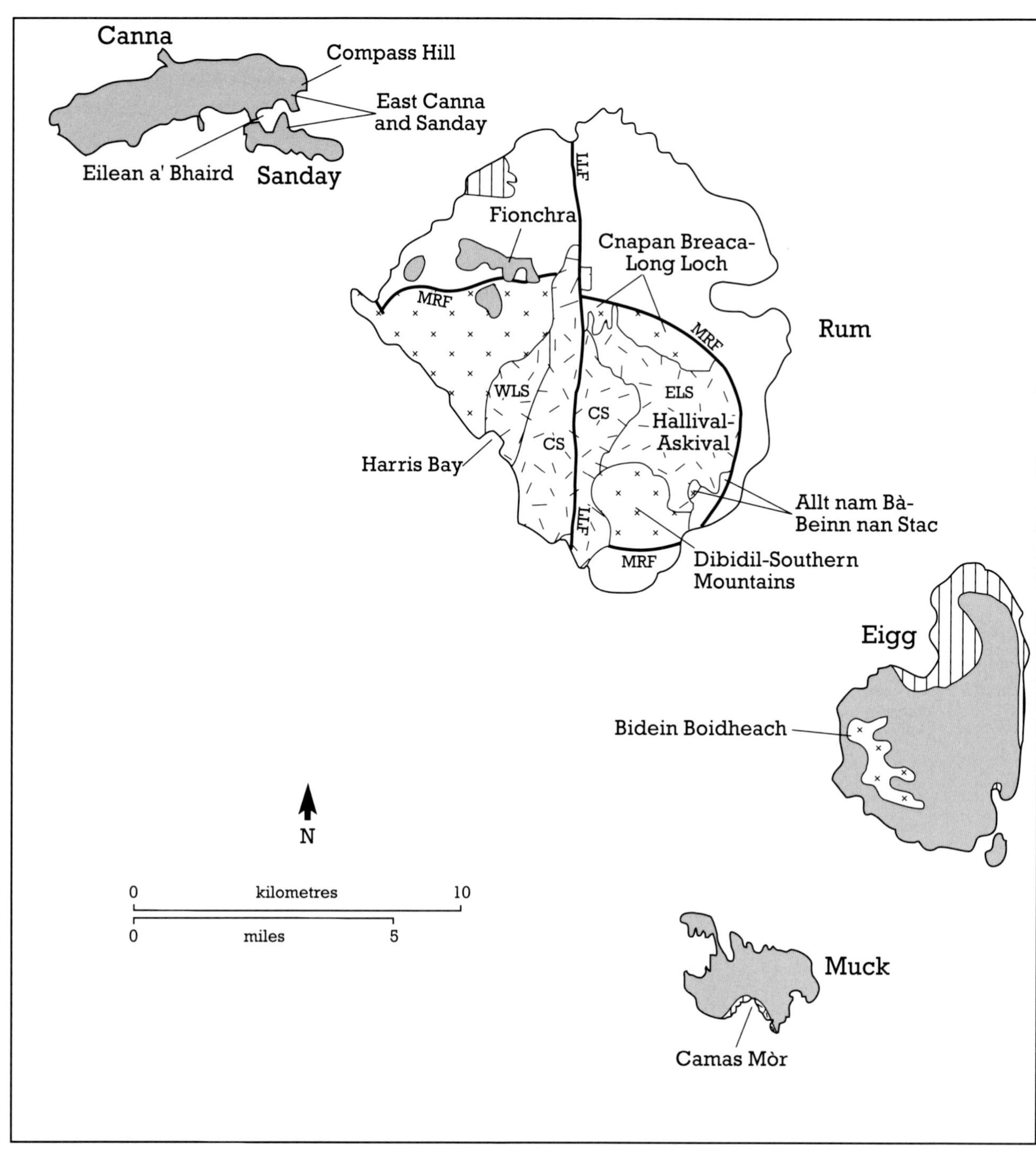

Tertiary

- Lavas, mainly basaltic
- Ultrabasic rocks and gabbro
- Granophyre, felsite, pitchstone, volcanic breccias
- Mesozoic strata
- Pre-Mesozoic rocks

On Rum

MRF Main Ring Fault
LLF Long Loch Fault

CS Central Series
WLS Western Layered Series
ELS Eastern Layered Series
} Ultrabasic rocks and gabbros

Table 3.1 Summary of the Palaeocene igneous geology of Rum and the Small Isles (based on Emeleus and Forster, 1979, table 1, with later amendments)

Valley-filling pitchstone of the Sgurr of Eigg, and associated conglomerates

Dolerite dykes

Lavas and fluviatile sediments of north-west Rum and Canna–Sanday, olivine basalts, hawaiites, mugearite (on Canna), including also tholeiitic basaltic andesite, icelandite (on Rum)

—— Period of profound erosion during which the Rum ——
central igneous complex was unroofed and eroded

The Rum Layered Igneous Complex:

Central Series: feldspathic peridotites, including breccias and some layered allivalites and peridotites

Western Layered Series (WLS): feldspathic peridotites and gabbroic rocks at Harris

Eastern Layered Series (ELS): layered feldspathic peridotite and allivalite, also gabbroic and ultrabasic intrusive bodies
(The WLS and ELS above may be coeval)

Dolerite and basalt dykes (some also post-date the Layered Igneous Complex)

Dolerite and basalt cone-sheets on Rum

Early phase of acid igneous activity:

Western Granite, also granite at Papadil and Long Loch

Porphyritic felsite (ignimbrites, in caldera, and intrusions)

Tuffisites (some may post-date porphyritic felsite)

Volcaniclastic breccias – probably a mixture of explosion breccias and breccias formed by caldera wall collapse

Dolerite and basalt dykes (some intruded after breccias and prior to felsites)

Initiation of the Main Ring Fault System: movement on this system of arcuate faults probably continued at least until emplacement of the ELS/WLS and was a major tectonic feature during the early acid phase of igneous activity.

Lavas of Eigg and Muck, and those involved in the Main Ring Fault on Rum. Principally olivine basalts, feldspar-phyric olivine basalts and mugearites on Eigg. The dykes cutting these lavas belong to the main post-felsite and granite phase of dyke intrusion on Rum. Thin sedimentary layers occur in the Eigg and Muck successions.

marized in the Reserve handbook (Black, 1974); an outline of the Palaeocene igneous sequence is given in Table 3.1.

The first comprehensive account of the geology of Rum is contained in the Geological Survey's Memoir on the Small Isles of Inverness-shire (Harker, 1908). Many of the views advanced by Harker were subsequently considerably amended in a key paper by Bailey (1945). These contributions, together with the results of investigations between about 1950 and 1966 have been reviewed by Dunham and Emeleus (1967).

The earliest Palaeocene igneous activity on Rum was the accumulation of basalt lavas, probably part of the Eigg and Muck lava fields. Initiation of the central igneous complex was probably preceded by intrusion of numerous gabbroic plugs followed by doming and the formation of an arcuate fault system. This faulting was recognized by Bailey (1945), who showed that gneisses and basal Torridonian sediments contained within it had been uplifted by possibly as much as 2000 m. Acid magmatism led to the

Figure 3.2 Map of the Small Isles showing localities mentioned in the text.

formation of volcanic breccias, tuffisites, bodies of porphyritic felsite and granites. Initially, the breccias were attributed to explosive volcanism associated with the acid magmas which formed the felsites (Bailey, 1945; Hughes, 1960a; Dunham, 1968) and which were considered to be intrusive. Subsequently, Williams (1985) showed that the felsite frequently exhibited typical eutaxitic structures and some, at least, were probably formed as ignimbrites, possibly ponded within a caldera (Emeleus *et al.*, 1985). Recent work on the chaotic breccias of Dibidil and Coire Dubh strongly suggests that much of the fragmented material may be due to catastrophic collapse, from time to time, of the walls of a caldera formed during subsidence on the Main Ring Fault (M. Errington, private communication; observations of the authors and B.R. Bell). The argument for the presence of a caldera was strengthened when it was found that Lower Lias fossiliferous sediments and basalt lavas similar to those on Eigg occurred within the Main Ring Fault system (Smith, 1985), implying that there must have been significant subsidence (1–2 km?) along this fault system after the uplift demonstrated by Bailey. It was also shown that the period of subsidence was followed by further central uplift during which Torridonian strata were brought up over the Mesozoic sediments and later lavas (Smith, 1987; Emeleus *et al.*, 1985). Furthermore, the subsequent emplacement of at least the Eastern Layered Series peridotites and allivalites was clearly guided by these arcuate faults. The complex interplay between acid magmatism, major doming and ring-faulting, intrusion and the extrusion of pyroclastic flows, caldera formation and the development of chaotic breccias, are noteworthy features of the igneous geology of Rum.

The layered ultrabasic rocks and associated gabbros were considered in some detail by Wager and Brown (1968) in their classic work on layered igneous intrusions. Subsequently, these rocks have figured prominently in the development of theories concerning the compositions of the magmas responsible for mafic bodies such as the Rum Layered Series. The frequent close association between the layered ultrabasic rocks and gabbros makes basaltic magmas attractive parent material. However, there is a growing body of opinion which advocates high-temperature picritic basalt or magmas of feldspathic peridotite composition as being parental to the Rum layered ultrabasic rocks and other similar masses. The existence of these high-temperature magmas at high crustal levels was advanced by Drever and Johnston (1958) and Wyllie and Drever (1963) from examination of the minor ultrabasic intrusions about the Cuillin gabbros (Cuillin Hills) and on the Isle of Soay, south of Skye. The close similarities between the minor ultrabasic intrusions and layered ultrabasic rocks on both Skye and Rum suggested to Gibb (1976) that the ultrabasic rocks had been formed from parental magmas consisting of a suspension of olivine crystals in ultrabasic ('eucritic') liquid. Donaldson (1975) subsequently investigated the harrisites and ultrabasic breccias of south-west Rum (Harris Bay) and postulated that the rocks had formed from (possibly hydrous) feldspathic peridotite liquids. Further supporting evidence for the presence of intrusive ultrabasic magmas came from McClurg's (1982) discovery of quenched aphyric ultrabasic dykes intruding the layered rocks, the recognition that at least some of the feldspathic peridotites in the layered succession of ultrabasic rocks were intrusive, sill-like sheets (Renner and Palacz, 1987; Bédard *et al.*, 1988), and the presence of quenched ultrabasic margins to the layered rocks at Beinn nan Stac and Harris Bay (Greenwood, 1987; Greenwood *et al.*, 1990). It thus appears inescapable, from the evidence of the Rum sites (Allt nam Ba-Beinn nan Stac; Askival–Hallival; Harris Bay) and elsewhere (for example, Skye Cuillins), that conditions in the larger central complexes of the BTVP and their immediate surroundings sometimes favoured the rise of hot, dense, picritic liquids to within a short distance (*c.* 1 km?) of the Earth's surface. Although not unique by any means, these examples are unusual and must have involved special conditions involving initial rapid, strongly localized throughput of hot basaltic magmas, thus providing preheated pathways along which the ultrabasic liquids were able to rise to high structural levels before crystallizing and congealing.

Brown (1956) and Wager and Brown (1968) attributed the prominent layering seen in the Rum Eastern Layered Series ultrabasic rocks to accumulation of early high-temperature crystals (mainly olivine, followed by plagioclase) from successive batches of fresh basaltic magma. The model of a frequently replenished magma chamber is generally accepted, but, as outlined earlier, many now consider that ultrabasic magmas, with or without contemporaneous basaltic liquids, were responsible for the large-scale layering.

Young *et al.* (1988) consider that there was a stratified picrite-basalt magma chamber in which both magmas were crystallizing simultaneously. Investigations of layers high in the Eastern Layered Series led Renner and Palacz (1987) to suggest that Unit 14 (Brown, 1956) records replenishments of the Rum magma chamber by fresh batches of both picritic and basaltic magmas, and their field observations show that some of the subsidiary peridotite layers are actually sill-like bodies, intruded into near solid or solid troctolitic (allivalitic) rocks. Bédard *et al.* (1988) also examined layers high in the Eastern Layered Series and concluded that it was highly likely that peridotite layers in this succession formed from thick sills of picrite magma intruded into partly consolidated troctolites, thus coming close to the original explanation of the layering put forward in the Memoir (Harker, 1908). The distinctive textures of the ultrabasic rocks provided type examples of cumulate textures (Wager *et al.*, 1960); early-formed olivine and/or plagioclase were thought to have been cemented by the crystallization of trapped, contemporaneous liquid to give well-formed (early) crystals enclosed by poikilitic (late) crystals. Subsequent studies on Rum and elsewhere (for example, Sparks *et al.*, 1985; Irvine 1987) have shown that there may have been substantial migration of the trapped liquids, which modified the early 'cumulate' minerals; it has also been realized that many of the textures of igneous rocks hitherto considered to have formed when the rocks first consolidated may, in fact, have been considerably modified. The ultrabasic rocks of Rum have been extensively referred to in such studies (for example, Hunter, 1987).

A field guide to the Tertiary igneous rocks of Rum has been published by the Nature Conservancy Council (Emeleus and Forster, 1979) and a compilation map of the solid geology has been published as one of the Nature Conservancy Council's 1:20 000 series on the island (Emeleus, 1980). A special issue of the *Geological Magazine* (Volume **122**, Part 5, 1985) contains a wide variety of papers entirely devoted to the Tertiary

Figure 3.3 The Nose, east end of Sgùrr of Eigg. Massive Eocene pitchstone flow overlies eroded Palaeocene basalt lavas. The pitchstone fills a steep-sided valley, columnar jointing is developed in the pitchstone perpendicular to the valley side (slopes top right to bottom left), but gives way to fine-scale, near-vertical jointing at higher levels. The individual lava flows cut out against base of pitchstone (bottom right side). South-west Eigg site. (Photo: A.P. McKirdy.)

igneous geology of Rum. Much of the recent research has concentrated on the layered ultrabasic rocks; this work has been reviewed and summarized by Emeleus (1987); the pre-layered rocks igneous geology has also been surveyed (Emeleus *et al.*, 1985).

Tertiary igneous rocks also crop out extensively on the islands of Eigg, Muck, Canna and Sanday which have been described by Harker (1908), Ridley (1971, 1973) and Allwright (1980).

On Eigg and Muck, subaerial basaltic lava flows overlie Jurassic sediments and are cut by a dense NW–SE-trending basaltic dyke swarm which continues north-west to Rum where it is cut by layered ultrabasic rocks; the lavas on these two islands thus pre-date the Rum centre. The Sgurr of Eigg forms a conspicuous feature at the south end of Eigg formed by a pitchstone floor filling a valley system eroded into the basalt lavas and the dyke swarm (Fig. 3.3). Radiometric dating of the pitchstone shows it to be one of the youngest igneous events in the British Tertiary Volcanic Province (52 Ma, Dickin and Jones, 1983).

The islands of Canna and Sanday consist entirely of Palaeocene lavas, volcaniclastic rocks and sediments. Fluviatile sediments are intimately associated with the products of the volcanic activity and clasts in inter-lava sediments provide evidence that the lavas on Canna and Sanday are closely linked with those of north-west Rum. Canna and Sanday are important links in a chain of sites that enable a relative dating of the Rum central complex and the later Skye Cuillin central complex in the study of the evolution of the British Tertiary Volcanic Province (Meighan *et al.*, 1981; Mussett, 1984).

FIONCHRA

Highlights

The site is of particular importance since it contains the only example of lavas clearly post-dating a central complex in the British Tertiary Volcanic Province. It also excellently demonstrates the interaction between the accumulation of lavas and the development of a contemporaneous fluviatile system.

Introduction

Four outliers of Tertiary volcanic rocks and associated fluviatile conglomerates and lacustrine sediments form the summits of Fionchra, West Minishal, Orval and Bloodstone Hill in north-west Rum (Figs 3.4 and 3.5). The site is unique within the Tertiary Igneous Province as it contains the only known occurrence of lavas which are demonstrably younger than the nearby central intrusive complex.

The lavas and associated sediments were first described by MacCulloch (1819). Judd (1874) correctly deduced their age relative to the plutonic rocks. However, Geikie (1897) and Harker (1908) both considered them to be relicts of a once extensive plateau embracing all the Small Isles subsequently intruded by the Rum Central Complex. Harker also frequently misinterpreted the massive flow interiors as intrusive sheets or sills (see site descriptions for northern Skye). Tomkeieff (1942) reinterpreted Harker's sills as trachyandesitic lava flows and, together with Bailey (1945), considered them to be of a pre-granophyre age.

Black (1952a) established beyond doubt that the lavas post-dated the Western Granophyre on Orval and had infilled a series of valleys carved into the granophyre and Torridonian country rocks. Through the work of Black and the more recent detailed studies of Dunham and Emeleus (1967), Emeleus and Forster (1979) and Emeleus (1985), the stratigraphy and petrology of these unique lavas are known in considerable detail.

Description

The lavas of the four outliers are essentially flat-lying or dipping gently to the west. They rest unconformably on Torridonian sediments and the Western Granophyre, overlapping the Main Ring Fault. The unconformity is irregular and the lavas appear to have filled an evolving river valley system. The most complete lava succession occurs on Fionchra where Black (1952b) estimated a total maximum thickness of over 500 m; however, not all members are represented here.

Emeleus (1985) remapped the Tertiary lavas of Rum and proposed a fourfold subdivision:

1. Lower Fionchra Formation. This is the oldest lava formation; it fills a river valley cut into

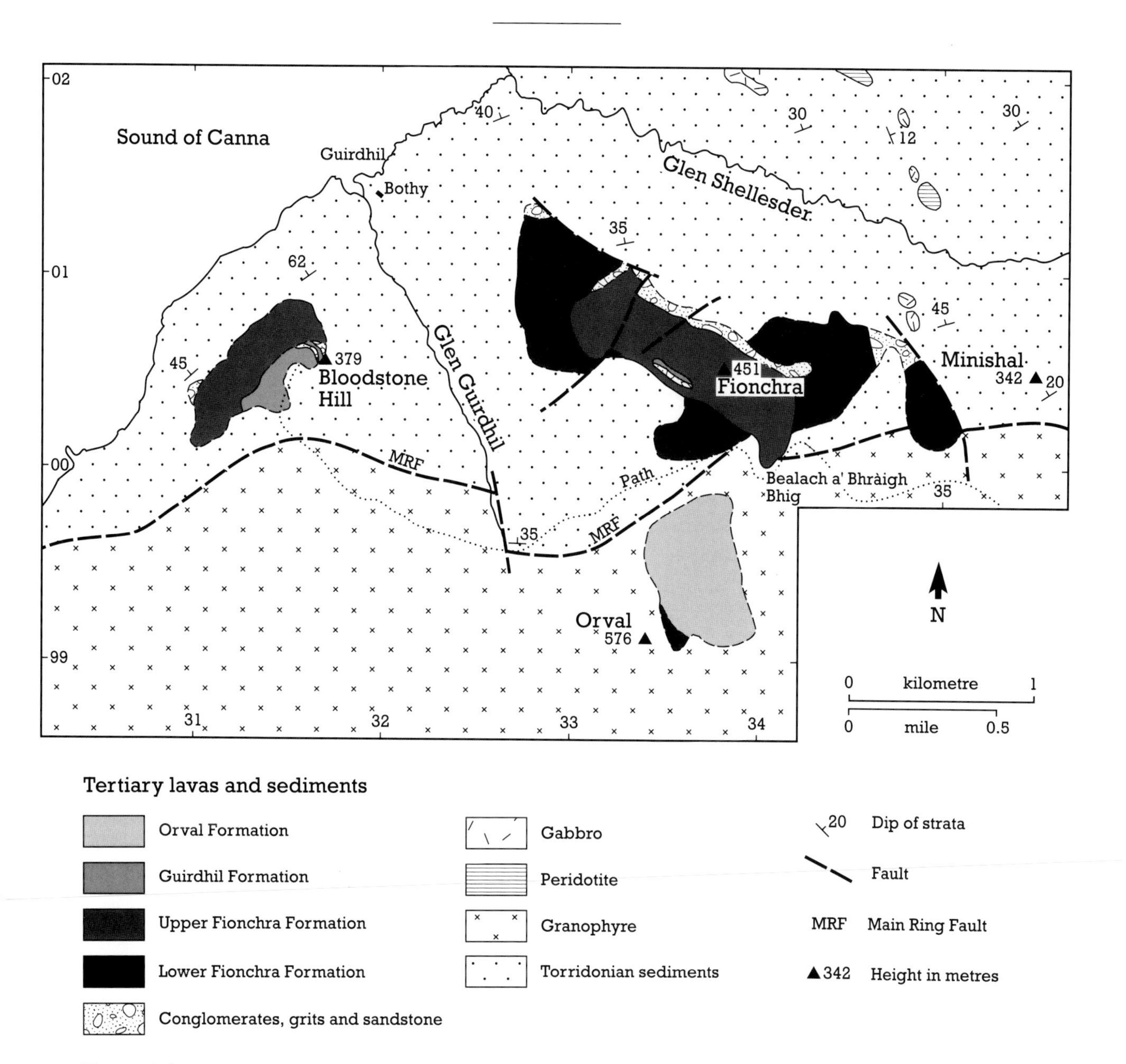

Figure 3.4 Geological map of the Fionchra site, Rum (after Emeleus, 1980).

Torridonian sediments and lined with Tertiary conglomerates. On the south side of West Minishal (NG 348 003), the lavas pass undisturbed across the Main Ring Fault to overlie the granophyre. The lavas are alkali-olivine basalts, typical of many Hebridean lavas, but less alkaline than those of Skye. Most contain phenocrysts of olivine and plagioclase. More fractionated, hawaiitic flows occur in the west of the Orval outlier where Black (1952a) demonstrated that they rest on an eroded surface of weathered granophyre. North of Minishal the lavas rest on thick fluviatile conglomerates which contain clasts derived from rocks now exposed in the Rum central complex (granophyre, porphyritic felsite, gabbro, explosion breccia, allivalite) and the country rocks (gneisses and Torridonian sandstones). There are also clasts of tholeiitic basalt, frequently vesicular, and dolerite. The basalts are compositionally unlike any now exposed on Rum or the adjoining islands.

2. Upper Fionchra Formation. This subdivision contains lavas termed 'mugearites' by earlier workers. Chemically, the lavas are tholeiitic

and contain clinopyroxene and plagioclase phenocrysts; in modern terminology they are tholeiitic basaltic andesites. Some flows low in the Bloodstone Hill outlier and the lowest flows on Fionchra are strongly feldspar-phyric and are superficially similar to some of the porphyritic upper members of the Skye composite flows (Harker, 1904; Kennedy, 1931b). The lavas rest on a slightly irregular surface of Torridonian arkose and Western Granite and are observed to overlap the Main Ring Fault at Bealach a'Bhraigh Bhig (NM 339 999). On the western side of Bloodstone Hill (NG 317 006) and along the western and northern faces of Fionchra (NG 332 009–337 006), a series of fluvial conglomerates, commonly containing porphyritic felsite and granophyre clasts, and silty beds separate the Upper Fionchra Formation from the underlying Lower Fionchra Formation lavas. These silty, lacustrine sediments contain plant remains such as leaf impressions. The sediments appear to fill a valley system partly excavated in the lavas of the Lower Fionchra Formation. The sediments frequently pass up into pillow breccias (for example, at NM 3367 0070) with the appearance of broken, angular or glassy lobate fragments of basalt in a finer basaltic breccia. Ponding of the lavas where they appear to have flowed into a small lake is indicated by the unusual thickness of the lowest flow and the occurrence of pillow breccias in its lower portions; the basal few metres of the flow are strongly columnar at Coire na Loigh (NG 332 009).

3. Guirdhil Formation. Rocks recognized by Ridley (1971, 1973) as icelandites comprise this formation. Their phenocryst assemblage consists of clinopyroxene, orthopyroxene, plagioclase and opaque oxide. The lavas of this formation occupy a twice-excavated, conglomerate-filled valley carved into the Upper Fionchra Formation and Torridonian sediments at Bloodstone Hill (NG 3165 0055). At Bealach a'Bhraigh Bhig (NG 340 002), a small valley fill of icelandite lava rests directly on rubbly granophyre about 150 m north of the col, and lava overlying conglomerate adheres to the south side of

Figure 3.5 Post-Central Complex basic lavas resting on an irregular surface of Torridonian sandstone. Fionchra site, Rum. (Photo: C.H. Emeleus.)

Fionchra (NG 335 005; Emeleus and Forster, 1979).

4. Orval Formation. These lavas are olivine- and feldspar-phyric hawaiites and basaltic hawaiites, petrographically similar to the lavas of the Lower Fionchra Formation. However, the Orval Formation lavas tend to be coarser-grained and contain biotite as overgrowths on opaque oxides and as discrete crystals. In addition, alkali-feldspar mantles strongly normal-zoned plagioclase phenocrysts. The flows are massive and the formation probably contains three or four flows, although these are not easily distinguished. High on Orval, just east of the summit, the formation cuts across strongly terraced flows belonging to the upper part of the Lower Fionchra Formation showing the younger age of the Orval Formation lavas. Elsewhere, the lavas of this formation lie directly on granophyre with no intervening conglomerate.

The chronological sequence of the formations is fairly clear; the oldest is the Lower Fionchra Formation which is overlain unconformably by both the Upper Fionchra Formation and the Orval Formation. The Upper Fionchra Formation is unconformably overlain by the Guirdhil Formation but there is no direct evidence for the relative ages of the Guirdhil and Orval Formations. However, the absence of clasts derived from the Orval Formation in the conglomerates associated with the Guirdhil Formation suggests that the Orval Formation is the younger.

No intrusions of compositions comparable with the lavas in the Fionchra site have been found on Rum, implying that the source(s) must have been outside the present area of the island. A possible source may have been offshore of eastern Canna, since coarse volcaniclastic rocks, probably of very local origin, are interbedded with the lavas of Compass Hill (East Canna and Sanday). However, this problem has not been fully resolved. These lavas are grouped with those of Canna and Sanday in the Canna Lava Formation, the individual 'formations' identified above are now termed 'members' (Emeleus, in preparation).

Interpretation

The Fionchra site contains an incomplete sequence of Palaeocene lavas which record the last major igneous event on Rum. The lavas and associated sediments are interpreted as having accumulated in a succession of hilly landscapes, filling valleys orientated parallel to the Main Ring Fault which was probably a zone of weakness exploited by weathering. River systems cut valleys into Tertiary granophyre and Torridonian rocks which occasionally became the sites of shallow lakes in which were deposited fine-grained sediments; the coarser conglomeratic sediments of the sequence probably had a fluviatile origin. Plant remains in the lacustrine sediments suggest a warm, temperate climate. Erosion occurred during periods of volcanic quiescence when the drainage system was re-established. The Tertiary sediments are considered by Emeleus (1985) to result from the erosion of a terrane consisting of Palaeocene lavas and igneous intrusive rocks as well as Lewisian and Torridonian basement rocks.

The site provides unequivocal evidence for the post-Central Complex age of the lavas. This comes from:

1. the presence of clasts of gabbro, allivalite, peridotite, felsite, granophyre/microgranite, tuffisite and volcanic breccia in the interlava conglomerates (Fig. 3.6);
2. the occurrence of lava flows resting on a weathered surface of granophyre; and
3. lavas overlapping the Main Ring Fault which separates granophyre and Torridonian sediments.

The Central Complex was clearly unroofed prior to accumulation of the lavas and it must have formed high ground which continued to undergo active subaerial erosion throughout the period of lava effusion.

The lavas and some of the sediments are comparable with those on Canna and Sanday (see below) to which they are related. Somewhat similar sediments also occur in south-west Skye (Allt Geodh a'Ghamhna). Taken together with these sites, Fionchra is a major link in a line of evidence which clearly indicates that the Rum central complex predated the central complex on Skye (for example, Meighan *et al.*, 1981; Dagley and Mussett, 1986).

The Rum lavas varied considerably in composition with time. The Lower Fionchra Formation flows are alkali basalts but they were preceded by tholeiitic flows, now only represented by clasts in underlying conglomerates. The succeeding Upper Fionchra and Guirdhil Formations are tholeiitic and show progressively more evolved, or frac-

Figure 3.6 Boulder conglomerate underlying flow-banded icelandite lava flow. The conglomerate contains granophyre, felsite and allivalite clasts derived from the weathering of the Rum Central Complex. South side of Fionchra. Fionchra site, Rum. (Photo: C.H. Emeleus.)

tionated compositions up the sequence; however, the final flows of the Orval Formation, are fractionated alkali basalts and hawaiites. The Rum flows were probably derived from at least two source magmas, one alkali basaltic in character, the other tholeiitic. There are only limited age data available (about 58 Ma; Dagley and Mussett, 1986), but all flows sampled, to date, gave reversed magnetic polarities and it is likely that they were erupted in a short space of time (Mussett *et al.*, 1988, figure 2; see also Chapter 1, Table 1.1).

Conclusions

The final phase of igneous activity on Rum resulted in the accumulation of lavas and associated sediments now exposed only in the Fionchra site. The site provides an excellent opportunity to study the interaction between the emplacement of lava flows and a palaeo-fluvial system.

The Tertiary lavas of Rum were possibly erupted from a centre or centres north-west of Rum and these filled valleys in an incised topography. Four lava members are recognized, each separated by unconformable contacts, conglomerates and some fine-grained lacustrine sediments. Clast populations in the intra-flow conglomerates show that the lavas post-date the emplacement of the Central Complex which was unroofed by the time that the first lavas erupted. Preliminary geochemical investigations show the presence of lavas of both tholeiitic and alkaline affinities.

ALLT NAM BÀ–BEINN NAN STAC

Highlights

The juxtaposition of Lewisian gneisses and basal Torridonian rocks with Jurassic sediments and Palaeocene lavas gives this site particular importance since it demonstrates both subsidence and uplift on the Main Ring Fault which bounds the Rum central complex. The site also contains one of the few examples in the British Tertiary Volcanic Province of a chilled margin to a gabbro–ultrabasic complex.

Allt nam Bà–Beinn nan Stac

Introduction

The site (Fig. 3.7 and inset) is characterized by a tectonic collage of fault-bounded slivers of metamorphosed Jurassic limestones, sandstones and shales, Palaeocene lavas, basal Torridonian sediments and Lewisian gneiss. These components occur within a fault zone coincident with the Main Ring Fault and provide a record of tectonic activity during the emplacement of the central complex.

Marble and calc-silicate rocks were discovered close to Allt nam Bà by Hughes (1960b) who proposed them to be of Lewisian age and recognized that their presence, together with Lewisian gneiss, could be attributed to movement along the Main Ring Fault. Subsequently, the discovery of poorly preserved fossil bivalves led to the recognition that the calc-silicate rocks were of Mesozoic age (Dunham and Emeleus, 1967). Smith (1985, 1987), in a detailed study of the fault zone, recorded fairly well-preserved fossils of bivalves, corals and belemnites of Lower Lias age in these rocks. These and other investigations have stimulated fresh interpretations of the Tertiary tectonics of Rum (Emeleus *et al.*, 1985).

Description

A variety of lithologies of strikingly contrasting ages occur as fault-bounded slivers and intrusives in a wide lensoid fault zone coincident with the Main Ring Fault at Allt nam Bà and extending to the south-east slopes of Beinn nan Stac (Fig. 3.7).

In the northern part of the fault zone a wedge of Mesozoic strata crops out. Metamorphosed calc-silicates have long been recognized at Allt nam Bà (Hughes, 1960b) and Smith (1985, 1987) has mapped associated fossiliferous sandstones, sandy limestones and shales of early Jurassic age. The succession of Mesozoic strata is up to 35 m thick and is highly inclined to the west and inverted. Palaeontological evidence suggests that they are probably an extension of the Lower Lias Broadford Beds (Smith, 1985).

Highly metamorphosed calc-silicate rocks within the marginal Layered Series are found in the Allt nam Bà valley (Fig. 3.7 inset; NM 4060 7945). They consist of various assemblages of calcite, grossularite, diopside, vesuvianite, leucoxene and tilleyite. Where they are cut by a narrow syenite vein they additionally contain wollastonite and pyrite (Hughes, 1960b). A further narrow strip of limestone, about 100 m to the south, is separated from the northern outcrop by gabbro and a minor intrusion of hybrid acid rocks. Two hundred metres to the south along the line of the same fault zone, a homogeneous, dull, coarse-grained marble is composed almost entirely of calcite (NM 404 942) and contains poorly preserved belemnite and bivalve remains (Emeleus and Forster, 1979). These calc-silicates and related sediments, together with limestones recently rediscovered by Smith on the south-western slopes of Beinn nan Stac (Smith, 1987) are the only Jurassic outcrops known on Rum. They are of early Jurassic age, comparable with the Broadford Beds on Skye (Smith, 1985). Younger Jurassic rocks are, however, well exposed on Eigg less than 10 km to the south-east.

The Main Ring Fault zone crosses to the eastern side of Beinn nan Stac about 1 km south-east of the summit where it is over 100 m wide. On the inner side it is bounded by Torridonian Basal Grit and Bagh na h-Uamha Shale (see Black and Welsh, 1961, for divisions of the Torridonian strata). On the outer, south-eastern side a thin strip of Rudha na Roinne Grit overlying Bagh na h-Uamha Shale occurs and dips steeply towards the fault zone. Within the fault zone, a variety of rocks are found including an elongate lens of Lewisian gneiss, a small area of Basal Grit and a strip of flinty, sheared amygdaloidal basalt. Smith (1985) has shown the basalts to be far more extensive than previously established; they are fault-bounded to the east and south-east with an unconformable contact between basalt and Mesozoic strata to the north. These lavas predate those at Fionchra in northern Rum and may be a faulted wedge of more extensive Tertiary lava fields which now cover much of Eigg and Muck and with which they have geochemical similarities (Smith, 1987).

Undeformed porphyritic felsite and an explosion breccia are also found within the fault zone and have probably exploited the lines of weakness along the faults.

The prominent cliff feature which extends from near the summit of Beinn nan Stac SSE along the crest of the ridge towards Allt nam Bà is formed by baked, resistant Torridonian sediments (Bagh na h-Uamha Shales) and probably owes its origin to the contact effects of the Marginal Gabbro of the ultrabasic/basic complex. At a locality about 450 m ESE of the summit (NM 4007 9403), the marginal olivine gabbro becomes fine grained and contains skeletal olivine

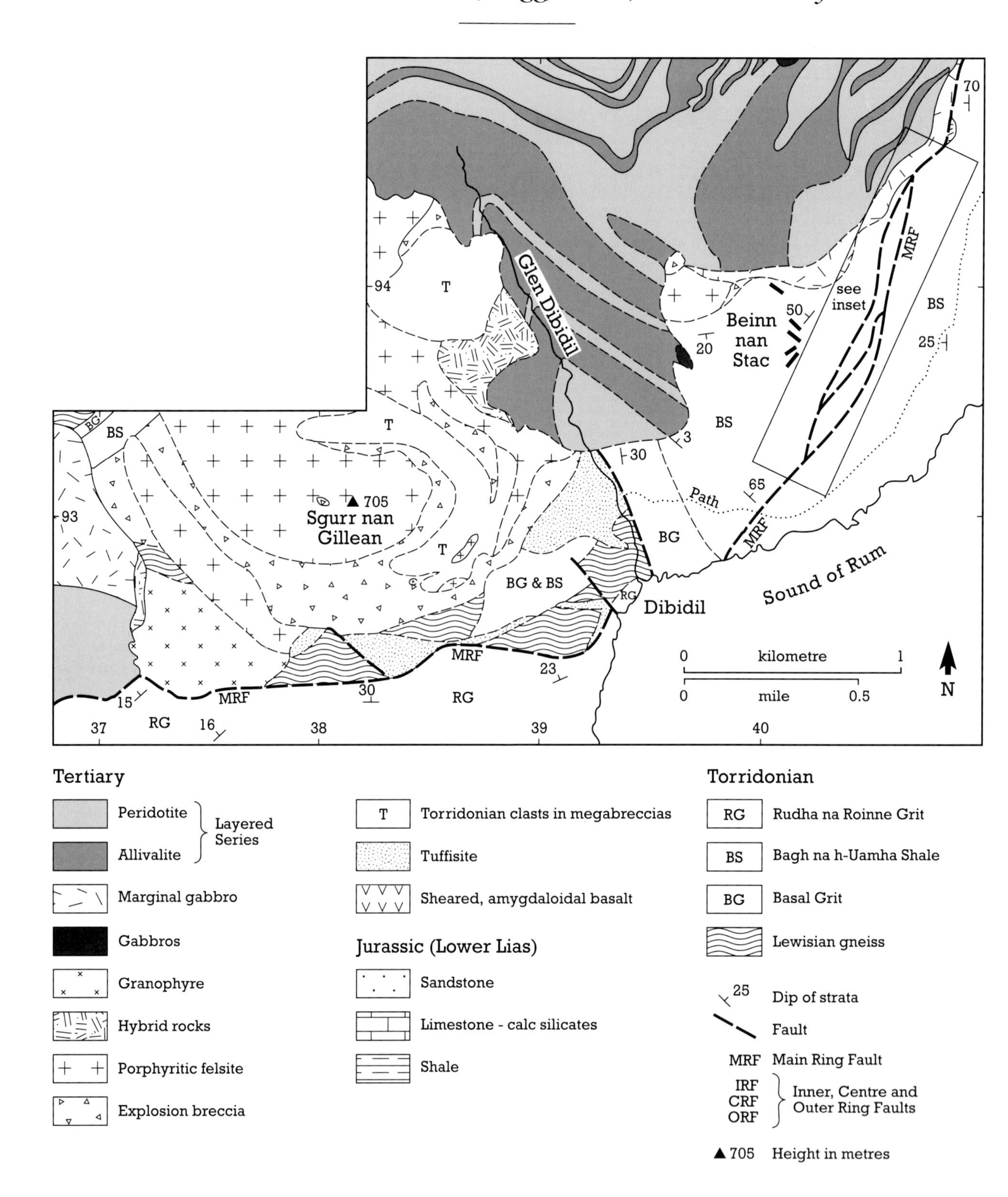

Figure 3.7 Geological map of the Dibidil–Southern Mountains and Allt nam Bà–Beinn nan Stac sites, Rum. Inset (on opposite page) shows detail to the south of Allt nam Bà. Main figure after Emeleus (1980) with subsequent modifications (Greenwood, 1987). Inset after Smith (1985, fig. 1).

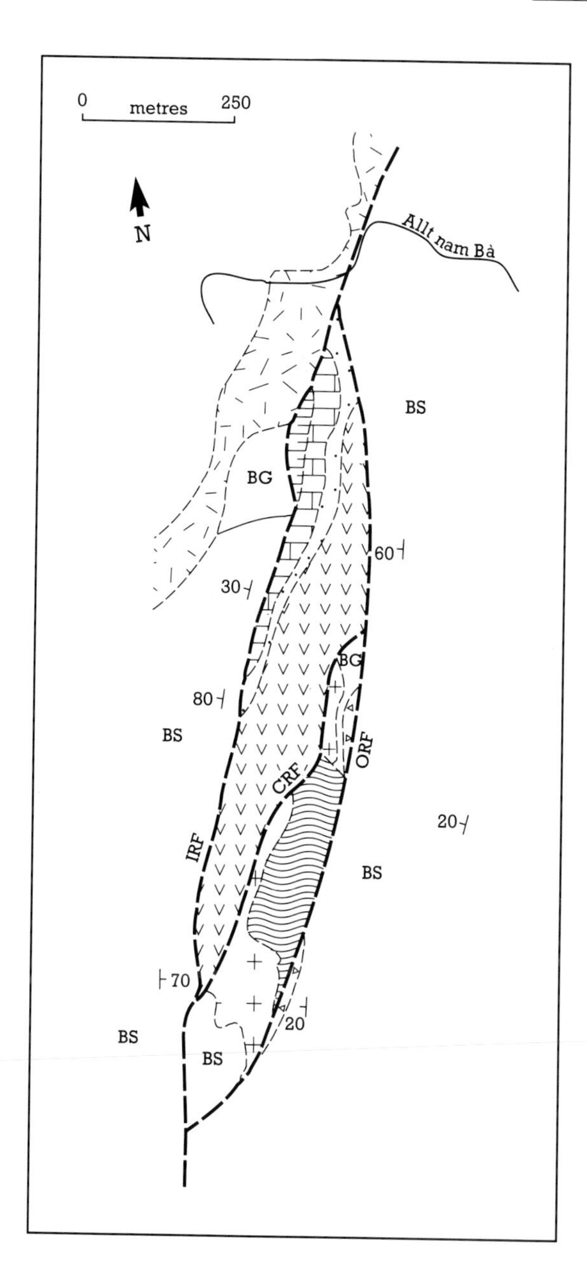

phenocrysts in a variolitic matrix. This is one of the few places on Rum where there is a clear indication of a chilled facies to the rocks rimming the ultrabasic–gabbro complex (Greenwood, 1987; Greenwood *et al.*, 1990). At this locality it is overlain by up to 7 m of hybrid rock which is in turn overlain by baked shale. The field relationships here suggest that the baked shales, explosion breccias and felsite of Beinn nan Stac form a SE-dipping roof to the later gabbros and ultrabasic rocks, similar to the roof seen at the east end of Cnapan Breaca (see below).

Interpretation

Smith (1985, 1987) has shown that the Jurassic sediments on Rum are closely comparable with the middle Broadford Beds of Skye and do not correlate with the nearby Great Estuarine sediments of Eigg. Eigg and Rum are separated by the southern extension of the Camasunary Fault (Binns *et al.*, 1974) which on Skye shows a considerable pre-Palaeocene downthrow to the east (Peach *et al.*, 1910). On Skye, only the lowermost Jurassic beds occur west of the Camasunary Fault and it is likely that this situation pertained on Rum at the start of Tertiary volcanism.

The wide fault zone in south-east Rum contains several individual, distinct, fault planes related to different stages of movement along the Main Ring Fault. Smith (1985) mapped the following faults which are shown on Fig. 3.7 inset.

Outer Ring Fault (ORF) – easternmost boundary fault.

Centre Ring Fault (CRF) – separating sheared Palaeocene basalts from felsites and gneisses.

Inner Ring Fault (IRF) – westernmost fault; responsible for the juxtaposition of stratigraphically low-level Jurassic sediments and Palaeocene basalts against Torridonian sediments.

The proposed model for the movement of the Main Ring Fault is discussed for the Cnapan Breaca and Dibidil sites. According to Smith (1985, 1987) the initial diapiric uplift and ring fracturing is thought to have occurred along the ORF, bringing Lewisian and basal Torridonian to higher stratigraphic levels. The ensuing caldera subsidence occurred along the CRF, evidence for which is the presence of Mesozoic sediments downfaulted against Torridonian and Lewisian rocks. Renewed uplift of about 2 km along the IRF is required to bring stratigraphically low Torridonian to the structural level that it now occupies within the Main Ring Fault. A more detailed discussion of the proposed movements within the fault zone is presented in Smith (1985). The occurrence of chilled picritic rocks at the edge of the layered ultrabasic rocks provides evidence for the existence of ultrabasic, or strongly picritic basaltic, liquids in the complex (Greenwood *et al.*, 1990).

Conclusions

The juxtaposition of basement Lewisian and Torridonian rocks and stratigraphically high Jurassic sedimentary rocks and Palaeocene basalts within the fault zone can be explained by movement along the different fault planes; the presence of Mesozoic strata provides crucial evidence for early subsidence during which the felsite and explosion breccia probably formed.

The exposed Main Ring Fault complex in south-east Rum is therefore of considerable importance in the study of the tectonic evolution of the Rum Complex in providing a comprehensive record of different stages of movement during its emplacement. Recent studies within this site have made an important contribution to the understanding of the Tertiary geology of Rum, in the overall context of the British Tertiary Volcanic Province. It is now evident that marginal complexes, such as that present here, merit further careful scrutiny both on Rum and elsewhere.

A chilled contact facies of the Layered Ultrabasic rocks provides evidence supporting the view that ultrabasic liquids played a role in the formation of the Central Complex.

ASKIVAL–HALLIVAL

Highlights

The site contains the thickest, unbroken succession of layered ultrabasic rocks in Great Britain. The large- and small-scale layered structures, the petrography and geochemistry of the layered rocks and their emplacement mechanisms have been studied in great detail and have contributed significantly to theories relating to the origins of igneous layering. The site is of international importance for these reasons.

Introduction

The Askival–Hallival site is a unique, internationally significant location for large-scale, cyclic layering in igneous rocks (Fig. 3.1). The site provides excellent exposure of repeated layered units comprising alternating olivine- and plagioclase-rich rocks which are type examples of igneous cumulates. A total thickness of about 700 m of layered rocks is exposed between *c.* 160 m OD and the summits of Askival and Hallival, which belong to the Eastern Layered Series in the core to the Rum central complex.

The earliest systematic investigation of the Rum Central Complex was carried out by Harker (1908), who interpreted the layering as multiple sill-like injections of contrasting magma types. In subsequent work, Brown (1956) favoured a genetic model which combined repeated magmatic replenishment of tholeiitic basalt magma and crystal accumulation processes, a model which is broadly accepted today, although the parental magmas are now argued to have been of a picritic composition (evidence summarized by Emeleus, 1987 and Young *et al.*, 1988).

Description

The rocks of the Eastern Layered Series in the Askival–Hallival site (Fig. 3.8) comprise a repeated succession of large-scale layered ultrabasic units. Each unit is composed of a basal feldspathic peridotite rich in cumulus olivine (and sometimes pyroxene) which is overlain by, and sometimes gradational into, a plagioclase-rich, troctolitic cumulate termed allivalite. Less commonly, extreme plagioclase cumulates, or anorthosites, may occur at the very top of a unit. Brown (1956) distinguished fifteen such units in the Askival–Hallival area ranging in thickness from under 10 m to over 80 m. The units are generally considered to be an upward-younging stratigraphic sequence with a primary easterly dip of about 20°. The terraced topography which characterizes the slopes of Askival and Hallival (Fig. 3.1) results from the contrasting weathering properties of peridotite and allivalite; the allivalites form prominent, resistant escarpments, while the peridotites erode more easily forming

Figure 3.8 Geological map of the Askival–Hallival site, Rum (after Emeleus, 1980).

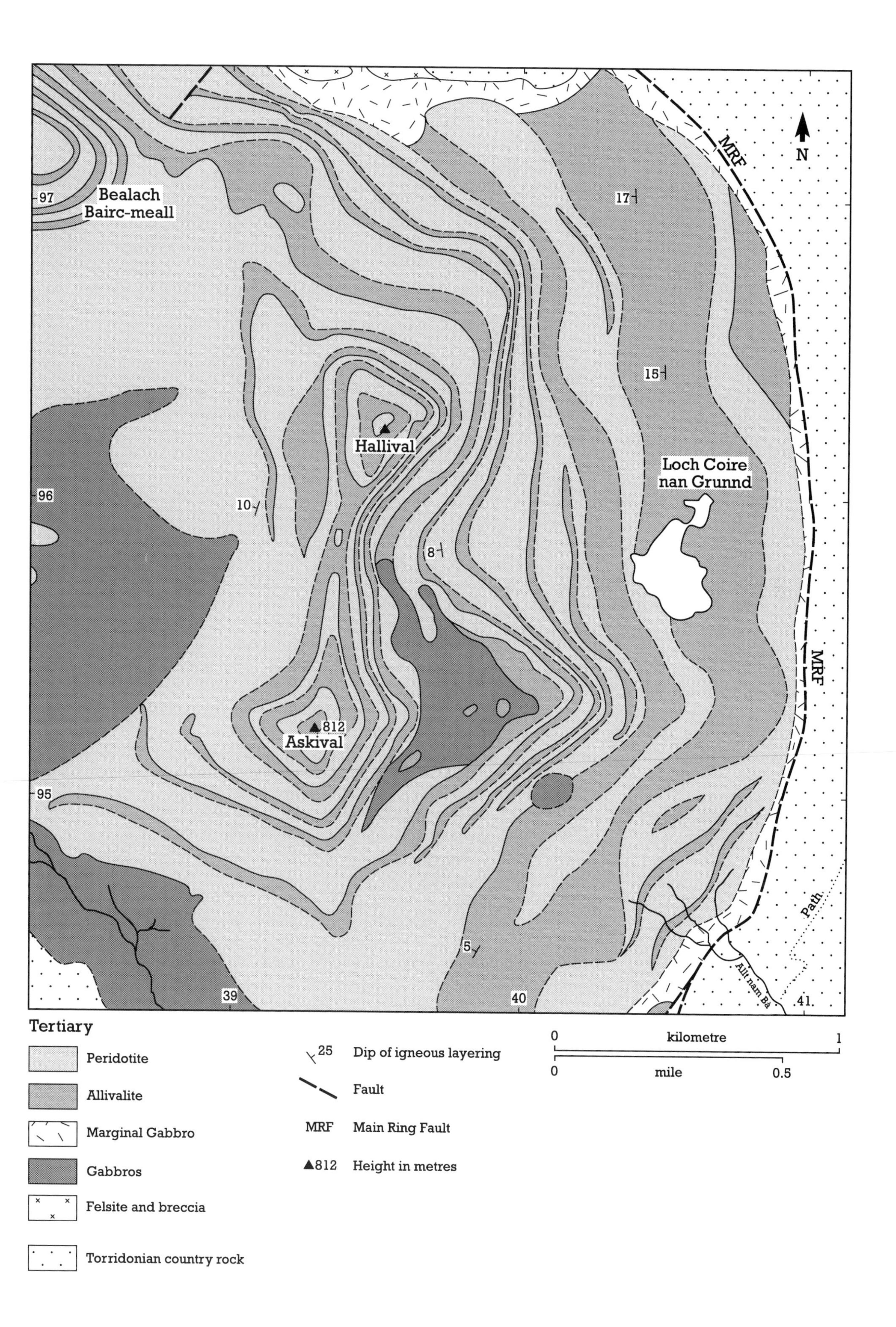

Bealach
Bairc-meall
97
96
95
Hallival
Askival
▲812
Loch Coire
nan Grunnd
MRF
MRF
N
17
15
10
8
5
39
40
41
Allt nam Ba
Path
Tertiary
Peridotite
Allivalite
Marginal Gabbro
Gabbros
Felsite and breccia
Torridonian country rock
25 Dip of igneous layering
Fault
MRF Main Ring Fault
▲812 Height in metres
0 kilometre 1
0 mile 0.5

the grass-covered terraces which provide the nesting sites for Rum's manx shearwater colony.

The allivalites and peridotites are mineralogically simple, containing variable proportions of Mg-rich olivine, diopsidic pyroxene, calcic plagioclase and chrome-spinel. The petrographic textures reflect cumulate processes (Wager *et al.*, 1960) and all minerals occur as cumulus phases; plagioclase in particular defines a strong lamination parallel to layering in allivalites. Plagioclase, pyroxene and, very occasionally, olivine are also intercumulus phases forming large poikilitic crystals up to 20 mm in diameter enclosing the cumulus phases. The resistance to weathering of plagioclase poikilocrysts is responsible for the characteristic honeycombed weathering surface of peridotite. Cumulus chrome-spinel, commonly an accessory mineral in peridotite, is frequently concentrated in seams several millimetres thick at the very base of peridotite layers marking the abrupt unit boundaries. Weak harrisitic textures are sometimes exhibited by olivine crystals in the peridotites, but they are not as well developed as those in the Harris Bay site (see below). While there is little overall cryptic variation through the layered units in eastern Rum, individual units show slight variability, becoming more fractionated upwards, with a return to less fractionated compositions in the basal feldspathic peridotite of the overlying unit (Dunham and Wadsworth, 1978). This corroborates the suggestion that each major unit in the Eastern Layered Series represents a fresh unit of unfractionated magma (Brown, 1956).

Allivalites commonly are rhythmically layered (Fig. 3.9) and frequently contain zones of slump deformational structures; a particularly good example is exposed on the eastern face of Askival in Unit 14 (Brown, 1956; Fig. 3.10). Such structures are of a similar nature to those observed in unlithified sediments and suggest the accumulation of considerable thicknesses of poorly consolidated crystal cumulates on, or near to, the floor of the magma chamber. Slumping of the unstable magmatic sediments may have been triggered by tectonic activity.

In addition to the layered peridotites and allivalites, distinctive, dark-grey, cumulate troctolitic gabbros are exposed as conformable sheets in the lower levels of the complex to the east and north-east of Hallival and on the Askival Plateau (Brown, 1956). These gabbro sheets have recently been interpreted to be an integral part of the Layered Series by Faithfull (1985). The gabbros often contain numerous subangular to rounded clasts of granular basic hornfels, probably derived from Tertiary basalts in the roof zone to the complex. Their presence suggests that fragments off the roof and upper wall country rocks were incorporated into the chamber and were sealed off as the intrusion solidified inwards and downwards from the walls and roof.

Interpretation

The spectacular large-scale cyclic layering exposed in this site has been subject to intensive investigation concerning its origins. Recently, interest has been focused on detailed petrographic and geochemical studies and the refinement of theories relating to the formation of such rocks. Consequently, the site has achieved significant petrological importance in the theory of layering in igneous rocks world-wide.

The layered rocks of eastern Rum are interpreted to be the result of successive pulses of magma injected into a shallow-level magma chamber (Wager and Brown, 1951; Brown, 1956; Wadsworth, 1961). This magma is currently argued to have been of a picritic composition (for example, Volker, 1983; Emeleus, 1987; Greenwood *et al.*, 1990). Each pulse, or replenishment of new magma, is envisaged to have produced a single unit by high-temperature crystal accumulation on the floor of the magma chamber (Wager and Brown, 1951). Successive pulses of new magma crystallized in this way to build up the Layered Series. Recent studies (Huppert and Sparks, 1980; Faithfull, 1985; Tait, 1985) have proposed that pulses of new picritic magma ponded at the base of the chamber beneath cooler, less-dense residual magma. Initial olivine crystallization in the picritic liquid lowered its density, and it was cooled against the older magma above. The resulting temperature and density differences are thought to have caused strong convection to develop in the picritic layer culminating in the mixing of the two magmas when their physical properties equilibrated following the fractionation of a peridotite layer. With further cooling, the resulting magmas formed allivalitic and other troctolitic gabbroic rocks until the next pulse of picritic magma was injected and the cycle repeated.

A wealth of new information has recently been published from detailed textural, mineralogical and geochemical studies on the Layered Series

Figure 3.9 Fine-scale layering in allivalite, west side of Hallival. Askival–Hallival site, Rum. (Photo: A.P. McKirdy.)

Figure 3.10 Slumped folding in allivalite, near Askival summit, Rum. Askival–Hallival site, Rum. (Photo: A.P. McKirdy.)

heralding a new era of current interest in the site. This work includes that of Faithfull (1985) who related cryptic variation in the lower Eastern Layered Series to post-cumulus effects; Tait (1985) produced detailed crystallization models from geochemical and isotope studies; Palacz and Tait (1985) investigated contamination of the magmas using isotope evidence; Butcher *et al.* (1985) described upwards-growing peridotite 'finger' structures and interpreted them as modal metasomatic replacement of allivalite by peridotite; Butcher (1985) discussed channelled metasomatism by intercumulus liquids within late-stage veins; and the cumulate rocks have been cited as examples exhibiting textural equilibrium (Hunter, 1987).

The Rum complex differs from other classic layered intrusions such as the Skaergaard (Wager and Brown, 1968) in that there is little or nothing of a marginal border group. Instead, the layered ultrabasic rocks are often bounded by rather variable gabbroic rocks which usually, but not always, separate them from the earlier Tertiary intrusions and Torridonian sediments. Marginal gabbros are exposed in the stream sections of Allt na h-Uamha (NM 409 968) and Allt Mor na h-Uamha (NM 405 973) in the east of the site. The gabbros have intruded along the line of the older Main Ring Fault near these streams and south to Allt nam Bà. Elsewhere, as on Beinn nan Stac and Cnapan Breaca, the relationships strongly suggest that the Marginal Gabbro and the layered ultrabasic rocks underlie earlier felsites and associated rocks which form an outward-dipping roof to the mafic rocks. In an investigation of the margin of the ultrabasic–gabbroic complex, Greenwood (1987) suggested, on the basis of detailed mineralogical and geochemical studies as well as a consideration of the field relations, that the Marginal Gabbro was probably not a distinct, separate body from the main succession of ultrabasic rocks but that it represented various degrees of modification of the more feldspathic ultrabasic rocks through reaction with country rocks. Extensive intrusion breccias are commonly developed at the contacts with earlier, more acid rocks (for example, Torridonian near Allt na Uamha); rheomorphic acid magmas generated at these contacts may well have reacted with partially crystallized mafic magmas to give the very variable gabbroic rocks that characterize the Marginal Gabbro zone of earlier workers.

Both Brown (1956) and Wadsworth (1961) postulated that the ultrabasic complex was emplaced as a solid mass along the Main Ring Fault lubricated by basaltic magma which subsequently formed the Marginal Gabbro. However, the lack of disturbance of layering right up to the edges of the intrusion and the roof-like contacts, give grounds for questioning this mode of emplacement. Recent publications argue that the layered rocks formed *in situ* from picritic magmas, with the possibility that some of the peridotites are in fact sill-like bodies intruded conformably into the layered allivalite rock (Emeleus, 1987; Renner and Palacz, 1987; Bédard *et al.*, 1988; Young *et al.*, 1988).

Conclusions

The excellent vertically and laterally extensive exposures of large-scale rhythmic layering in ultrabasic and gabbroic rocks on Hallival and Askival are unique within the British Isles. Unlike the other Tertiary centres of Mull and Skye, the Rum layered rocks formed as the last major event in the central complex and are thus virtually free from the overprinting effects of any subsequent igneous or tectonic events. Owing to this, they are particularly suited to the development of models concerning the origins of the layering, the accompanying rock textures and the variations found in their mineralogy and chemistry. The large- and small-scale structures and textures in the layered rocks are in many respects very similar to those developed in clastic sediments and invite interpretation in terms of the sedimentation from mafic magmas of successive crops of crystals of different densities. These features are common in peridotitic, gabbroic and other igneous rocks world-wide and, because of the exceptional clarity of the Rum examples, the theories developed here have strongly influenced petrogenetic thought regarding the crystallization of high-level magma chambers (for example, Wager and Brown, 1968; various articles in Parsons, 1987; Renner and Palacz, 1987; Bédard *et al.*, 1988).

HARRIS BAY

Highlights

The site is the type area for harrisite, a rock containing spectacular growths of olivine crystals. It contains arguably the best-exposed contact of layered gabbroic rocks against (granitic) country rocks in the BTVP, which shows, firstly, the intrusive nature of the gabbros and, secondly, evidence for extensive partial melting of the earlier granite.

Introduction

The Harris Bay site (Fig. 3.11) contains exposures of ultrabasic rocks belonging to the Western Layered Series and the Central Series of the Rum Central Complex. Unusual cumulate features are well-developed in the rocks of this site including 'harrisitic', or crescumulate, olivine textures and poikilo-macrospherulitic plagioclase textures. In addition, the site is also of special significance because it contains excellent exposures of the contact between the ultrabasic rocks and the earlier granitic rocks of western Rum; evidence for partial melting of the granite by the mafic rocks is particularly convincing (Greenwood, 1987).

Wadsworth (1961) identified at least four petrographically distinct stratigraphic units in the layered ultrabasic rocks within the site. At Harris Bay these are underlain by unusual eucritic cumulates which were termed the Harris Bay Series. Later investigations within the site by McClurg (1982) and Volker (1983) assigned part of Wadsworth's Western Layered Series to the later intrusive Central Series. The view that harrisitic olivines represent local coral-like growth from the floor of the magma chamber (Wadsworth, 1961) has been modified by Donaldson (1982).

Description

The Western Layered Series, which occupies most of the site, comprises the Harris Bay, Transitional and Ard Mheall series (Wadsworth, 1961). Eucritic gabbros of the Harris Bay Series form the lower part of the succession, above which a 50 m thick gradational unit, the Transitional Series, passes up into feldspathic peridotites of the Ard Mheall Series (*c.* 380 m thick). Unlike the Eastern Layered Series, layering is on a scale of a few metres to centimetres, dipping gently to a centre just east of An Dornabac, but nearly flat lying in Harris Bay. The layering is frequently defined by texture rather than modal layering. The later intrusive Central Series, recognized by McClurg (1982) and verified by Volker (1983), occupies a north–south strip up to 2 km wide from the south coast to the Long Loch and Minishal. Central Series rocks crop out on the eastern edge of the site and include rocks previously designated by Wadsworth (1961) as the Dornabac and Ruinsival series. The Central Series is characterized by predominant feldspathic peridotite and peridotite breccia, together with minor dunite, allivalite and gabbro.

The characteristics and subdivisions of the ultrabasic and basic layered rocks within the Harris Bay site are summarized in Table 3.2 (after Wadsworth, 1961).

The majority of ultrabasic rocks within the Western Layered Series are feldspathic peridotites containing varying proportions of olivine, plagioclase, chrome-spinel and diopsidic augite. There is a great variety of cumulate igneous textures in these rocks, signifying the interest of the site. Harrisitic textures (Wager and Brown, 1951), or crescumulate textures (Wager and Brown, 1968), were first described from Harris Bay by Harker (1908). Unusual poikilo-macrospherulitic textures are also found here which take the form of massive radial, braid-like growths of plagioclase (Donaldson *et al.*, 1973).

The harrisitic (olivine crescumulate) textures characterize the rock that was termed harrisite by Harker (1908). They were formed by large skeletal growth of forsteritic olivine. The individual crystals, or bundles of crystals, may be up to 2 m in length and often form coral-like growths from the planar surface provided by an underlying normal olivine cumulate (cf. Wadsworth, 1961, plate 4). Good examples of this type of harrisitic texture can be seen in the Transitional Series close to the road near Harris (NM 340 964) and at many places on Ard Mheall in the overlying Ard Mheall Series. Radiating, and sometimes apparently isolated pods of harrisitic-textured rock also occur, particularly within the eucritic Harris Bay Series in coastal exposures near Harris Lodge (NM 337 957).

Massive radial and braid-like growths of poikilitic plagioclase crystals enclosing numerous olivines occur within igneous breccias a short way north-east of Loch an Dornabac, just north of the site at NM 357 977. These individual radial growth structures range in size from 0.15 m to almost a metre in diameter and up to fifteen radiating crystals have been found growing out from a single nucleus. The term poikilo-macrospherulitic has been applied to these structures (Donaldson *et al.*, 1973). In some growths there is a radial symmetry. The feldspar textures resemble the 'lace' or 'honeycomb' textures described by Brown (1956) from the Eastern Layered Series. The feldspars involved in these structures have grown *in situ* and support current views that 'diagenetic' processes are of considerable importance in the crystallization of igneous cumulates (for example, Irvine, 1987).

Extensive zones of peridotite breccia occur within the Central Series ultrabasic rocks of the site. The breccias contain rounded, irregular and angular blocks of feldspathic peridotite lying randomly in a more feldspathic ultrabasic matrix, often approximating the mineralogy of allivalite. In places the breccias contain blocks which themselves display layering or even slumping. Clearly, the source for the blocks was a well-lithified cumulate. Wadsworth (1961) mapped an extensive zone of breccia, the Lag Sleitir Breccia, transgressing the Harris Bay, Transitional and Ard Mheall series. These breccias are well exposed in the stream sections where the Allt Lag Sleitir and the Abhainn Rangail join (NM 346 956). Wadsworth demonstrated that many of the blocks were derived from the Ard Mheall Series and considered that the breccias formed along a fault scarp within the magma chamber. An extensive zone of breccias cutting the Ruinsival Central Series was also thought to have such an origin.

Numerous other occurrences of breccia have been recorded from the same general area (Donaldson, 1975; Volker, 1983) and to the north around the Long Loch and Barkeval (McClurg, 1982). Structures within these breccias suggest formation by either intraformational slides, brittle deformation or plastic deformation. The transgressive nature of the Lag Sleitir Breccia towards the Western Layered Series is one of the stronger pieces of evidence which justifies the establishment of the Central Series of ultrabasic rocks and gabbros (McClurg, 1982; Volker, 1983).

The ultrabasic rocks are cut by numerous intrusions of gabbro. These are clearly later than the host rock which they vein but, as their margins show no signs of chilling, they would appear to have been intruded before the surrounding rocks had cooled. Wadsworth (1961) suggested that they formed from a magma closely related to that from which the ultrabasic rocks formed. The most unusual of these intrusions is the Glen Duian Gabbro, an assemblage of numerous thin gabbroic sheets which intrude the layered eucrites of the Harris Bay Series with

Figure 3.11 Geological map of the Harris Bay site, Rum (after Wadsworth, 1961, figure 2; Volker, 1983).

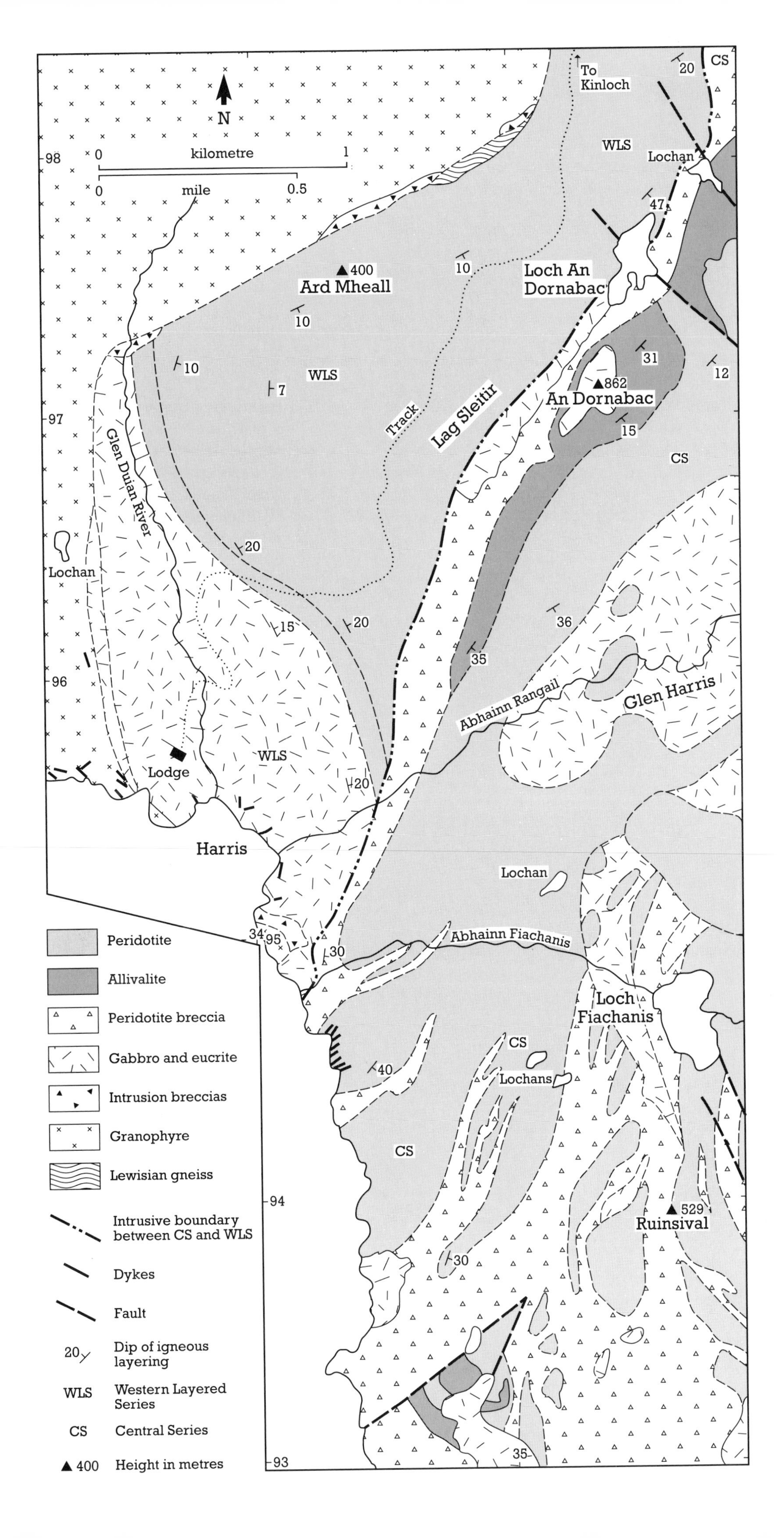

N
kilometre
0
1
mile
0
0.5
98
97
96
94
93
To Kinloch
CS
WLS
Lochan
Loch An Dornabac
Ard Mheall
400
862
An Dornabac
Track
Lag Sleitir
Glen Duian River
Lochan
Lodge
Harris
Abhainn Rangail
Glen Harris
Lochan
Abhainn Fiachanis
Loch Fiachanis
Lochans
Ruinsival
529
Peridotite
Allivalite
Peridotite breccia
Gabbro and eucrite
Intrusion breccias
Granophyre
Lewisian gneiss
Intrusive boundary between CS and WLS
Dykes
Fault
Dip of igneous layering
WLS Western Layered Series
CS Central Series
400 Height in metres

Table 3.2 Harris Bay: subdivisions of the ultrabasic and basic layered rocks (modified from Wadsworth, 1961, table 1, with amended Western Layered Series).

		Thickness	Distinctive features
PART OF CENTRAL SERIES	Upper Ruinsival Series	~ 330 m	Both Ruinsival series show an upwards gradation from olivine cumulates often with feldspar to feldspar–olivine cumulates often with pyroxene.
	Lower Ruinsival Series*	~ 500 m	Exposure is generally poor and the sequence is complicated by transgressive later intrusions, zones of igneous breccia and structural disturbances. In places, gravity stratification, rhythmic layering and slump structures occur.
	Transition Layer	~ 0.5 m	Olivine–feldspar cumulate. Variable dips (5°–50°) in all directions but predominantly in general easterly direction.
	Dornabac Series	~ 130 m	Olivine–feldspar and feldspar olivine cumulates often with streaky or rhythmic layering and frequently with slump structures and evidence of gravity stratification. Layering dips at 35° to 40° to the east and south-east. The rocks show similarities to the allivalites of the Hallival–Askival area. Feldspathic peridotite breccia at the base of the Central Series cuts transgressively across all Western Layered Series units.
AMENDED (1982) WESTERN LAYERED SERIES	Ard Mheall Series	~ 400 m	Olivine and olivine-feldspar cumulates with rhythmic layering throughout. Harrisitic cumulates are intimately associated with normal cumulates and are very prominent within the lower half to two-thirds of the sequence and they are also locally important higher in the series. The layering has a general dip of 5° to 10° (exceptionally 15°) to the south-east or east.
	Transition Series	~ 50–60 m	Olivine-feldspar cumulates, often with pyroxene, of both harrisitic and normal types. Olivine is more abundant than in the Harris Bay Series, while the content of feldspar is higher than in the Ard Mheall Series.

(Table 3.2 contd)

	Thickness	Distinctive features
Harris Bay Series	~130–140 m	Essentially eucritic mesocumulates in texture with olivine, feldspar and ubiquitous pyroxene as cumulus phases. Olivine is the most abundant phase and forms distinctive tabular crystals exhibiting igneous lamination in the normal cumulates. Intercalations of generally thin harrisitic cumulates (crescumulates) richer in feldspar and pyroxene than those of the Ard Mheall Series occur. Layering dips at low angles (5–10°) to the north-east.

* Now termed the Long Loch Group (of Volker and Upton, 1990).

surprisingly constant conformity to the layering. These are well exposed in the low cliffs bordering Glen Duian Burn close to the road bridge at Harris (NM 338 960).

The Harris Bay Series is in contact with the Western Granite at Harris Bay. Along the coastal section from the Mausoleum (NM 336 956) to Gualain na Pairce, the pinkish-weathering granophyre is well exposed and material from this section has been used for radiometric age determinations, giving a date of 59 Ma (Dagley and Mussett, 1981). The granophyre is cut by numerous basaltic sheets and dykes striking more-or-less parallel to the coast. Near the contact the granophyre becomes very tough and takes on a bluish-grey colour. In the cove 200 m south-west of Harris Lodge, the dull, altered granophyre is separated from the Harris Bay Series by a zone, of variable width (2–10 m), of acid hybrid rock containing acicular amphibole and plagioclase crystals. The Harris Bay Series eucritic rocks are layered to within a metre of the contact with hybridized rocks and there is little or no disturbance of the layering at the contact. The contact exposures at the south-east end of Harris Bay provide outstanding examples of intrusion breccias where the basic/ultrabasic rocks have come into contact with acidic rocks; this is a common phenomenon on Rum and it provides a particularly clear example of melting, or partial melting of acid rock by a later mafic intrusion (Fig. 3.12). The intrusion breccia zone is several metres wide on the east of Harris Bay but less than a metre in width on the west where layered ultrabasic rocks are virtually in contact with baked granophyre on the coast close to the Mausoleum (Emeleus and Forster, 1979).

Although not strictly part of the Tertiary geology, a final feature of this site merits a mention. Harris Bay is backed by a series of fine raised beaches at about 30 m. These are Quaternary storm beaches which consist almost exclusively of cobbles and boulders of granophyre and arkose; the paucity of basic and ultrabasic rocks is very striking and indicates their poor resistance under conditions of prolonged mechanical weathering. This may, in part, explain their relative scarcity in the basal and intralava conglomerates of the Fionchra site.

Interpretation

The formation of the Western Layered Series, partly exposed in the Harris Bay site, was probably contemporaneous with the Eastern Layered Series (Emeleus, 1987), the intrusion of the picritic magmas being partly controlled by the Main Ring Fault. However, in the Harris Bay site, the Western Layered Series has intruded and cut the western granophyre mass causing a zone of hybridization and brecciation. The granophyre–Layered Series contact in Western Rum has all the characteristics of an original igneous feature and not a fault as previously suggested (Wadsworth, 1961). This feature, together with

Figure 3.12 Intrusion breccia at the contact of ultrabasic rocks with earlier granite. Eastern end of Harris Bay. Harris Bay site, Rum. (Photo: C.H. Emeleus.)

the roof contacts to the ultrabasic intrusion which are exposed in the Cnapan Breaca and Dibidil sites and on Ard Nev north of this site, and the continuation of the undisturbed layering up to the margins of the complex, suggest that the layered ultrabasic rocks crystallized essentially *in situ* in relation to these contacts.

Unlike the Eastern Layered Series of the Askival–Hallival site, the Western Layered Series and the Central Series contain predominantly feldspathic peridotites and layering involving allivalite is only sporadically developed. It has been suggested that feeders for the successive batches of new magma may have been sited in the western part of the complex, spreading magmas across the chamber floor to the east (Emeleus, 1987), causing peridotite rocks to develop near the feeder(s) and allivalites at the furthest extremities.

The gabbroic cumulates of the Harris Bay Series were formed from a magma considerably less basic than the overlying feldspathic peridotite cumulates. Wadsworth (1961) considered that this could reflect primary differences in magma composition, contamination of more basic magma with granophyre, or the result of strong crystal fractionation in a very thick unit of which the Harris Bay Series is the top. The overlying Transitional Series may represent, according to Wadsworth, gradual replacement of the basaltic magma of the Harris Bay Series with the more mafic magma of the Ard Mheall Series.

The contrasted thicknesses of intrusion breccia along the intrusion contact either side of Harris Bay may be explained by the movement of rheomorphic acid magma, together with inclusions, along the contact. Partially brecciated dykes in granophyre at the eastern exposure indicate that little movement occurred apart from shattering of the edge of the mafic pluton by the acid magma. On the western exposures, acid magma is seen intruding back into the mafic complex in cliffs near the Mausoleum; this probably represents an injection of low-density rheomorphic magma formed at the contact zone and which subsequently moved up and away from its source (cf. Greenwood, 1987).

The layered peridotites and breccias of the Central Series were intruded after the Western Layered Series. They cut across the Transitional and Ard Mheall members and also transgress the Main Ring Fault on its northern and southern margins. The long axis of the Central Series parallels the Long Loch Fault which probably had a strong influence on the emplacement of the Central Series (McClurg, 1982) and possibly controlled the feeder conduits to the whole Layered Series.

The formation of harrisites has been conventionally explained by the upward growth of olivines from the crystal mush on the floor of the magma chamber during periods of quiescence. Wager *et al.* (1960) regarded this texture as a special kind of crescumulate. Donaldson (1982) agreed that most harrisites have such an origin, but argued that some unusual occurrences must have crystallized within the crystal mush. This concept was supported by occurrences of discontinuous, lensoid layers of harrisite; cumulate layers terminating against harrisite; isolated lensoid masses of crescumulate; tongues/lobes of harrisite protruding up into overlying cumulate layers; coarse, randomly orientated hopper and branching olivines growing upwards and downwards in the centres of harrisite lenses. Donaldson postulated that a process of filter-pressing concentrated a differentiate of upwards-migrating intercumulus melt which collected beneath layers of low permeability in the crystal mush. These accumulations of melt were thought to be capable of propagating laterally as sill-like injections and crystallizing as harrisites. The poikilomacrospherulitic plagioclase growths (Donaldson *et al.*, 1973) also probably have a similar, post-depositional 'diagenetic' origin.

Donaldson (1977b) examined the morphology of the olivines in the harrisitic rocks and found that their features could be reproduced in the laboratory. These experiments indicated that crystallization of the branching olivines in harrisites involved crystallization with 30–50° undercooling of the magma. The skeletal and dendritic olivine crystals in the harrisites were found to show notable similarities to the spinifex-textured olivines of Archaean ultramafic rocks (Donaldson, 1974).

Conclusions

The ultrabasic/gabbroic layered rocks of Harris Bay belong to the Western Layered Series and they comprise a lower gabbroic unit which grades into overlying feldspathic peridotite. The intrusion of the later Central Series, which contains spectacular ultrabasic breccias, is also demonstrated by this site. The ultrabasic rocks can be shown to have intruded against the Western Granophyre, the contact being an igneous feature and not a fault as previously proposed. This contact is one of the best exposed in the British Tertiary Volcanic Province between low-melting-point country rocks and later high-temperature mafic intrusives. The site is, however, of special interest as the ultrabasic and gabbroic rocks show many textural characteristics which are as yet unknown in, or not as well-developed in, other layered intrusions. The site is the type locality for harrisite, displaying spectacular crescumulate olivines of both cumulate and post-cumulate origin. Pioneering work on the origin of these rocks has been carried out at this site.

CNAPAN BREACA–LONG LOCH AND DIBIDIL–SOUTHERN MOUNTAINS

Highlights

These sites contain clear evidence for substantial uplift along the Main Ring Fault of Rum. The well-developed felsite/explosion breccia/tuffisite association is now known to be a combination of subaerial ignimbrites, tuffs and caldera breccias. At the roof-like contacts with the gabbros and ultrabasic rocks, there is excellent evidence for hybridization between the gabbroic rocks and remelted felsite.

Introduction

The Cnapan Breaca–Long Loch and Dibidil–Southern Mountains sites (Figs 3.7 and 3.13) demonstrate the earlier Tertiary igneous rocks in the northern and southern margins of the Rum Central Complex, respectively. Both sites contain closely related felsite, granophyre, explosion breccia, tuffisite and bedded tuff in association with the Main Ring Fault. They record an early acidic phase in the evolution of the Rum complex. The Dibidil–Southern Mountains site (Figs 3.7) is of special significance in this respect because of the size and excellent exposure of the acid igneous bodies. The relationships between these early intrusives and Lewisian/Torridonian basement and the later Cenozoic ultrabasic and basic intrusives are well exposed in both sites. Together, the sites provide crucial information in the study of the magmatic and tectonic evolution of the Rum complex.

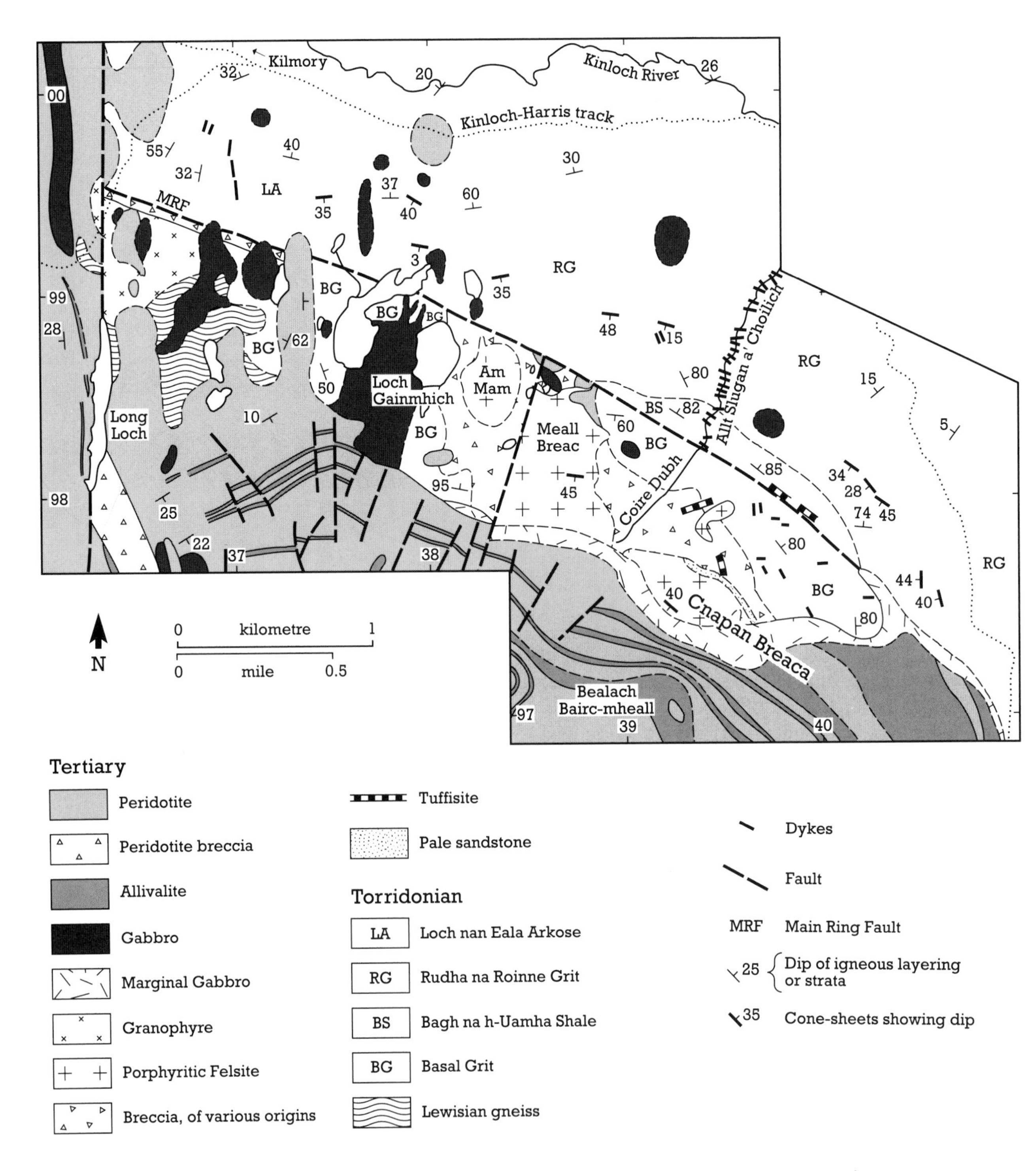

Figure 3.13 Geological map of the Cnapan Breaca–Long Loch site, Rum (after Emeleus, 1980).

Dunham (1962, 1964, 1965a, 1968) carried out the first comprehensive investigation of the felsites and associated rocks in the Cnapan Breaca–Long Loch site. The felsites and granophyres were related to a common parent magma generated by the fusion of Lewisian country rock. The degassing of this magma was considered to be responsible for the occurrence of explosion breccia and tuffisite. However, Williams (1985) reinterpreted some of the felsites to be of sub-aerial, pyroclastic origin. The Dibidil–Southern Mountains area was first mapped in detail by Hughes (1960a), who termed it the Southern Mountains Igneous Complex. Hughes regarded the complex as being bounded by a minor ring fracture, within and partly coincident with the Main Ring Fault. Current ideas, however, suggest a rather different structural setting; the Main Ring Fault clearly bounds the acidic Tertiary igneous rocks, but elsewhere they terminate against intrusions belonging to the later phase of emplacement of the ultrabasic and associated basic rocks.

Description

Cnapan Breaca–Long Loch

The northern part of the site (Fig. 3.13) is underlain by Torridonian arkoses and shales to the north of the Main Ring Fault of eastern Rum which bounds the plutonic complex. Away from the complex, the Torridonian strata have a fairly uniform dip of 10°–20° to the north-west but become progressively disturbed towards the Ring Fault near which they strike east–west and dip north at angle up to 70° and more. The lowermost Torridonian sediments, the Basal Grit and Bagh na h-Uamha Shale groups (Black and Welsh, 1961), occur within the Ring Fault and are well exposed in the area between Cnapan Breaca and the northern end of Meall Breac. They have been highly disturbed by movement along the fault and are intricately folded.

Patches of granodioritic, dioritic, amphibolitic and feldspathic Lewisian gneisses are exposed in a wide area around the Priomh Lochs (NM 368 986), either side of the Long Loch Fault and adjacent to the Main Ring Fault north of Meall Breac (NM 386 988). The presence of these rocks within the Main Ring Fault was used by Bailey (1945) as key evidence for an uplift of at least a thousand metres of the central, fault-bounded block.

Contacts of Lewisian and Torridonian rocks are exposed to the east and north of the Priomh Lochs (NM 372 988). Originally virtually planar, they have been affected by later movements and now dip north-eastwards at between 30° and 85°. The junction is generally faulted, although movement on the fault may not be large. An apparently unconformable relationship is seen north of the lochs (at *c.* NM 369 990), where gneiss is overlain by a coarse sedimentary breccia which passes up into bedded, gritty, coarse sandstone.

In the Cnapan Breaca–Long Loch and Dibidil–Southern Mountains sites, large masses of porphyritic felsite are closely associated with highly brecciated country rock. A felsite sheet caps Cnapan Breaca, dipping 35° to the south-west; a pipe-like mass with steep contacts is found in the east of Coire Dubh; a partly sheet-like, steep-walled mass occurs in western Coire Dubh and on Meall Breac; and on Am Mam, a felsite body with steep northern margins becomes sheet-like in the south. Small lenses of felsite also occur in an east–west-trending mass of explosion breccia along the Main Ring Fault to the north of Long Loch; unlike the other felsites, these are demonstrably older than the breccia which also contains fragments of the felsite as well as gabbro and arkosic sandstone.

The grey, weathered felsites contain conspicuous glomeroporphyritic aggregates of plagioclase, augite and opaque oxides, together with separate phenocrysts of quartz set in a holocrystalline groundmass of quartz and alkali feldspar. In addition, Williams (1985) has recognized some of the felsites at the base of the Cnapan Breaca sheet to be of pyroclastic origin. Typical features of eutaxitic welded tuffs are described by Williams from areas within the felsite, these are: a strong planar fabric, formed by collapsed, attenuated pumice fragments (*fiamme*) and Y-shaped flattened glass shards; rounded Torridonian clasts also occur. A subaerial origin, as opposed to a shallow intrusive origin as suggested by Dunham (1968), for at least some of the felsites is therefore invoked from such evidence. Similar features occur south-west of Meall Breac (NM 384 981). Some of the felsites are ignimbritic.

The felsites are closely associated with coarse breccias and tuffisites which occur almost wholly within the Main Ring Fault. Largely unbedded breccias have a wide outcrop north of Cnapan Breaca, in Coire Dubh around Meall Breac and around Three Lochs Hill (NM 373 987) and form an E–W-trending strip along the line of the Ring Fault to the north of Long Loch. The breccias contain predominantly angular Torridonian sedimentary rock clasts which range from a few centimetres up to several metres in size. Basic igneous and Lewisian gneiss fragments are also occasionally present. Dunham (1968) distinguished two main types of breccia, one made up almost entirely of subangular, rounded blocks derived from the basal Torridonian set in a matrix composed of finely comminuted Torridonian, and the other containing Lewisian and basic igneous fragments set in a very fine-grained matrix of Lewisian. The first type occurs in Coire Dubh and west of Meall Breac, while the second type is found to the north of both Meall Breac and Three Lochs Hill (Am Mam). The east–west strip of breccia to the north of Long Loch differs from the other outcrops in that it contains fragments of rounded basal Torridonian, gabbro, angular felsite and gneiss set in a dark, almost glassy, comminuted matrix derived from all rock types present as clasts except felsite; it is spatially closely associated with the Main Ring Fault and may have resulted from explosive activity localized along the fault.

The occurrence of blocks of coarse gabbro in the second type of breccia recognized by Dunham (1968) is important since it indicates that there were plutonic gabbroic intrusions in existence before emplacement of the felsites and other acidic bodies. This view is reinforced by the discovery of rare blocks of feldspathic peridotite in these breccias on the north end of Meall Breac (Emeleus, in preparation). Some of the coarse gabbros show the effects of crushing, possibly produced during movement of the Main Ring Fault; quite extensive areas of uncrushed gabbro crop out east of Loch Bealach Mhic Neill (NM 376 990) and plugs of petrographically similar gabbro occur on both sides of the Main Ring Fault, for example, north of Loch Gainmich (NM 380 988) and between Kinloch and Coire Dubh (see Emeleus, 1980).

Thin, intrusive tuffisite sheets crop out in the eastern part of the site in close association with felsite and explosion breccia. The petrography of this unusual rock type is described below in the Dibidil–Southern Mountains description; there extensive sheets of tuffisite occur. On Cnapan Breaca, the tuffisites can be generally shown to be younger than the explosion breccia; however, they are also sometimes demonstrably older than the felsites. Dunham (1968) records several tuffisite bodies cutting the Torridonian, both within and outside the Ring Fault, which show no apparent association with either felsite or explosion breccia.

From the southern side of Meall Breac, an olivine gabbro lying between the explosion breccia/felsite and the layered ultrabasic rock extends south-eastwards with widening outcrop. The mass, which is generally poorly exposed, is dyke-like in form and identical to the gabbro which crosses Cnapan Breaca and extends to the Main Ring Fault further to the east. This gabbro is the Marginal Gabbro (Brown, 1956) which has been postulated to have provided a 'lubricant' during the solid emplacement of the layered ultrabasic rocks, although recent evidence questions this interpretation (Greenwood, 1987).

The emplacement of the later gabbro against the felsites to the south of Meall Breac and Cnapan Breaca caused partial fusion of the felsite, resulting in extensive back-veining of the basic rocks by acidic material. Dunham (1964) reported that the remelted felsite back-veined the solid chilled margin of the gabbro, caused some acidification of partially solidified gabbro and then mixed with the still-liquid gabbroic magma in the interior of the intrusion to produce hybrid rocks.

In the west of the site, three tongues of ultrabasic rock extend northward from the main ultrabasic body, cutting through Lewisian and Torridonian country rock, explosion breccias, granophyre and the Main Ring Fault. McClurg (1982) recorded several similarities in these peridotites with the Layered Series peridotites and the peridotite matrices of the ultrabasic breccias. Consequently, McClurg considered the emplacement of the tongues to be contemporaneous with the tectonic disturbances responsible for the formation of the intra-magmatic ultrabasic breccias found elsewhere in this site.

Small-scale (1–3 cm thick) banded structures occur in the tongue peridotites, and in small ultrabasic and gabbroic intrusions elsewhere in the Province (for example, Rubha Hunish, Skye and Camas Mòr, Muck). On Rum, these structures reflect variation in the modal proportions of interstitial clinopyroxene and plagioclase, with the proportion of modal olivine varying very little; this has been termed 'matrix banding' by Dunham (1965b).

Superb examples of layered peridotites of the Central Series occur on the low ridge immediately west of Long Loch and south of the Kinloch–Harris road (NM 363 991; see Fig. 3.14). This is one of the most accessible localities for examination of the varied layered structures in allivalite and feldspathic peridotite: for example, small-scale phase layering, density-stratified layering, layers with size grading, erosional surfaces and a variety of structures due to slumping. Excellent examples of distorted allivalite layering adjacent to slumped peridotite blocks occur just north of the Harris track (NM 363 994), and spectacular peridotite breccias, with subsided blocks up to 3 m in diameter and highly distorted layering, are beautifully displayed 300 m SSE of the south end of Long Loch (McClurg, 1982;

Figure 3.14 Gravity stratified rhythmic layering in allivalite, west of Long Loch, Rum. (Photo: C.H. Emeleus.)

Emeleus, 1987, Fig. 11). Numerous basaltic cone-sheets are exposed in the Allt Slugain a'Choilich which dip either towards a centre in upper Glen Harris or to one somewhat further west. These tholeiitic sheets, which both cut and are cut by basaltic and doleritic dykes, have been studied by Forster (1980).

The major north–south Long Loch Fault, one of the principal features of the geology of Rum, occurs at the western edge of the site. This fault has a considerable zone of crushing, up to 50 m wide in places, involving ultrabasic and earlier rocks. The other spectacular fault on Rum, the Main Ring Fault, is well exposed within the site in Coire Dubh near to the intake of the hydroelectric pipeline (NM 393 983) and close to the deer fence gate. To the east of Cnapan Breaca, the marginal gabbro joins but is unaffected by the Ring Fault and the two are coincident for some distance to the south. West of this point, recent mapping of the margin of this gabbro suggests an upper surface dipping north at a low to moderate angle. Thus, the pre-Marginal Gabbro rocks of this site, the felsites, breccias and associated Torridonian rocks of the northern marginal complex, probably form a roof to these later mafic rocks which may continue beneath this roof as far north as the Main Ring Fault.

Dibidil–Southern Mountains

Lewisian gneisses crop out in a series of elongated, partly fault-bounded blocks for about 2 km west from Dibidil Bay (Fig. 3.7). Here, as on the south-east flanks of Beinn nan Stac, the distribution of gneiss is obviously closely linked with the Main Ring Fault. Torridonian rocks are also present as isolated masses within the igneous rocks and as a more substantial mass in the lower east side of Dibidil within the ring fault. Torridonian country rock outside the ring fault shows signs of disturbance near to this structure on the southern edge of the site.

Extensive sheets of porphyritic felsite are closely associated with coarse breccias and Torridonian sediments at several levels on Sgurr nan Gillean. The felsites in this site have the same general characteristics as those described from Cnapan Breaca–Long Loch. The largest sheet covers the summit area of Sgurr nan Gillean and forms the high ridge extending north out of the site to Ainshval (NM 377 944). At a lower level, on the east side of the hill and about 600 m north-west of the bothy (NM 393 929), a steep-sided felsite intrusion appears to connect the uppermost and lower sheets extending down towards the Dibidil River. This felsite was regarded as a feeder for the felsite sheets (Hughes, 1960a).

The close relationship between the felsite and breccia is clearly demonstrated on Sgurr nan Gillean, where the felsite sheets are generally bordered by breccias. The breccia has the form of flat-lying sheets, sometimes sandwiched between undisturbed Torridonian sediments. This relationship suggests that the breccia formed well below the land surface, however, like some of the Cnapan Breaca felsites in the north, the felsites show evidence in places for a subaerial, ignimbritic origin.

This site is of particular note in that it contains the most extensive developments of tuffisite in any of the Hebridean Tertiary central complexes. Hughes (1960a) first described the rocks as intrusive tuffs, and they were later recognized to be tuffisites by Dunham (1968), following the terminology of Reynolds (1954). Several tuffisite masses have been mapped along the southern edge of the site, closely connected with the Main Ring Fault. Excellent exposures of tuffisite occur near the ford in lower Dibidil (NM 393 931), where slabs in the Dibidil River expose a network of veins and stringers of dark-coloured, fine-grained rock charged with crystals of feldspar and quartz and numerous fragments of Lewisian country rock. Inclusions of porphyritic felsite also occur within the tuffisites of lower Dibidil and, since there is evidence that felsite cuts tuffisite on the southern slopes of Sgurr nan Gillean, there must be a time overlap in felsite and tuffisite formation.

Coarse, acidified, hybrid gabbros are found in several places in the Rum central complex, usually around the periphery of the ultrabasic/gabbroic complex. Hybrid rocks containing conspicuous elongate plagioclase and amphibole (after orthopyroxene) crystals up to 30 mm in length are found in sharp contact with later gabbro to the west of the Dibidil River (NM 3398 9373); further extensive exposures of finer-grained hybrid rocks occur on slabs and rock surfaces for some distance west of this locality. These hybrid rocks appear to be of late formation as they are not cut by the numerous dykes and inclined sheets which are so abundant in the felsite immediately to the west (for example, NM 385 935). The gabbros in Glen Dibidil have recently been recognized to belong to the layered ultrabasic series and are not later intrusive bodies as previously thought (Greenwood, 1987).

The small area of granophyre to the east of Papadil (NM 374 924) is not well exposed. It probably cuts felsite and Lewisian gneiss to the east and north, being fault bounded to the south where it is crushed near to the Main Ring Fault. It is cut by the gabbro which margins Central Series ultrabasic rocks against which it has been recrystallized.

Interpretation

The exposures, within both sites, of porphyritic felsite, explosion breccia, tuffisite and granophyre along the margins of the central complex record an early phase of acidic magmatism and associated tectonism along the Main Ring Fault. Walker (1975) has proposed that acidic magmatism is a characteristic feature of all British Tertiary igneous centres and the record of this early event is particularly well seen on Rum.

The acidic magmatism on Rum was closely associated with the development of the Main Ring Fault (Bailey, 1945). Emeleus *et al.* (1985) have recently proposed that rising acidic magma caused initial doming of Lewisian and Torridonian country rock which ultimately led to ring fracturing. Basal Torridonian and Lewisian rocks were uplifted within the ring fault system. Subsequent relaxation of magmatic pressures caused the central uplifted block to subside along the ring fracture. This is attested by the presence of Mesozoic sediments and Cenozoic lavas (see Allt nam Bà) juxtaposed against faulted slivers of basal Torridonian and Lewisian gneiss which had been elevated by the initial uplift and left stranded at high structural levels along the Main Ring Fault (Emeleus *et al.*, 1985). Large masses of basal Torridonian and Lewisian gneiss on the southern lower slopes of Sgurr nan Gillean and in the Cnapan Breaca–Long Loch site, represent relict roof rocks to the Central Complex.

The caldera-like subsidence within the Main Ring Fault appears to have been clearly associated with a reduction of magmatic pressures caused by escaping magma and volatiles along fault systems. The explosion breccia provides evidence for violent release of volatiles from the acid magma which shattered the country rock along lines of weakness such as faults and bedding planes (Hughes, 1960a). However, Williams (1985) has argued that they may represent vent breccias in a deeply eroded edifice through which felsic magma rose resulting in either shallow, sill-like and steep-sided intrusions or subaerial pyroclastic extrusion; evidence for the latter occurs on Cnapan Breaca. Williams also postulated that the close association of breccias with felsites on the inner margin of the Main Ring Fault could indicate an origin for the breccias by caldera-wall collapse; evidence for this interpretation is good in Dibidil (M. Errington, pers. comm.).

Using petrographic and geochemical evidence, Hughes (1960a) demonstrated that the felsite and granophyre crystallized from the same magma derived from the fusion of Lewisian basement. The granophyre probably represents a thick ring-dyke intruded along the line of the ring fault at deeper levels than the felsite and it thus did not suffer degassing and associated explosive activity. Contemporaneous emplacement of tuffisite along the ring fracture system also occurred, representing fluidized, high-pressure injections of shattered country rock (Hughes, 1960a) mixed with fragment porphyritic felsite magma.

The northern and southern sites provide valuable information as to the nature of the margin to the later ultrabasic/basic complex. The contact between the latter and the felsites and associated rocks has been shown generally to dip outwards at both sites at angles as low as 40°, representing a roof-like contact (Emeleus, 1987). The ultrabasic layering is undisturbed right up to these contacts and extends beneath the overlying rocks where hybridization of basic/ultrabasic intrusives with felsite is observed. It is probable that large parts of the northern marginal complex and Southern Mountains marginal complex are immediately underlain by the layered ultrabasic rocks and that the contact represents the original roof to the mafic complex. If the roof contacts are projected upwards to the centre of the complex, the vertical extent of the Eastern Layered Series is limited to a few hundred metres above the present-day peaks. The layered ultrabasic rocks are, therefore, not considered to have been emplaced as a solid, upfaulted block, as previously suggested. The ultrabasic magma may have intruded upwards, causing further uplift along the Main Ring Fault involving the felsites and associated rocks (see Emeleus, 1987 for discussion), the layered series crystallizing essentially *in situ* beneath them. The emplacement of the Layered Series, however, still presents many difficulties and work is currently in progress which will hopefully resolve these problems.

Conclusions

The Dibidil–Southern Mountains and Cnapan Breaca–Long Loch sites are important localities exposing the margin of the igneous complex and allow investigation of the early magmatic and tectonic evolution of the Rum centre. The felsite–granophyre explosion breccia–tuffisite association can be related to major caldera-like subsidence along the line of the Main Ring Fault, with contemporaneous acidic magmatism. The roof contacts between these rocks and the underlying layered series are of particular importance in these sites since they provide evidence that the ultrabasic/basic complex crystallized *in situ* in relation to these rocks, although further uplift along the Main Ring Fault may have occurred during the emplacement of the ultrabasic magmas/rocks.

The precise nature of the origin of the felsites and associated rocks, and the structural complexities of the sites, have received little attention since the work of Dunham (1968). The areas merit reassessment in view of the reinterpretation of the Cnapan Breaca felsite–explosion breccia association (Emeleus *et al.*, 1985; Williams, 1985) and work is currently in progress. Early acidic magmatism and the presence of welded tuffaceous felsitic rocks, such as those observed on the northern margin of the Rum complex, are common to many Tertiary igneous centres. The opportunity for a comprehensive understanding of well exposed acidic rocks in these sites on Rum will provide valuable information on the early magmatic and tectonic evolution of the British Tertiary Igneous Volcanic Province as a whole (cf. Bell and Emeleus, 1988).

SOUTH-WEST EIGG

Highlights

The Sgùrr of Eigg pitchstone lava flow dominates the site. The flow fills a valley system, floored with fluviatile sediments, which was carved into Tertiary basaltic lavas. The pitchstone is one of

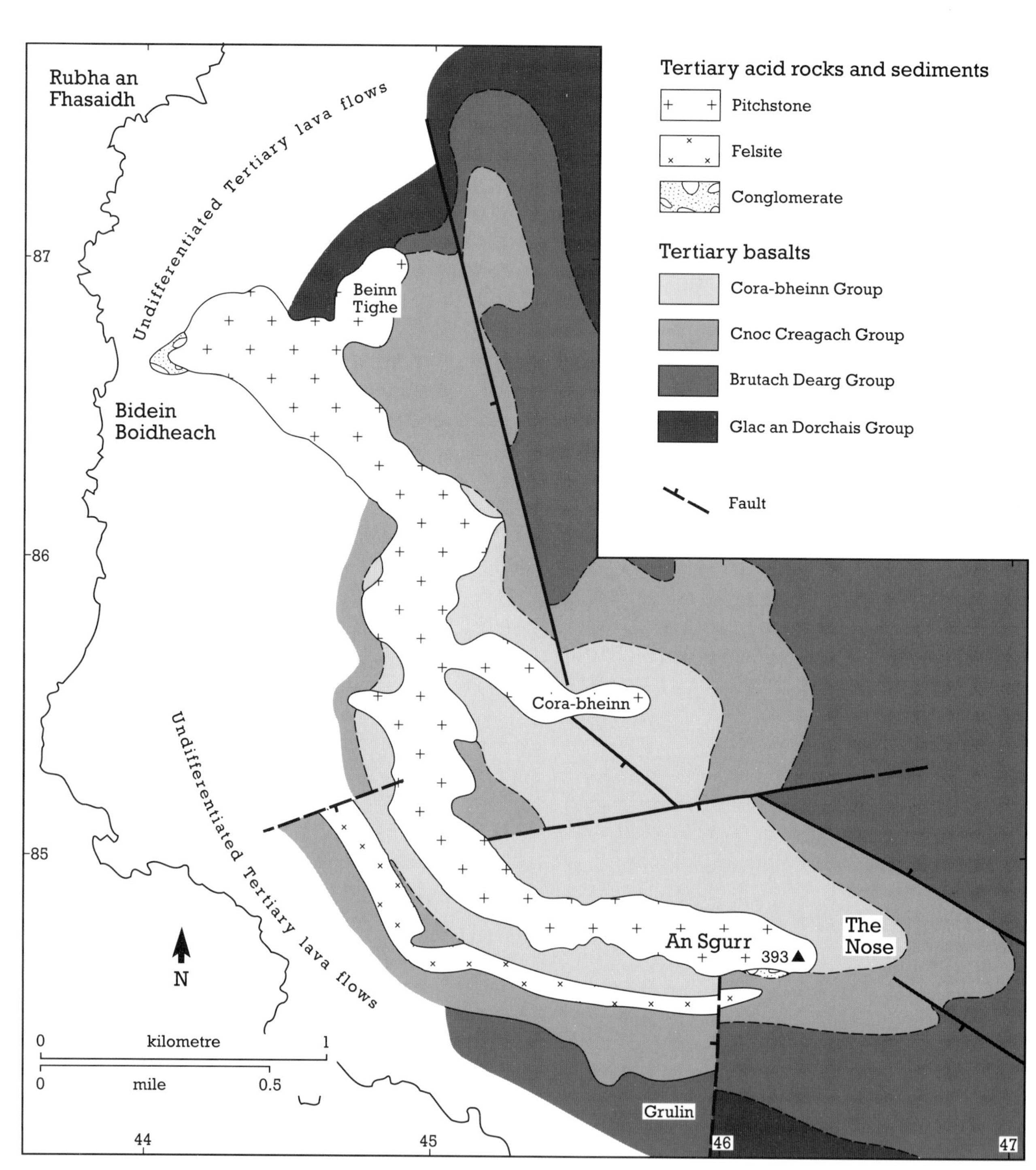

Figure 3.15 Geological map of south-west Eigg (after Allwright, 1980, fig. 2.3.2).

the youngest igneous rocks in the British Tertiary Volcanic Province.

Introduction

The south-western part of Eigg lying between Rubha an Fhasaidh and An Sgùrr provides good exposure through a Tertiary lava succession (Fig. 3.15). The site is dominated by the unique columnar pitchstone outcrop forming the Sgùrr (Fig. 3.16) and also demonstrates intercalated sediments in the lava pile, together with both acid and basic minor intrusions.

On Eigg, the basaltic lavas show slight differences from those on Skye, but vary little in detail from those on Mull. The most striking and

Figure 3.16 Ridge of the Sgùrr of Eigg, formed by an Eocene pitchstone flow filling a valley eroded from Palaeocene basalt lavas. South-west Eigg site. (Photo: C.H. Emeleus.)

controversial geological feature is the relatively young pitchstone forming An Sgùrr and adjoining hills. Geikie (1897) regarded the pitchstone as a subaerial lava flow which had occupied a system of small river valleys eroded into the underlying lavas, but Harker (1908) reinterpreted it as an intrusion. Bailey (1914) subsequently supported Geikie's view but invoked auto-intrusion to explain some features. Ridley (1973) supplied new mineralogical and geochemical data from both An Sgùrr and the earlier lavas but remained uncommitted as to the nature of the pitchstone. Recent research, however, favours an extrusive origin for the Sgùrr pitchstone (Allwright and Hudson, 1982).

Description

From Laig cliffs (NM 463 880) to the outcrops of pitchstone at Beannan Breca (NM 448 865) or Cora-bheinn (NM 457 856), the exposure in the crags displays a full sequence, at least 200 m in thickness, through the lavas of western Eigg. The flows are predominantly alkali to transitional olivine-phyric or olivine-rich basalts. The only exceptions are several flows of feldspar-phyric basalts near to the top of the sequence.

The lavas are typical Palaeocene basalts containing olivine phenocrysts in a groundmass of labradorite, clinopyroxene and titanomagnetite. In hand specimen, they commonly weather with rusty crusts. In contrast, the pitchstone is black and lustrous and carries alkali-feldspar phenocrysts and glomeroporphyritic aggregates of clino- and orthopyroxene, titanomagnetite and alkali feldspar in a pale-brown, glassy matrix (Ridley, 1973; Allwright, 1980).

A thick pitchstone sheet forms the ridge from An Sgùrr westwards to Beinn Tighe and rests, often with spectacular unconformity, upon the basaltic lavas. At several localities along its base, the pitchstone is seen to be brecciated, flow banded, and to contain possible flattened shards (*fiamme*). It lies upon a conglomerate possibly produced by fluvial reworking of agglomerate. Fragments of wood and other plant remains have been found hereabouts. A thick lens of fluviatile conglomerate underlies the pitchstone at Bidein Boidheach (Fig. 3.17; NM 441 867).

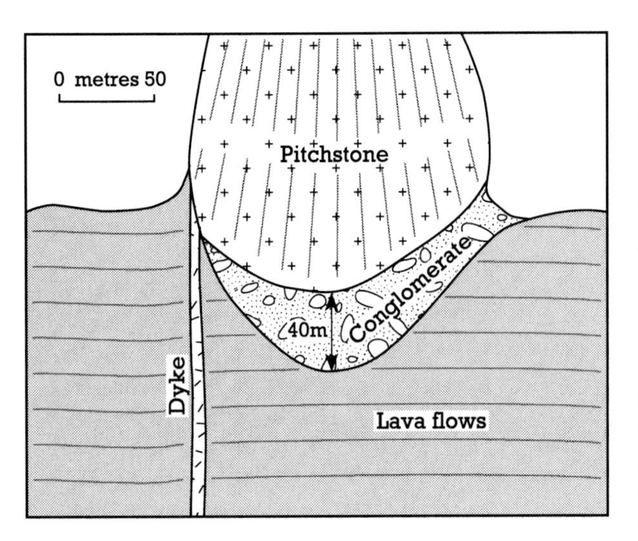

Figure 3.17. Section through pitchstone and lava flows, near Bidein Boidheach, south-west Eigg (after Allwright, 1980, figure 6.4b).

Along the south face of An Sgùrr (NM 460 846) several felsite sheets intrude the pitchstone. These are interpreted as being due to the back-injection of residual acid magma since they are mineralogically identical to the pitchstone, differing only in their well-crystallized matrices (Ridley, 1973; Allwright, 1980). Below the ridge of An Sgùrr another major felsite–the Grulin Felsite – intrudes the basaltic lavas and is demonstrably later than some of the basic dykes. The Grulin Felsite has been regarded as the feeder for the Sgùrr pitchstone, although this view is now discounted (for example, Dickin and Jones, 1983). Petrographically, this rock is a quartz microsyenite.

The pitchstone post-dates the NW-trending basic dykes of Eigg as can be seen at the east end of An Sgùrr (NM 464 847) and in the cliff section near Bidean Boidheach. Radiometric age determinations on the pitchstone indicate an age of *c.* 52 Ma; it is thus one of the youngest igneous rocks in the British Tertiary Volcanic Province (Dickin and Jones, 1983).

Interpretation

The basaltic lavas of this site are cut by an extensive swarm of NW-trending basaltic dykes (Speight *et al.*, 1982; Allwright, 1980). The probable continuation of the dyke swarm appears in south-east Rum where it is cut by the layered complex. The Eigg (and Muck, see below) lavas thus predate the Rum central complex and their equivalents appear to have been caught up in the Rum Main Ring Fault (Smith, 1985, 1987).

The site illustrates how it is possible for eminent geologists to come to diametrically opposed interpretations despite excellent exposure. Harker (1908) held firmly to his view of the intrusive origin for the Sgùrr pitchstone despite the earlier work of Geikie (1897) and a vigorous defence of Geikie's views by Bailey (1914). The careful work of Allwright (1980) leaves no doubt as to the correctness of Geikie's and Bailey's general interpretations; at Bidean Boidheach, the site contains an excellent example of a valley system which has been filled by a pitchstone flow. The fluviatile conglomerates exposed at the base of the pitchstone at Bidean Boidheach have yielded clasts of arkosic and other sandstones of Torridonian age. Since the Eigg Tertiary lavas overlie a thick Jurassic succession it is likely that the provenance of these sediments was high ground west of the Camasunary Fault's southern continuation (Binns *et al.*, 1974). In this connection, it is of interest that a petrographically identical pitchstone is present well to the west of this fault, forming the islets of Oigh-sgeir (or Hyskeir; NM 156 963). The possible connection between these two pitchstones requires further investigation, as does the possibility that the pitchstones may be welded ash flows.

Conclusions

The lavas on Eigg represent some of the earliest volcanic activity in the BTVP, predating the nearby Rum central complex. As in the Fionchra site on Rum, the pitchstones, which are of special interest in this site, have interacted with a fluvial system and provide an excellently exposed example of the filling of a valley system which was carved into the earlier Tertiary basaltic lavas which now cover much of Eigg. An Eocene age (52 Ma) has been obtained from the pitchstone which is one of the youngest igneous rocks of the province.

CAMAS MÒR, ISLE OF MUCK

Highlights

This locality shows a well exposed sequence of interbedded tuffs, clastic sediments and lavas at the base of the Tertiary succession. It also shows that a 'giant' dyke of gabbro has produced a suite of high-temperature calc-silicate minerals by the thermal metamorphism of Jurassic sediments.

Introduction

The base of the Palaeocene lava succession overlies Jurassic sediments at Camas Mòr and these rocks are intruded by a dense swarm of NW-trending basalt dykes and by a large gabbro dyke. The general geology of the site has been described by Harker (1908) and the Camas Mòr gabbro and associated metamorphism of Jurassic sediments was the subject of a detailed investigation by Tilley (1947). The dyke swarm was included in the survey by Speight *et al.* (1982) and the Tertiary lavas, tuffs and sediments have been mapped and described by Allwright (1980).

Description

The exposures on the shores of Camas Mòr show an interesting sequence of basal Tertiary tuffs and sediments at the junction of the Tertiary lavas with Jurassic sediments (Fig. 3.18). Two localities are of special interest, the first is among the boulders on the foreshore below the An Stac cliffs, a few metres east of Sgorr nan Loagh where two horizons of bedded, water-laid tuffs occur. The exposures are terminated to the east by a dyke which has apparently intruded along a fault which throws the basal Palaeocene beds down against the Jurassic limestones. The second locality, at the eastern end of the An Stac cliffs (NM 403 789), is figured by Harker (1908) and has been reinvestigated by Allwright (1980) who produced the following succession:

		Approximate thickness
10.	Coarse plagioclase-rich basaltic lava in discontinuous flow units.	8.5 m
9.	Red laminated tuff infilling fissures in reddened scoriaceous lava.	0.75 m
8.	Plagiophyric basaltic lava.	7.10 m
7.	Finely brecciated flow top.	1.10 m
6.	Aphyric, very fine-grained, mugearite lava.	5.80 m
5.	Coarsely bedded, greyish-pink sediment enclosing numerous small basalt fragments.	0.95 m
4.	Gap (possibly the thin lava flow of Harker, 1904).	0.60 m
3.	Brown–red mudstone.	0.80 m
2.	Purplish-red, sandy tuffs often distinctly banded.	1.10 m
1.	Jurassic limestone with thin shales and concretionary sandstones.	

The Palaeocene sediments are predominantly air-fall deposits and the bedded units most probably originated through the deposition in shallow water of airborne basaltic ash and lapilli from adjacent fissures, perhaps by lava fountaining.

A complete sequence through the lavas of Muck is provided by a gully at the west end of An Stac and exposures above the cliff tops on the hillside towards Beinn Aireinn (NM 403 792). The lavas are transitional in composition between alkaline and tholeiitic basalts, they are mostly olivine-phyric, but with plagioclase-rich porphyritic types and a mugearite towards the base.

The cliffs and the beach on the eastern side of Camas Mòr, especially between NM 416 782 and NM 412 781, expose lavas and Mesozoic sediments in contact with a major olivine gabbro dyke trending NW–SE. The dyke rock is a relatively iron-rich, hypersthene-normative gabbro consisting of large titaniferous augites enclosing laths of zoned labradorite, irregular olivines and anhedra of titanomagnetite. This central facies locally develops centimetre-scale diffuse mafic and felsic layering, somewhat similar to that found in the Trotternish sills of Skye (Rubha Hunish). There is also a finer-grained marginal facies to the gabbro which is mildly nepheline normative. The dyke, which in places attains 50 m–100 m in width, has incorporated and partly assimilated some of the adjacent Mesozoic limestones, resulting in the formation of contact assemblages as reported by Tilley (1947); the assimilation is believed responsible for the undersaturated, nepheline-bearing nature of the marginal facies of the dyke (Tilley, 1952; Ridley, 1973).

Minerals recorded by Tilley from the contact zone include calcite, grossularite, wollastonite, monticellite, melilite (gehlenite), spurrite, merwinite, larnite, rankinite, cuspidine, tilleyite, periclase, brucite, spinel and perovskite and from the skarn zone where the limestones were soaked in solutions from the gabbro he noted

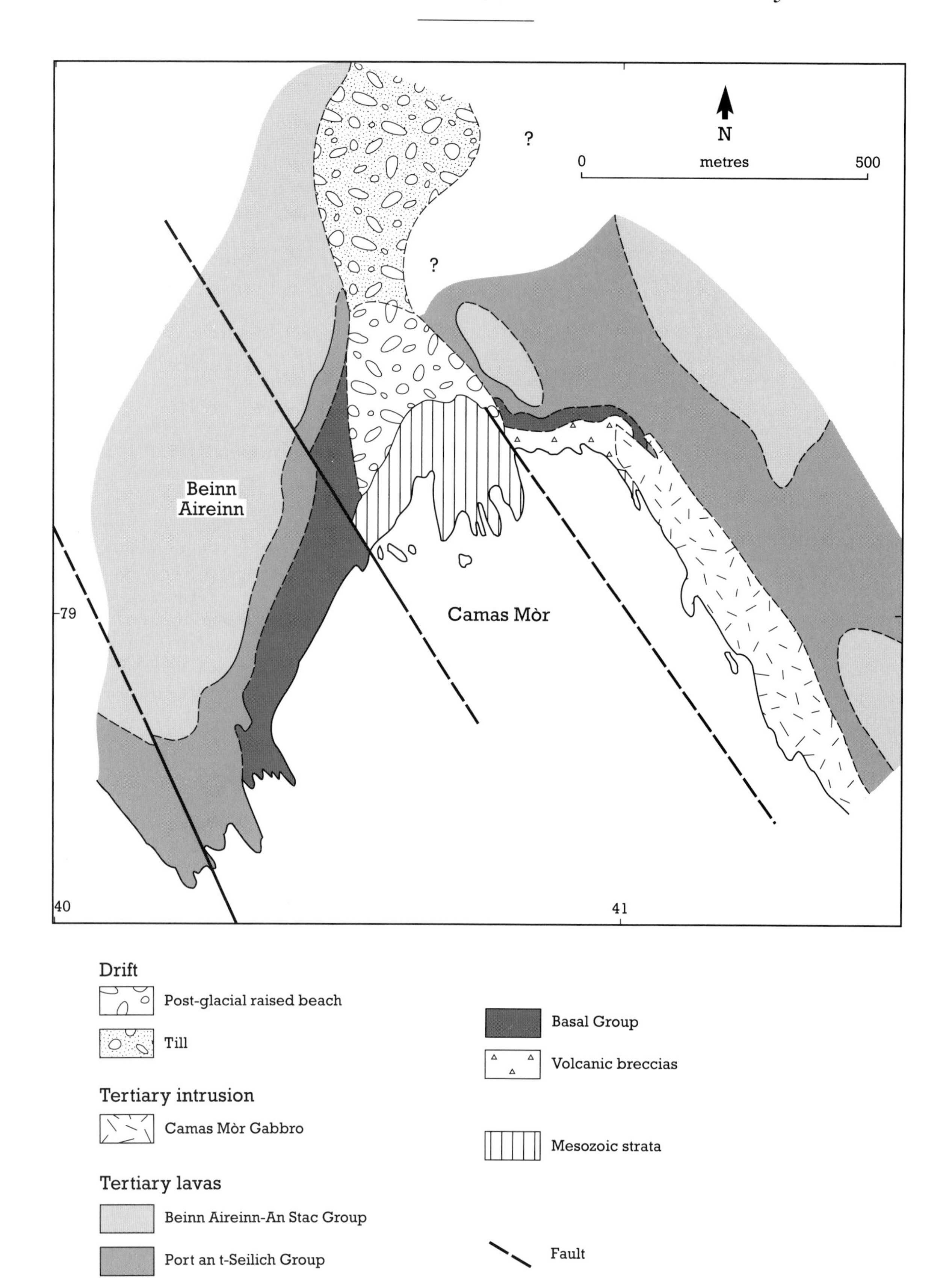

Figure 3.18 Geological map of the Camas Mòr site, Muck (after Allwright, 1980, fig. 2.2.2).

clinopyroxene, analcite, soda-sanidine and titanite (sphene). The gabbro is modified for a few centimetres from its contact with the sediments; pyroxenite is followed away from the contact by theralite (nepheline gabbro) with wollastonite, melilite and soda-sanidine and iron-rich olivine in segregations.

Immediately north of the dyke, the Mesozoic sediments are found to be extensively brecciated. Blocks of limestone, sandstone and black shale up to 1 m in diameter are set in a comminuted matrix of these rocks which is, apparently, free of any igneous material. Unbrecciated Palaeocene lavas margin the breccias but do not themselves show any disturbance. The origin of this breccia is uncertain; it may represent an early explosive vent immediately prior to lava effusion. It predates intrusion of the olivine gabbro dyke.

Muck is characterized by a dense swarm of NW-trending basic dykes which are less alkaline than the lavas they intrude (Allwright, 1980). Harker (1908) recorded at least 40 dykes along the south coast between Camas Mor and Port Mòr with an aggregate width of about 60 m.

Interpretation

The Palaeocene lava successions in the BTVP frequently provide glimpses of interbedded sediments and thin pyroclastic deposits (Anderson and Dunham, 1966). This site clearly exposes a succession of water-laid sediments, fine-grained air-fall tuffs, possible volcanic breccias and basaltic lava flows which mark the onset of the Palaeocene volcanism. These rocks are cut by numerous basaltic dykes which represent a crustal dilation of *c.* 6%, not including the very thick gabbroic dyke at Camas Mòr. Tilley's (1947) examination of this dyke showed that, in addition to producing a suite of high-temperature, calc-silicate minerals in the country rocks, there had been a reaction between the hypersthene-normative basaltic magma of the dyke and the calcareous sediments to form a limited, marginal zone of nepheline normative, critically undersaturated rocks rather similar in nature to that described by Tilley from Scawt Hill, Co. Antrim (Tilley and Harwood, 1931). Ages of about 63 Ma obtained from the Muck and Eigg lavas are among the oldest in the BTVP (Mussett *et al.*, 1988).

Conclusions

Camas Mòr contains particularly good examples of basaltic lavas with interbedded sediments of volcanic origin and a dense swarm of basaltic dykes. A very thick gabbro dyke cuts the lavas and Jurassic sediments, altering them to high-temperature hornfelses with distinctive calc-silicate minerals. Reaction between the gabbro and calcareous sediments has resulted in a distinctive marginal zone of nepheline-normative rock and skarn mineral assemblages. The lavas are among the oldest in the Province, which accords with their pre-Rum Central Complex age.

EAST CANNA AND SANDAY

Highlights

The inter-lava fluvial sediments and pyroclastics exposed here are the best developed in the British Tertiary Volcanic Province, providing an essential link in a chain of sites in the study of the Tertiary volcanic history of the region. Derived clasts suggest that the Rum Complex is appreciably older than the Skye Cuillin centre.

Introduction

Agglomerates, fluvial conglomerates and other sediments closely associated with the volcanic succession of Canna, occur within this site and are the best developed and most extensive examples within the BTVP. These formations were the first such deposits within the Province to be described in detail. Geikie (1897) concluded that the conglomerates had filled a river channel cut into the lavas. The islands were later mapped by Harker (1908) who accepted the fluvial origin. Allwright (1980) has more recently interpreted coarse conglomerates on Compass Hill in terms of a nearby source of coarse pyroclastic debris which was possibly connected with a volcano which fed some of the lava flows.

Description

The spectacular sea cliffs, stacks and intertidal areas of south-eastern Sanday, Eilean a'Bhaird and eastern Canna contain good exposures of the interlava clastic sediments. In and near to the

Figure 3.19 Stack of Dùn Mòr, Sanday, formed of basalt lavas with interbedded coarse conglomerates. The conglomerates contain rare granite pebbles from the Rum Central Complex. Canna–Sanday vicinity. (Photo: A.P. McKirdy.)

stacks of Dùn Mòr (NG 2877 0374; Fig. 3.19) and Dùn Beag (NG 2888 0375; Fig. 3.20), conglomerates occur beneath a thick, columnar basalt lava. On Dùn Mòr and in the adjacent cliffs, the conglomerates locally exceed 2 m in thickness and appear to blanket underlying vesicular basalts. On Dùn Beag, on the other hand, the conglomerates reach 15 m in thickness overlying amygdaloidal lavas and abutting directly against them. This important exposure was interpreted by Geikie as the wall of a river channel or gorge, against which the deposits were banked. The conglomerates of Dùn Mòr pass laterally into finer-grained tuffs and shales, the transition being accompanied by a marked reduction in overall thickness. Identical sediments and overlying lavas can be traced for some distance westwards along the cliffs of southern Sanday. The sediments are heterogeneous and comprise crudely bedded conglomerates (often exceptionally coarse, with boulders over a metre in diameter), thin-bedded, green/pink ashy shales, sandstones and tuffaceous mudstones. Carbonaceous streaks and organic remains have been reported by Geikie (1897) and Harker (1908). The clasts include Torridonian arkoses, Tertiary lavas, various types of schist, gneiss and granophyre, the last of these probably derived from the Rum western granite (Emeleus, 1973). The columnar basalt which overlies the sediments has a markedly chilled base; the occurrence of lobate structures with radial fractures and diffuse vesicles arranged concentrically, strongly suggests that the lava had flowed into and along channels occupied by waterlogged sediments. The sediments, lavas and conglomerates of south-east Sanday lie towards the base of the exposed volcanic succession (Harker, 1908; Stewart, 1965; Allwright, 1980).

The impressive hundred-metre-high cliffs and the coastal sections below Compass Hill (NG 280 063) in eastern Canna expose intercalations of basaltic lavas and pyroclastic rocks and the thickest development of predominantly water-laid sediments on the island (Fig. 3.21). These comprise conglomerates and sandstones derived largely from the erosion and transport of contemporaneous pyroclastic material. The sediments have been interpreted as deposits formed in fluvial and perhaps marginal lacustrine environments (Geikie, 1897; Harker, 1908; Allwright,

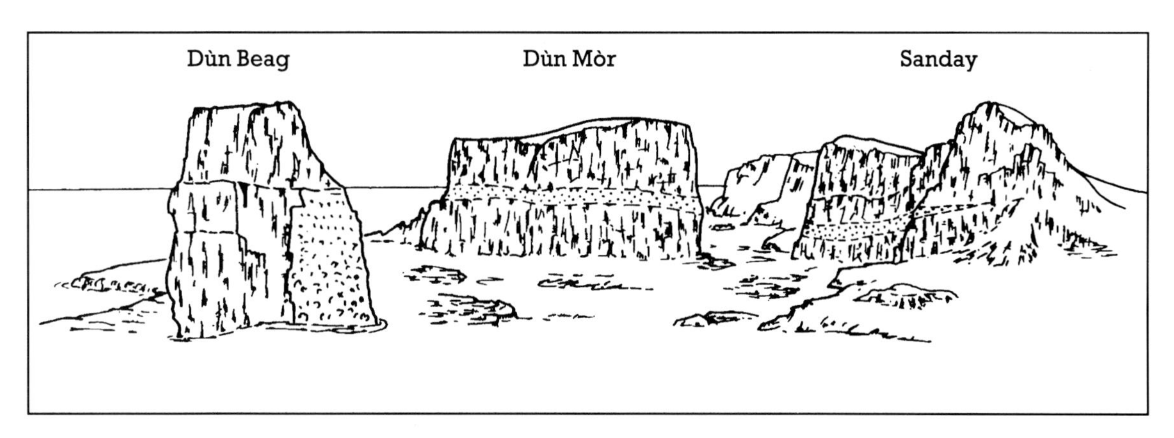

Figure 3.20 Sketch of Dùn Beag, Dùn Mòr and the cliffs of Sanday (after Harker, 1908, figure 12). Columnar basalt flows with interbedded conglomerate are seen on Dùn Mòr and the cliffs of Sanday. On Dùn Beag, lavas fill a steep-sided valley eroded in conglomerate.

1980). Compass Hill is significant in that it shows the intimate relationships between lavas, conglomerates and agglomerates. Lithologically, the conglomerates are similar to those on Dùn Mòr, no doubt owing to their closer proximity to agglomerates, and contain a higher percentage of basic igneous pebbles, whilst granophyre and gneiss pebbles are rare. Porphyritic felsite pebbles, similar to the Rum felsites, have been found here.

The sea stacks of Alman (NG 2805 0545) and Coroghon Mor (NG 2797 0554) expose confused masses of pebble conglomerates in basaltic lavas; the lavas have tended to nose their way into the sediments which must have been unconsolidated. No vent has been identified on Canna but the coarseness of the reworked conglomerates and 'agglomerates' suggest that one was nearby, possibly only a short distance offshore to the east, as indicated by the westwards thinning of the deposits. Large blocks of Torridonian arkose occur in the 'agglomerates' on the shore to the north-east of Compass Hill which indicates that the basal lavas may rest directly on Torridonian sediments.

Eilean a' Bhaird (NG 270 050) is a small island rising above the tidal flats of Canna Harbour. It contains the most evolved lava on Canna, a very fine-grained flow which is chemically of tholeiitic andesite rather than mugearite (Muir and Tilley, 1961; Ridley, 1973). The flow overlies a substantial thickness of very coarse conglomerates. They are identical to the others on Canna and appear to infill a small channel (Figs 3.22 and 3.23; Allwright, 1980).

Interpretation

The value of the Canna–Sanday site lies in the evidence that it provides for the development of extensive, vigorous river systems during the period of Palaeocene lava accumulation. Geikie (1897) placed particular emphasis on the palaeogeographical implications of the clasts in the fluviatile sediments; the occurrence of gneisses, schists and Torridonian clasts indicates erosion in the Palaeocene of a landmass containing lithologies similar to present-day western Scotland. The abundance of Torridonian fragments in the pyroclastic-derived deposits of Compass Hill might alternatively indicate a Torridonian basement to the Tertiary lavas, although it is surprising that no Mesozoic sedimentary fragments have been recognized, since Canna overlies a Mesozoic basin (Binns *et al.*, 1974). The position of the Eilean a'Bhaird flow within the Canna sequence is problematic. Allwright argues that it must be fairly low in the stratigraphic succession of lavas; however, it could be one of the latest lavas in the area, since it overlies conglomerates filling a valley eroded in lavas and no examples of a lava of similar composition are known to be overlain by other lavas on Canna or Sanday.

The similarity between some of the Canna–Sanday lavas and those of north-west Rum (Emeleus, 1985), and the occurrence of clasts apparently derived from the acid rocks in the Rum complex, establishes a close connection with Rum. The site is a vital link in the chain of sites from Muck and Eigg, through Rum, to south-

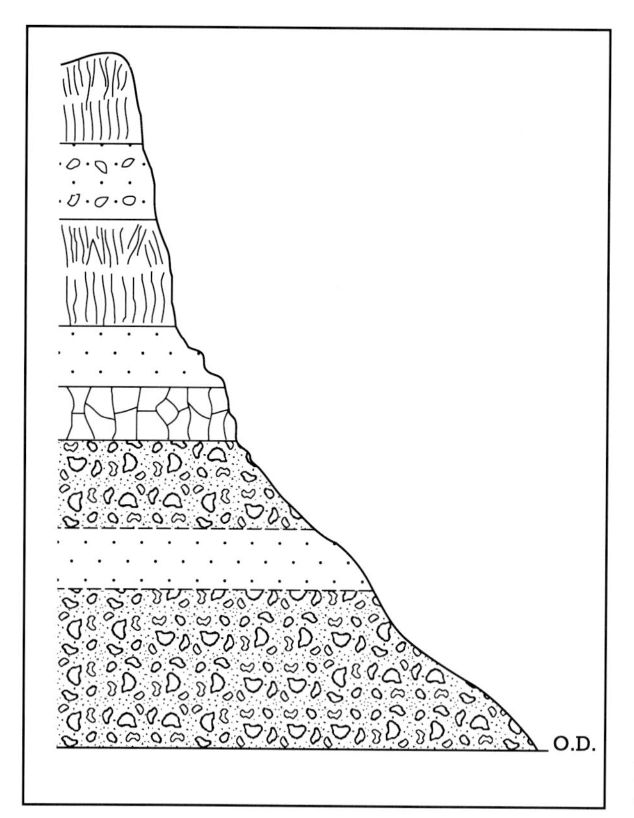

Figure 3.21 Cliff section at Compass Hill, Canna (after Harker, 1908, figure 8).

west Skye and Ardnamurchan which enables continuous, relative dating of igneous rocks across the British Tertiary Volcanic Province to be attempted, thus aiding assessment of the evolution of the Province (Meighan *et al.*, 1981; Dagley and Mussett, 1986).

Conclusions

The intra-lava sediments in the volcanic succession of east Canna and Sanday provide unequivocal evidence that a river system draining areas of gneiss, schist and probably Torridonian sediments, was established during periods when active effusion of basaltic and intermediate lavas was occurring. The rivers also drained Rum and probably extended to Skye, the distinctive pebbles laid down in the associated fluviatile deposits make it possible to conclude that the Skye Cuillin Central Complex almost certainly post-dated the Rum Central Complex.

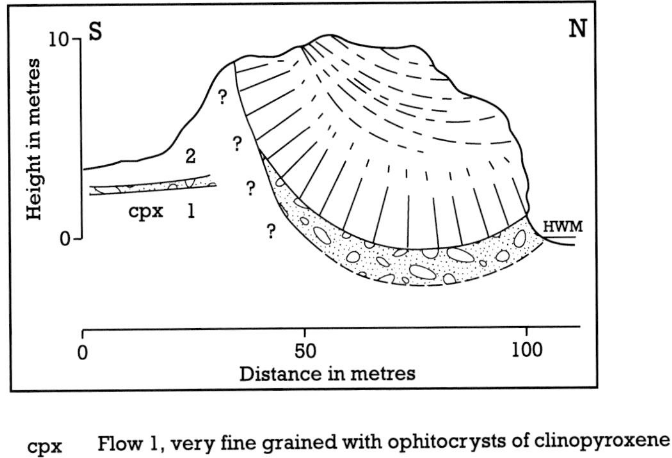

cpx Flow 1, very fine grained with ophitocrysts of clinopyroxene

2 Flow 2, fine grained aphyric lava

Upper part of valley-infill flow (horizontal jointing)

Lower part of valley-infill flow (columnar flow)

Conglomerate

Figure 3.22 Canna Harbour: Eilean a' Bhaird from the east (after Allwright, 1980, figure 2.4.13).

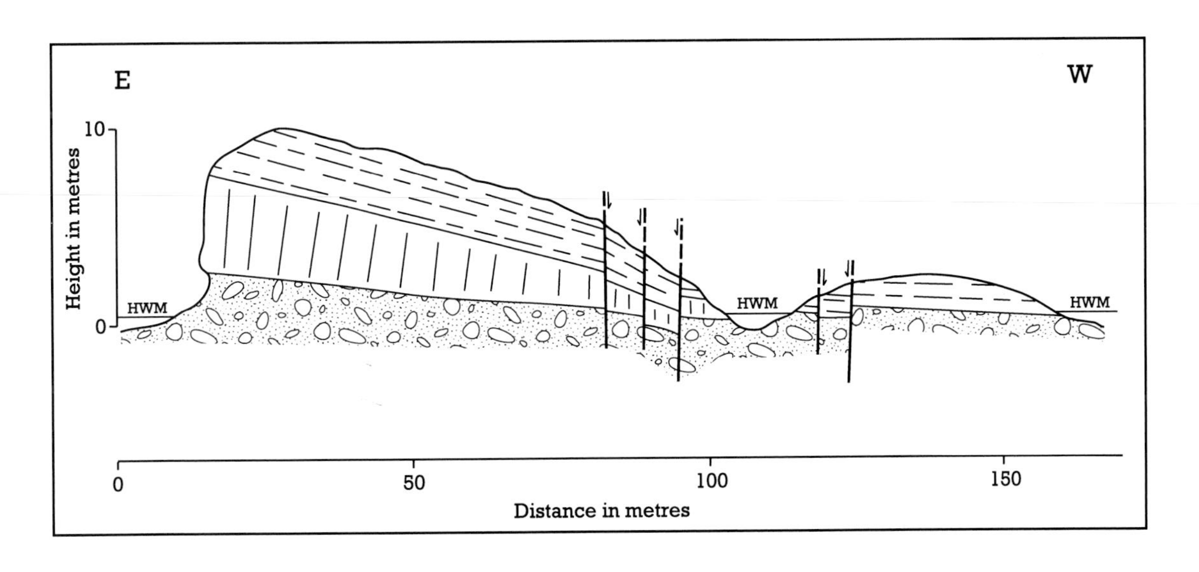

Upper part of valley-infill flow (horizontal jointing)

Lower part of valley-infill flow (columnar flow)

Conglomerate

Figure 3.23 Canna Harbour: Eilean a' Bhaird from the north (after Allwright, 1980, figure 2.4.14).

Chapter 4

Ardnamurchan

INTRODUCTION

The Ardnamurchan Peninsula is the site of a Tertiary igneous complex which was emplaced into Moine schists, Mesozoic sediments, and early Cenozoic volcanic rocks. It is much less rugged and mountainous than the other British Tertiary Volcanic Province central complexes, yet in many ways, it more strikingly – and often much more simply – demonstrates the salient geological features common to many of the centres.

The complex igneous geology of Ardnamurchan became recognized through the investigations of several geologists during the nineteenth century (for example, Judd, 1874, 1886; Geikie, 1888, 1897); however, it was not comprehensively mapped and examined until the official survey was undertaken by Richey and Thomas (1930). Despite appearing over half a century ago, the Ardnamurchan Memoir probably remains the most complete and widely used work on the area, although some of the interpretations and general conclusions reached in it have been questioned by subsequent investigators. The recent facsimile (1987) reprint of this Memoir is to be welcomed.

The Ardnamurchan Centre is renowned for its development of sets of cone-sheets and numerous arcuate, dominantly gabbroic intrusions which were, almost without exception, interpreted by the authors of the Memoir as ring-dykes. From the disposition of the ring-dykes and cone-sheets, Richey and Thomas (1930) recognized that these arcuate intrusions defined three separate centres of igneous activity (referred to henceforth as Centres 1, 2 and 3 respectively) and that the focus of the activity, as indicated by these centres, had shifted progressively with time (Table 4.1). Although there are many other features of interest in the igneous rocks of Ardnamurchan, it is these features in particular for which the area is renowned and upon which its claim to be of both national and international geological importance rests. The geology of the area has been the subject of several reviews and field guides (Richey, 1933; Richey, 1961; Stewart, 1965; Deer, 1969; Gribble *et al.*, 1976; Emeleus, 1982, 1983). It should, however, be mentioned that the status of the three independent but overlapping centres has been questioned (for example, Durrance, 1967; Green and Wright, 1969, 1974; Gribble *et al.*, 1976), as has the interpretation of many of the gabbroic and other intrusions as ring-dykes (for example, Wells, 1954a; Bradshaw, 1961; Skelhorn and Elwell, 1966). The Tertiary igneous geology of Ardnamurchan is represented by six SSSIs (Fig. 4.1).

Relicts of the earliest activity, which was dominantly volcanic and probably covered much of the peninsula, are now found in the Ben Hiant area (Table 4.1). Remnants of mildly alkaline plateau lavas, overlying a basal ashy mudstone, are cut by a series of vents and pitchstone lavas of intermediate composition which post-date the basalts. The relationships of these rocks, together with numerous later dolerite intrusions, including a set of cone-sheets, are all excellently displayed in the Ben Hiant site; coarse volcanic breccias, possibly representing vent infills, are particularly well developed at Maclean's Nose.

The painstaking field and laboratory investigations of Richey and Thomas (1930) demonstrated the presence of a complex series of suites of cone-sheets and ring-dykes. Representative sections across such intrusions, assigned in the Memoir to Centre 2, are present in the Beinn na Seilg–Beinn nan Ord and Ardnamurchan Point to Sanna sites, while the most complete major ring structures are exposed in Centre 3 representing the final focus of activity. Studies made on these intrusions since the appearance of the Memoir have revealed features which, in some instances, do not conform to the expected pattern of the classic ring-dyke. Variations in these patterns have been noted in the two outermost intrusions of Centre 2 (Richey, 1940; Wells, 1954a, 1954b; Skelhorn and Elwell, 1966, 1971; Wells and McRae, 1969; Butchins, 1973). In addition to field studies, the mineralogy and geochemistry of the two principal centres (Centres 2 and 3) have received considerable attention (Bradshaw, 1961; Smith, 1957; Walsh, 1971, 1975; Gribble, 1974; Walsh and Henderson, 1977). Other features such as the dolerites, xenolithic inclusions and associated granophyric bodies with their distinctive net-veined relationships in dolerites have been described by Gribble (1974), MacGregor (1931), Wells (1951), Brown (1954), Paithankar (1968) and Vogel (1982).

The mutual relationships between all three centres of activity, and the successive truncation of the margins of intrusions and internal megascopic structures by later intrusions, are demonstrated in the Glas Bheinn–Glebe Hill site, where remnants of the former volcanic cover of the peninsula are also present as a screen between two ring-dykes. Intense thermal metamorphism of country rocks adjoining the major intrusions has

Table 4.1 The geological succession in the Ardnamurchan Central Complex (based on Richey and Thomas, 1930, Chapter 7)

(youngest)
- Late NNW-trending dolerite dykes

Centre 3
- Quartz monzonite
- Tonalite
- Fluxion biotite gabbro of Glendrain
- Fluxion biotite gabbro of Sithean Mor
- Quartz–biotite gabbro
- Quartz dolerite, granophyre-veined
- Inner Eucrite
- Biotite eucrite
- Quartz gabbro, southern side of Meall an Tarmachain
- Quartz gabbro of Meall an Tarmachain summit
- Outer Eucrite
- Great Eucrite
- Cone-sheets of Centre 3 (sparse)
- Porphyritic gabbro of Meall nan Con screen
- Gabbro, south-east of Rudha Groulin
- Gabbro of Plochaig
- Fluxion gabbro of Faskadale
- Quartz gabbro of Faskadale

(Migration of focus of activity to Achnaha area)

Centre 2
- Felsite, south of Aodann
- Fluxion gabbro of Portuairk
- Younger quartz gabbro of Beinn Bhuidhe
- Quartz gabbro of Beinn na Seilg
- Quartz gabbro of Loch Caorach
- Eucrite of Beinn nan Ord
- Inner cone-sheets of Centre 2
- Quartz dolerite of Sgurr nam Meann
- Quartz gabbro of Aodann
- Older quartz gabbro of Beinn Bhuidhe
- Granophyre of Grigadale
- Quartz gabbro of Garbh-dhail
- Old Gabbro of Lochan an Aodainn
- Hypersthene gabbro of Ardnamurchan Point
- Glas Eilean vent
- Outer cone-sheets of Centre 2

(Migration of focus of activity to Aodann area [NM 453 664])

Centre 1 and the Ben Hiant vent*
- Cone-sheets of Centre 1 (penecontemporaneous with the quartz dolerite intrusion of Ben Hiant)
- Ben Hiant quartz dolerite
- Composite intrusion of Beinn an Leathaid
- Augite diorite of Camphouse
- Quartz dolerite of Camphouse

Table 4.1 *contd*

Porphyritic dolerite of Ben Hiant
Granophyre west of Faskadale
Quartz gabbro west of Faskadale
Old Gabbro of Meall nan Con
Porphyritic dolerite of Glas Bheinn
Agglomerates of Northern Vents
Tuffs, agglomerates and lavas of Ben Hiant vents
Trachyte plug

(Igneous activity localized at Ben Hiant and also centred on a focus *c.* 1.3 km west of Meall nan Con)

Palaeocene basalt lavas and thin sediments
Jurassic and Triassic sandstones, shales, limestones, conglomerates
Moine metasediments

(oldest)

*The relative ages of many of the units assigned to Centre 1 and Ben Hiant are uncertain. (From Emeleus, in Sutherland, 1982, table 29.5).

been noted at several localities. A classic example of hornfelsed, aluminous, iron-rich, sediments (bole?) occurs in the site at Glebe Hill (Richey and Thomas, 1930), while a complex suite of calc-silicate hornfelses containing the mineral kilchoanite, first identified on Ardnamurchan (Agrell, 1965), crops out a short distance to the east.

Many of the Tertiary central complexes in the British Isles contain examples of cone-sheets, but those belonging to the Ardnamurchan complex are perhaps the most obvious, well exposed, easily accessible and widely studied (Harker, 1917; Richey and Thomas, 1930; Anderson, 1936; Kuenen, 1937; Durrance, 1967; Le Bas, 1971; Holland and Brown, 1972; Phillips, 1974). The most extensive suites are those developed about Centre 2, and the Glas Eilean–Mingary Pier site has been selected as a standard reference section where the form, petrology, contact-metamorphic effects and, to some extent, the emplacement mechanism of these intrusions can be studied. In Richey and Thomas's classification, the cone-sheets in this site comprise part of the outer set of Centre 2; members of the inner set may be seen in the Beinn na Seilg–Beinn nan Ord site and the rather poorly developed suite attributed to Centre 3 may be studied on the south-east of the Centre 3 site (Table 4.1). The Glas Bheinn-Glebe Hill site contains cone-sheets attributed to Centre 1, but it has been suggested that all the cone-sheets belong to a single spiral suite (Durrance, 1967), and Holland and Brown (1972) were unable to discriminate geochemically between sheets assigned to Centres 1 and 2.

Ardnamurchan is the site of a positive Bouguer gravity anomaly (Bott and Tuson, 1973) but the anomaly is markedly less intense than those found on Mull, Skye or Rum which indicates a rather shallower body of mafic rock underlying the Ardnamurchan complex. Radiometric age determinations on rocks of the complex indicate that the activity took place in the Palaeocene (*c.* 60 Ma; Miller and Brown, 1965; Mitchell and Reen, 1973) but the data do not allow the different intrusions or centres of activity to be separated, since the duration of igneous activity at Ardnamurchan was probably of the order of one million years, comparable with the margin of error of the Ardnamurchan age determinations (cf. Table 1.1; Mussett *et al.*, 1988).

BEN HIANT

Highlights

The steep slopes of Ben Hiant provide an excellent cross-section of volcanic vents filled with agglomerate, ash and pitchstone lavas. The vent rocks are cut by large basic intrusions of complex origins.

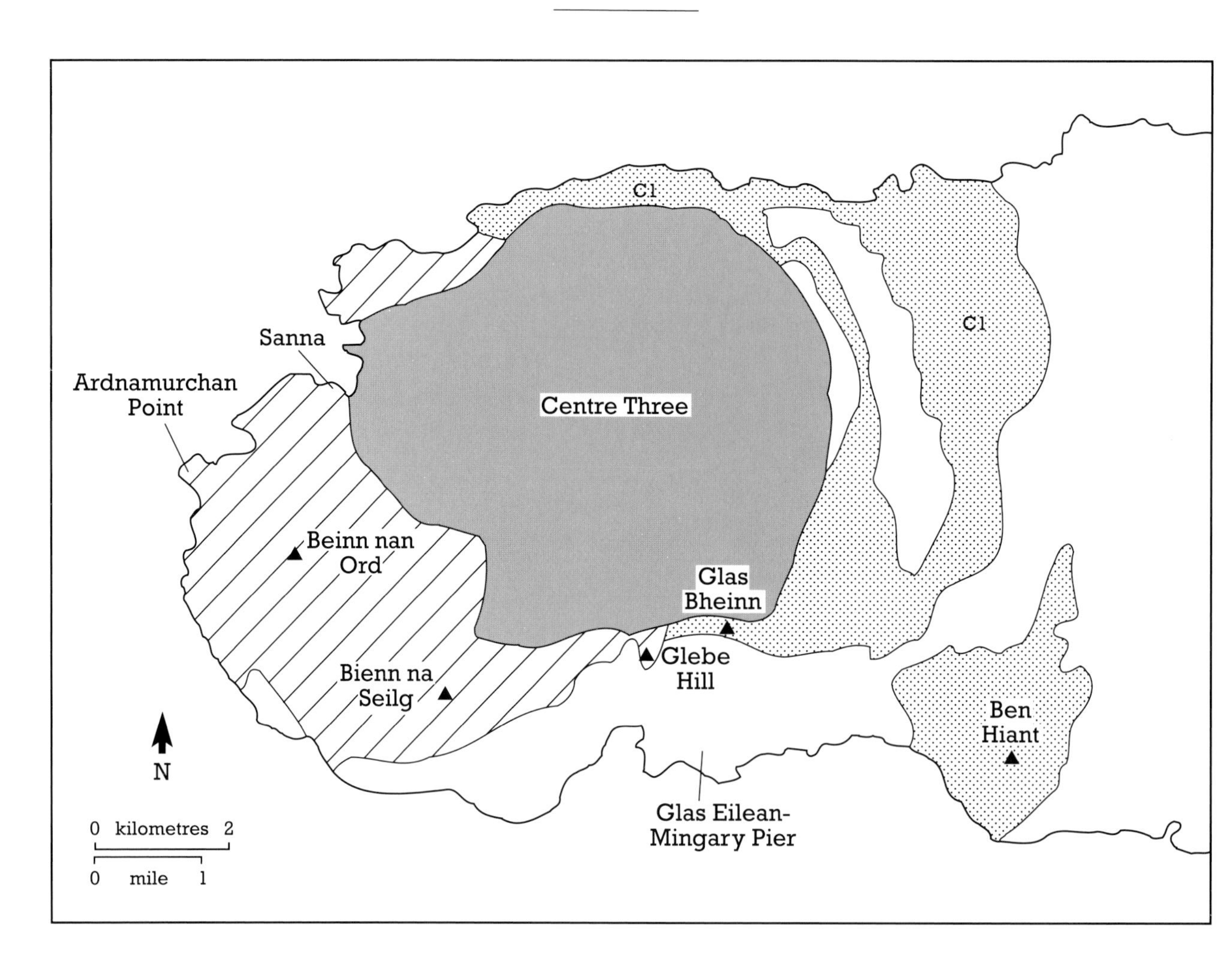

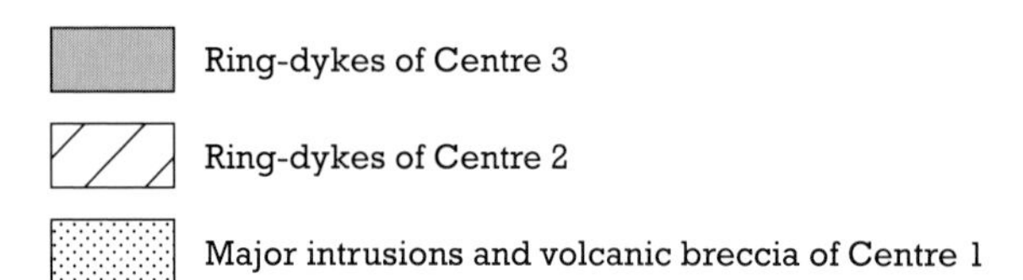

Figure 4.1 Map of the Ardnamurchan Peninsula showing localities mentioned in the text.

Introduction

The geological interest of Ben Hiant and the neighbouring hills of Beinn na h-Urchrach and Stallachan Dubha lies in the well developed assemblage of volcanic vents, associated lavas and major basic intrusions which are part of the early Centre 1 of the Ardnamurchan complex. The Ben Hiant quartz dolerite mass and vent agglomerates dominate the geology of the site and thin remnants of the earlier lava plateau of this region are also well represented (Figs 4.2 and 4.3).

Early research in Ardnamurchan centred largely around the Ben Hiant area (Judd, 1874, 1886, 1890 and Geikie, 1888, 1897), but the significance and complexity of the area was not revealed until the work of Richey and Thomas (1930) and Richey (1938). Subsequently, little research specifically related to the area has been published, although samples from Ben Hiant have been used in a petrological study of doleritic intrusions within the Ardnamurchan complex by Gribble (1974) and in an investigation of the radiometric ages of the rocks in the complex by

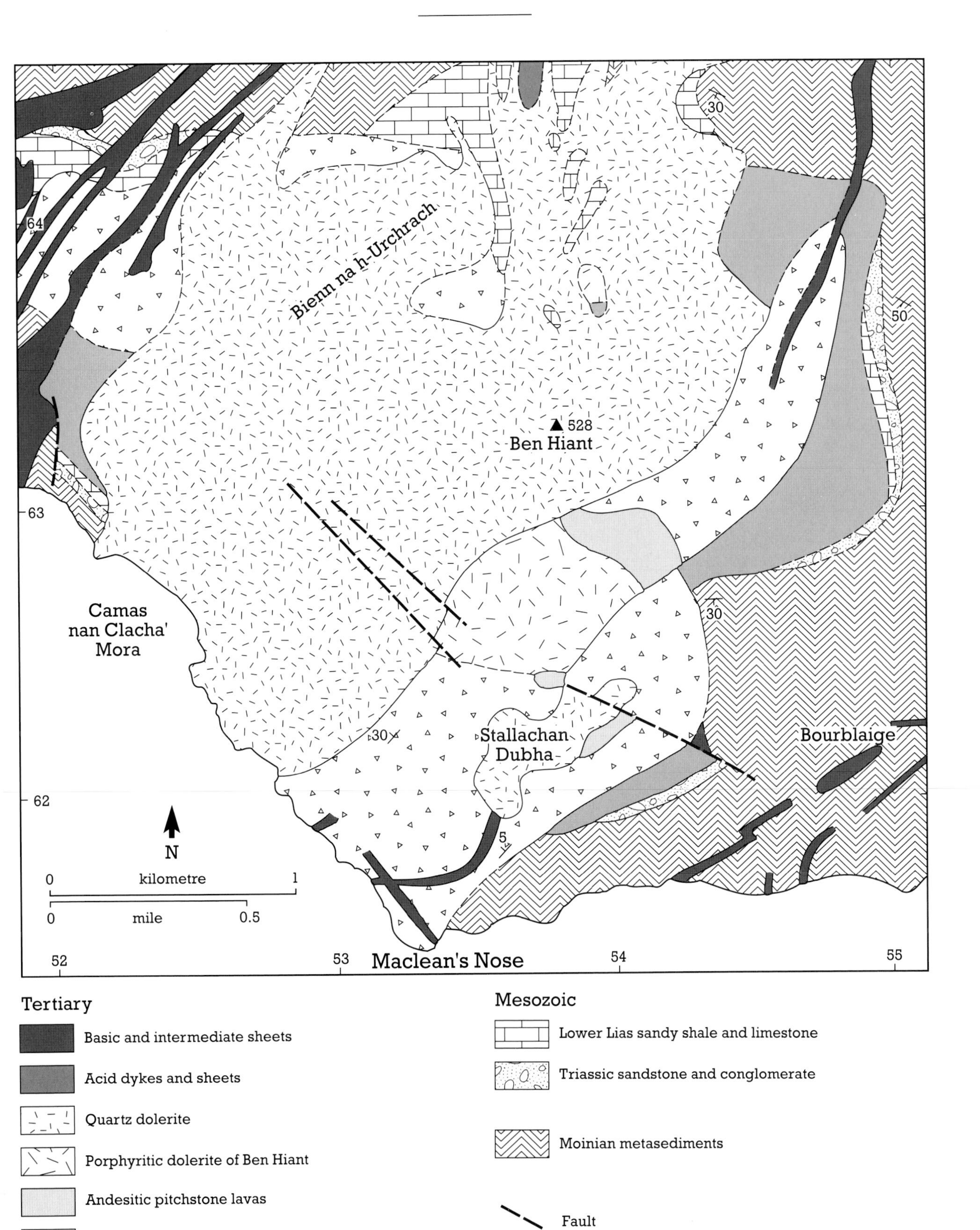

Figure 4.2 Geological map of the Ben Hiant site (after Gribble, 1976).

Figure 4.3 Ben Hiant from the east, showing terracing developed along the location of minor intrusions. The headland to the left is Maclean's Nose, formed by volcanic breccias. Ben Hiant site. (Photo: C.H. Emeleus.)

Mitchell and Reen (1973). In addition, the need for a reassessment of the status of Centre 1 has been suggested by Green and Wright (1969, 1974).

Description

The eastern slopes of Ben Hiant (Fig. 4.2) to the north and south-west of Bourblaige (NM 547 623) expose the most representative remnants of the early lava plateau which has been mostly obliterated by the central complex. The lavas are either non-porphyritic or microphyric, alkali-olivine basalts containing alkali-rich pegmatoid patches in which analcite and alkali feldspar are present, together with augite zoned to aegirine augite. Several thin flows can be distinguished which commonly exhibit spheroidal weathering.

In two places, the base of the lava sequence rests upon a basal mudstone of a similar nature to sediments beneath the Mull lavas (Bailey *et al.*, 1924). This deposit is underlain by Triassic sediments, themselves unconformably overlying Moine schists.

Two vents, infilled with agglomerates, tuffs, crater lavas and two major dolerite intrusions, cut through the remnants of the lava plateau and the underlying country rocks; they form the high terrain of Ben Hiant and Beinn na h-Urchrach. The volcaniclastic material within the vents consists of fragments up to 0.3 m in diameter and larger volcanic bombs several metres across. These deposits are both bedded and unbedded. The fragments are predominantly of trachytes and porphyritic basalt lavas, but acid and rare fine-grained basaltic rock types also occur. Clasts of Moine country rock contribute to the agglomerate near to the vent margins.

The well known exposures at Maclean's Nose (NM 533 616) show a sharply defined and near-vertical vent margin. Here the agglomerate contains several beds of tuffaceous material composed of comminuted basic rock, plus crystals of quartz and mica likely to have been derived from Moine schists. Large basaltic blocks have close petrological affinities to the adjacent lavas and appear to have fallen into the vent from the walls of the crater.

In the south-eastern vent, lavas are intercalated with the volcaniclastic rocks. The most noteworthy is a suite of andesitic pitchstone lavas seen at three localities, which are considered to have been erupted within a caldera. At least three flows interbedded with thin tuffs are recognized, each showing a triple-tiered structure comprising a slaggy amygdaloidal lava top overlying a layer of lava with small, subhorizontal columns and passing downwards into a layer characterized by larger vertical columns; such a flow structure is typical of lavas which have cooled slowly. There is some evidence in the pitchstone lavas to suggest at least limited auto-intrusion not dissimilar to that found in the pitchstone of the Sgùrr of Eigg (see above). The rocks are relatively fresh, dark-brown to black in colour with a fine-grained or glassy texture. Microphenocrysts of augite–ferroaugite, pigeonite–ferrous pigeonite (Emeleus *et al.*, 1971) and labradorite occur in a groundmass which shows different degrees of devitrification and consists of acid glass, oligoclase microlites, rare orthopyroxene and accessory iron–titanium oxides and apatite. The chemistry of the pitchstones is andesitic (being more basic than the Sgùrr of Eigg flow); they were termed augite andesites or inninmoreites by Richey and Thomas (1930) and are similar to rocks now termed icelandites (Carmichael, 1964).

Within the Ben Hiant vents there are two major dolerite intrusions. The smaller is a roughly circular mass of porphyritic dolerite intruded into the agglomerates and pitchstone lavas of the south-east vent, to the south of Ben Hiant. This dolerite is cut by the later Ben Hiant intrusion which contains a large xenolith of the porphyritic dolerite to the north of the Ben Hiant summit. A similar rock type, characterized by large, conspicuous labradorite/bytownite phenocrysts (up to 15 mm in length), forms the escarpment of Glas Bheinn (NM 495 648) near Kilchoan and is perhaps related in origin. The contacts of the porphyritic dolerite with the andesitic pitchstone lavas are of particular interest since the earlier, but not the later, lavas show contact alteration suggesting that intrusion of the dolerite was contemporaneous with the infilling of the vent.

The larger intrusion of Ben Hiant forms the hills of Ben Hiant, Beinn na h-Urchrach and Stallachan Dubha and reaches the coast at Camas nan Clacha' Mora (NM 524 626). The mass consists of a number of varieties of quartz dolerite and of less common olivine dolerite, all having strong tholeiitic characteristics. The intrusion is best described as a non-porphyritic, ophitic dolerite; however, larger crystals of altered olivine, augite, ilmenite and labradorite do occasionally occur. A mass of columnar-jointed variolitic rock, probably formed by chilling at the upper contact, overlies and grades into normal dolerite to the south-west of the Ben Hiant summit. Labradorite microphenocrysts lie in a typically feathery-textured, variolitic groundmass of acicular augite, magnetite and oligoclase/andesine feldspar. Rare glomeroporphyritic aggregates of augite and labradorite also occur. The margins of the Ben Hiant intrusion contain xenoliths of basic volcanic rocks and of schist; where the latter have suffered partial melting, hybrid rocks are in evidence.

Minor intrusions are scattered throughout the site and include basic and composite dykes and intermediate cone-sheets.

Interpretation

Vent deposits in eastern Ardnamurchan around Ben Hiant, and those outside the site forming an arcuate belt from Camphouse to Kilmory, represent the first manifestations of centralized volcanic activity in the region after the extrusion of the plateau lavas. The vents appear to be of a much later date than the lavas since they contain material of a very different composition and acidic vent lavas, interstratified with agglomerate and ash, are significantly different from the crater lavas on Mull. The two vents of Ben Hiant were probably active for a prolonged period, during which explosive activity predominated and plateau basalts were not erupted; the products of this activity infilling enormous craters (Richey and Thomas, 1930). As the greatest height of the vent deposits is now over 300 m above sea-level, Richey and Thomas have argued that the crater walls were at least this high; this assumes that there has been no tectonic uplift of the deposits. Where the vent margins are exposed in the Ben Hiant area, there is usually remarkably little brecciation of the country rock which is difficult to reconcile with vents characterized by violent explosive activity; the breccias may be partly the products of debris avalanches off the walls.

The form of the Ben Hiant intrusion is of particular interest and uniqueness. Judd (1890) erroneously interpreted the terrace- or scarp-like topographic features of the south-eastern parts of the outcrop as a succession of lava flows but Geikie (1888) correctly concluded that the dolerite was intrusive (Fig. 4.3). He speculated that the intrusion has the form of a suite of coalescing sills, an interpretation also favoured by Gribble (1974). Richey and Thomas (1930) interpreted the north-western part of the intrusion, which overlies agglomerate, as a lateral off shoot extending from a lower mass with vertical contacts, while the south-eastern margin was considered to be formed by a suite of sheets which coalesced to form a single intrusion. The sheets dip at angles slightly less than those shown by the Centre 1 cone-sheets with which they share a similar composition. The intrusion is therefore suggested to be a mushroom-shaped body in part and also an assemblage of coalesced sheets. The rocks which form the successive terraces vary slightly in chemistry (Gribble, 1974; Gribble *et al.*, 1976) and modal mineralogy (Gribble *et al.*, 1976, confirming Judd, 1890). Gribble (1974) has argued that the intrusion is not a single homogeneous body and it is possible that several pulses of magma were responsible for the intrusion, the problems that it poses not yet having been resolved.

In comparison with the other major late dolerites of Ardnamurchan (Richey and Thomas, 1930; Skelhorn and Elwell, 1966; Holland and Brown, 1972; Gribble, 1974), those of Ben Hiant are distinctly tholeiitic, showing only slight iron-enrichment trends. Gribble (1974) suggested that the magma which formed the Ben Hiant Intrusion was the parental magma for all the rocks of Centre 1 and the source from which the rocks of Centre 2 and 3 were ultimately derived having a composition similar to the non-porphyritic Central Magma Type of Bailey *et al.* (1924).

Conclusions

Early explosive volcanic activity and limited lava effusion produced the vents of Ben Hiant and eastern Ardnamurchan which represent the first manifestations of the central complex. A significant repose period had intervened between this activity and the earlier eruption of plateau basalts. The vents were active for a prolonged period and were later intruded by doleritic and quartz gabbro masses and the largest of these, the Ben Hiant intrusion, is probably largely a mass of coalesced cone-sheets. The magma which formed this intrusion may have been parental to all the rocks in Centre 1 and compositionally similar to the parent magmas for the intrusions in Centres 2 and 3.

GLAS EILEAN–MINGARY PIER

Highlights

One of the best sections through a cone-sheet swarm to be found in the British Tertiary Volcanic Province is exposed on the shore west of Mingary Pier. Numerous basic, rare acid and composite sheets enclose thin screens of country rock, some of which have developed distinctive thermal metamorphic minerals. A well exposed linear volcanic vent on Glas Eilean cuts the cone-sheets and is itself cut by a tuffisite dyke.

Introduction

The numerous cone-sheets of the Ardnamurchan complex are exceptional in their development and are of international importance. The 1.5 km stretch of coast between Glas Eilean on Kilchoan Bay and Mingary Pier provides a representative section through a large number of basic and less-abundant felsite sheets associated with Centre 2. Other features of interest within the site include remnants of the early lava plateau and deposits in a linear vent (Fig. 4.4).

The cone-sheets of Ardnamurchan were originally mapped by the Survey (Richey and Thomas, 1930) which led to the recognition of several distinct sets related to each of the intrusive centres. The origin of the intrusions has been subject to continuing debate since the interpretations of Bailey *et al.* (1924) and Richey and Thomas (1930) for the cone-sheets of Mull and Ardnamurchan, respectively. These workers, and Anderson (1936), attributed the sheets to the updoming of the country rock by rising magma which resulted in conical fractures along which the cone-sheets were emplaced. Durrance (1967), Phillips (1974) and Walker (1975) have subsequently presented modified interpretations on the mechanisms of cone-sheet emplacement. The geochemical characteristics of these intrusions have been studied by Richey and Thomas (1930) and, more recently, by Holland and Brown

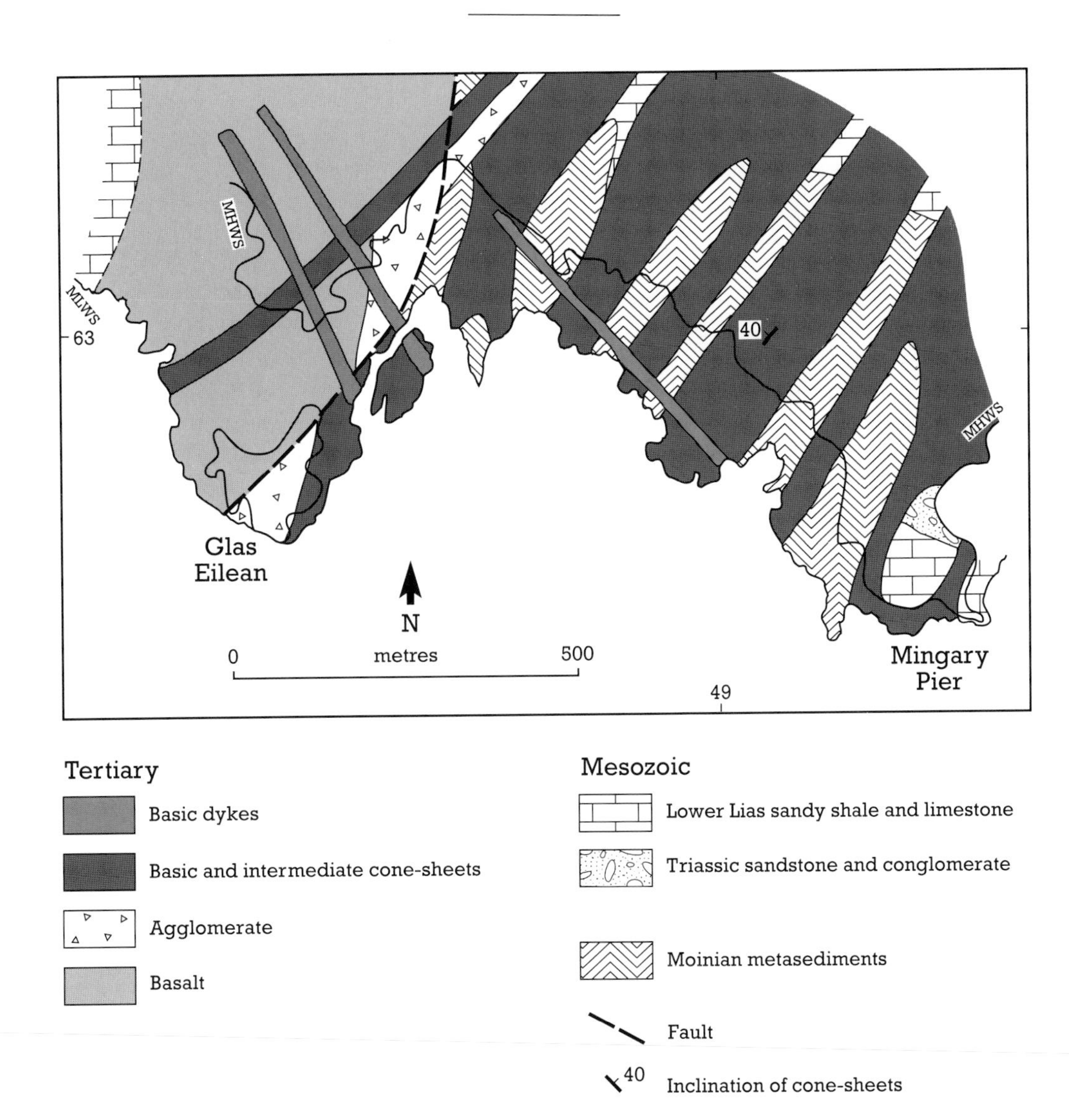

Figure 4.4 Geological map of the Glas Eilean–Mingary Pier site (after Gribble *et al.*, 1976).

(1972). Reynolds (1954) described and discussed features of the Glas Eilean linear vent.

Description

The predominantly intertidal exposures in the site (Fig. 4.4) demonstrate a profusion of cone-sheet intrusions which are so frequent that the proportion of igneous rocks often exceeds that of the country rocks they intrude. They belong to the outer set of cone-sheets associated with Centre 2 and include some of the most silicic compositions in the complex (Richey and Thomas, 1930; Holland and Brown, 1972), although being mostly of non-porphyritic quartz-dolerite, porphyritic dolerite or basalt. Individual cone-sheets vary greatly in inclination (20° to 75°) and range in thickness, from 0.5 to 15 metres; however, most are inclined at 35° to 45° towards the Aodainn–Achosnich area and are between 0.5 and 2 metres thick. In general, they cannot be traced for any substantial distance along strike and in many places the sheets interdigitate and form anastomosing masses several metres thick. Country rocks are biotite-grade Moine psammites and phyllites, Triassic breccias and conglomerates, Lias limestone–shale rhythmic sequences and Tertiary lavas, against all of which the cone-sheets are conspicuously chilled. Contact-metamorphic effects involving the formation of idocrase, sphene, garnet, clinopyroxene, tremolite and prehnite are commonly observed in the adjoining Lias calcareous shale rocks. The sheets are cut by numerous WNW- to NW-trending basic dykes and themselves cut earlier dykes and sills. The country rocks occur as thin septa, or screens, between the cone-sheets.

South of Mingary Pier (NM 494 627), a composite sill of quartz dolerite and granophyre is exposed and, a few metres to the west, there is a composite cone-sheet. Such composite sheets have margins of basic quartz dolerite and central portions of felsite (often xenolithic), granophyre, craignurite or acidified quartz dolerite. The internal contacts between the acid and basic members may be sharply defined or gradational.

At Glas Eilean (NM 484 628) a small remnant mass of gently westerly dipping lavas is exposed. At least two amygdaloidal flows are present, intruded by numerous cone-sheets. The original hydrothermal assemblage of the rocks has been almost totally obliterated by pneumatolytic alteration similar to that found around the Mull central complex. The rocks are deeply weathered to an earthy-red and green colour and are traversed by anastomosing veins of chlorite, albite and epidote, often with prehnite. Similar mineral assemblages occupy the altered amygdales. No fresh olivine has survived, although pseudomorphs are common.

The small tidal island of Glas Eilean, and the foreshore to the north-east, expose a linear vent cutting the lavas, cone-sheets and Moine schists. It is broken into two outcrops by a fault which, in places, throws the lavas against the schists. Fragments of quartz dolerite and tholeiitic basalt derived from the cone-sheets occur within the vent agglomerate, which also contains fragments of Moine schist, basalt lava, Jurassic sandstone, limestone and shale. The clasts lie in a devitrified, chloritic matrix containing spherulites of quartz and alkali feldspar. The agglomerate is locally intricately veined by an acid tuff composed of devitrified acid glass enclosing bodies of basic devitrified glass (Paithankar, 1968). A tuffisite dyke with flow-textured margins cuts the agglomerate. The dyke matrix, which closely resembles the agglomerate matrix, contains a variety of clasts, of which basalt is the dominant type. The clasts increase in size from the margins of the dyke towards its centre, where the concentration of coarse fragments may make it difficult to distinguish the rock from the normal vent agglomerate.

Interpretation

Cone-sheets are associated with many of the central intrusive complexes in the British Tertiary Volcanic Province but are relatively scarce elsewhere, implying unusual conditions for their formation (Walker, 1975). Their outcrops are concentric and they are inclined inwards having an inverted cone shape, each cone sharing a common apex. The apex probably marks the position of the roof to the magma chamber from which the intrusions originated. Cone-sheets are important in understanding the tectonic processes, mechanisms of intrusion and the stress fields associated with the emplacement of the British Tertiary central complexes. In addition, their ubiquity makes them useful time markers dividing periods of intrusion.

The cone-sheet swarms of Ardnamurchan were first interpreted by Richey and Thomas (1930) who, in accordance with the structural and stress analysis studies previously applied to the Mull centre (Anderson *in* Bailey *et al.*, 1924), concluded that upward pressure from the magma chamber produced conical fractures in its roof (cf. Anderson, 1936) and that the cone-sheets were intruded along these fractures. Two distinct sets of cone-sheets were recognized by Richey and Thomas, arranged around separate foci located east of Glendrain (Centre 1) and south of Achosnish (Centre 2). A partial set associated with Centre 3 was also recognized.

Durrance (1967) argued that the cone-sheets could not be divided into two independent suites but occupy one large conjugate spiral (as opposed to concentric) shear fracture system centred on Sanna. This focus is located at the periphery of the complex and corresponds closely with the gravity maximum of the Ardnamurchan complex (Bott and Tuson, 1973). This model postulated that a release of magma pressure and associated compressional stresses, coupled with a degree of torsion, resulted in the spiral fracture system along which the cone-sheets were subsequently injected when the magma pressure increased. Gribble *et al.* (1976) strengthened this model by noting that the cone-sheets originally attributed by Richey and Thomas (1930) to Centre 1, and to the outer set of Centre 2, occupy predominantly sinistral shear fractures. However, the inner set belonging to Centre 2 and the few Centre 3 cone-sheets, tend to have been intruded into dextral fractures. According to Durrance's model, the shear sense of these fractures would have been controlled by torsional stresses set up by the high-level emplacement of magma bodies.

Walker (1975), in presenting a new concept for the evolution of the Tertiary intrusive centres, suggested a radically different mechanism for cone-sheet formation. The cone-sheets in the British Tertiary Volcanic Province generally coincide with areas of updoming, presumed to have been caused by early uprising acid diapirs (but see Le Bas, 1971). Walker suggested that the sheets are emplaced passively along fractures produced by rising magma which tended to move in the direction of maximum hydrostatic pressure (the amount by which the fluid pressure of the magma exceeds the lithostatic pressure). Where basic magmas followed preferred pathways of increasing excess pressure, governed by the perturbations in the surfaces of equal excess fluid pressure near to a high-level acid diapir, cone-sheets instead of dykes were produced.

The Ardnamurchan cone-sheets vary considerably in lithology and texture, but chemically all of them appear to belong to a single tholeiitic lineage (Richey and Thomas, 1930; Holland and Brown, 1972). They range from plagioclase- and clinopyroxene-bearing quartz dolerites to felsites; composite examples are also known. Holland and Brown concluded that on the basis of 'mean' chemical analyses, there is no easily distinguishable difference in magma type between Centre 1 and Centre 2 cone-sheet swarms, and no evidence to support or oppose the hypothesis of Durrance (1967). On total alkali–silica plots, the cone-sheets cluster around the Hawaiian alkali-basalt–tholeiite field boundary (MacDonald and Katsura, 1964) and appear to be quite different from the regional Hebridean alkaline trends (Bailey *et al.*, 1924; Tilley and Muir, 1962; Thompson *et al.*, 1972).

The variation in clast size within the tuffisite dyke intruding the Glas Eilean vent must reflect a marked increase in gas velocity towards the centre of the dyke, where the entrained fragments are as much as 100 mm across compared with the 2–4 mm sizes of the marginal fragments.

Conclusions

Magma intruding a series of fractures centred beneath Sanna gave rise to the dense outer cone-sheet swarm associated with Centre 2, and possibly also to the cone-sheets associated with Centre 1. Diapirs of acid magma may have caused uprising basaltic magma to be diverted in the direction of maximum excess hydrostatic pressure to form the cone-sheets. The cone-sheets intruded cold country rocks against which individual intrusions were chilled; thermal metamorphic effects were limited and distinctive mineral assemblages were confined to the compositionally-favourable Jurassic calcareous shales. The Ardnamurchan cone-sheets have tholeiitic affinities and contrast with the regional alkaline basaltic lavas. The Glas Eilean linear vent, which cuts the cone-sheets, is the youngest major body of pyroclastic rocks in Ardnamurchan, the intrusive tuff, or tuffisite, cutting it demonstrates the ability of flowing gases to entrain and transport fragments of considerable size.

GLAS BHEINN–GLEBE HILL

Highlights

Several massive, arcuate, cross-cutting intrusions belonging to Ardnamurchan Centres 1, 2 and 3 intruded and metamorphosed pre-Tertiary country rocks and Tertiary plateau basalts. Centre 2 gabbros contain distinctive sapphire-bearing xenoliths derived by alteration of contemporaneous soil (bole) and weathered basalt; calcareous Jurassic sediments at the margins of the gabbros have been altered to high-temperature calc-silicate hornfelses which contain the rare mineral kilchoanite (type locality). The age relationships between Centre 2 cone-sheets and the major intrusions are well-illustrated here.

Introduction

The original survey of Ardnamurchan by Richey and Thomas (1930) identified three independent but overlapping centres of igneous activity and showed them to consist of volcanic vents, various types of arcuate intrusion and cone-sheet swarms. Within the relatively small area of the Glas Bheinn–Glebe Hill site (*c.* 3 km^2), various intrusions belonging to all three centres are represented, with a clear demonstration of their relative ages. The site also contains exposures of the contact between the intrusions and the country rocks which include Moine schists, Lower Lias sediments, Tertiary lavas and minor intrusions other than cone-sheets, a small remnant mass of

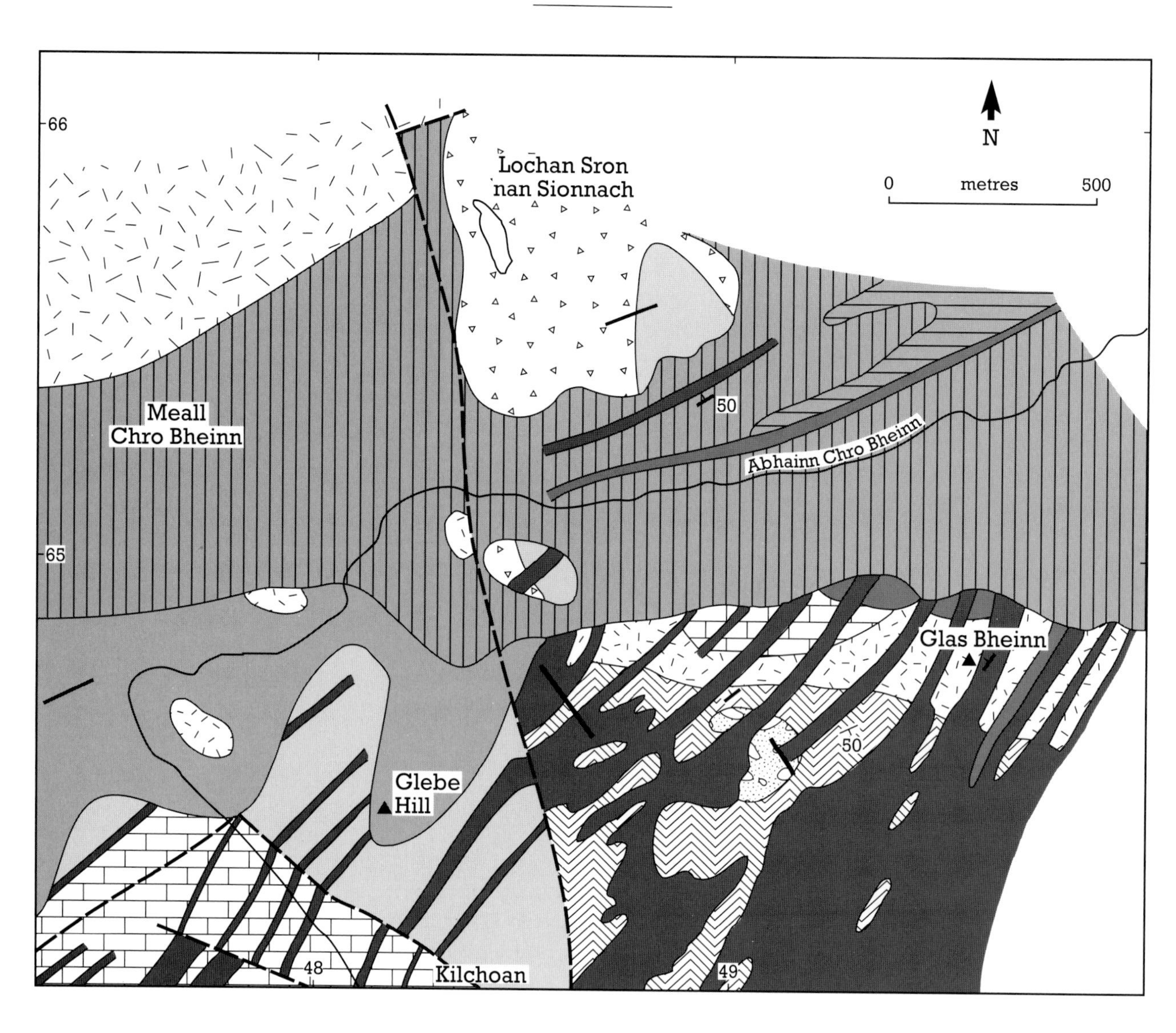

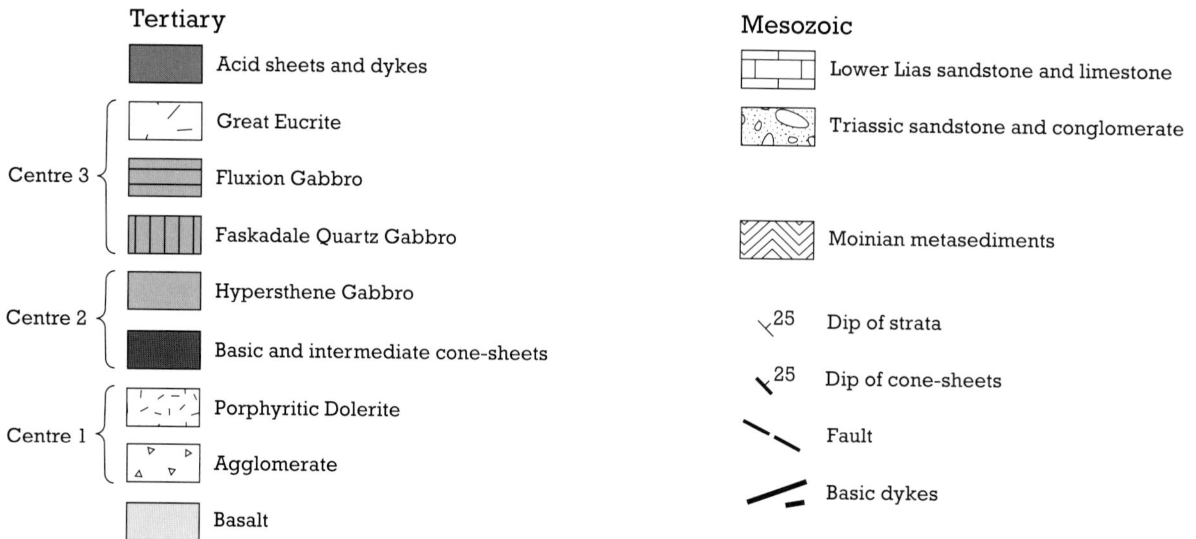

Figure 4.5 Geological map of the Glas Bheinn–Glebe Hill site (after Gribble *et al.*, 1976).

agglomerate and highly metamorphosed xenoliths in the outer gabbros of Centre 2 (Fig. 4.5). Agrell (1965) studied the thermally metamorphosed calcareous Lias sediments and identified a wide range of minerals, including kilchoanite ($Ca_3Si_2O_7$) west of Glas Bheinn.

Description

On Glas Bheinn (NM 493 648), a large, E–W-trending, dyke-like intrusion of plagiophyric dolerite (attributed to Centre 1) is poorly exposed. The intrusion appears to have near-vertical and often chilled, aphyric contacts with Moine psammites and Lias shaly sandstones. It is dissected by a swarm of non-porphyritic and (later) feldspar-phyric cone-sheets belonging to the outer set of Centre 2 and a later porphyritic sheet which is inclined towards the Centre 3 focus (Richey and Thomas, 1930). The main intrusion is a quartz dolerite or gabbro, bearing zoned labradorite phenocrysts in a matrix of acicular labradorite, augite and brown basaltic glass. Schistose xenoliths are commonplace and gabbroic xenoliths, of probable cognate origin, occur about 300 m west of the Glas Bheinn summit (NM 497 647). Near to the summit, the dolerite is intruded by a small mass of felsite or granophyre which, together with the cone-sheets and country rock, is almost entirely recrystallized owing to the thermal effects of the nearby Faskadale Quartz Gabbro of Centre 3.

The Faskadale Quartz Gabbro is poorly exposed within the site on the southern side of the Abhainn Chro Bheinn, and the form and intrusive relationships have been largely inferred rather than observed here. Cone-sheets do not cut this intrusion. Although typically a medium-grained quartz gabbro, the Faskadale intrusion varies from olivine eucrite to a basic granophyre especially towards the roof (Richey and Thomas, 1930; Gribble *et al.*, 1976). The occurrence of internal contacts between coarse- and fine-grained members implies that the mass probably has a composite form.

Glebe Hill (NM 480 647) is formed from baked and partially granulitized amygdaloidal basaltic lavas which are cut by thin, basic cone-sheets, a few minor felsites and the Hypersthene Gabbro. Most of the basalts are aphyric, containing olivine pseudomorphs and the amygdale assemblages have been altered to chlorite and plagioclase. The lavas are probably related to the relatively unaltered mildly alkaline, olivine-basalt lavas around Ben Hiant. At Glebe Hill, the lavas form a roof to the Hypersthene Gabbro intrusion, the contact varying between being virtually horizontal and steeply inclined. The Hypersthene Gabbro is generally a fine-grained, pyroxene-rich rock but, around Glebe Hill, coarse pegmatitic quartz gabbro and allivalitic–troctolitic varieties are prevalent. Xenoliths, which are frequently observed in the gabbro, contain equigranular plagioclase and dark-green hercynitic spinel accompanied by interstitial, colourless and blue corundum (including sapphire) and magnetite with exsolved ilmenite lamellae. These were derived from aluminous sediments, possibly bole horizons in the lavas.

The contact between the Hypersthene Gabbro and the Faskadale Quartz Gabbro trends east–west to the north of Glebe Hill. Partial exposure of this contact occurs in the bed of the Abhainn Chro Bheinn a few metres north of where the stream turns westwards before passing under the Kilchoan–Sanna road. The Hypersthene Gabbro is shattered, deeply weathered and veined by granophyre. The Quartz Gabbro is chilled towards the contact and locally a thin screen of amygdaloidal basalt intervenes. Elsewhere, the position of the contact can be accurately inferred although it is rarely exposed. On following the contact eastwards towards Glas Bheinn, the Hypershene Gabbro, Faskadale Quartz Gabbro and the Glas Bheinn Porphyritic Dolerite, with its numerous cone-sheets, can all be seen in close proximity to one another.

Around Lochan Sron nan Sionnach (NM 484 656), the slopes of Meall an Tarmachain expose a mass of basalt lava and agglomerate roofing the Faskadale Quartz Gabbro and the Great Eucrite. These are cut by sparse cone-sheets inclined towards the focus of Centre 3 and consist mostly of basaltic, felsitic and country rock fragments in a tuffaceous matrix. The lavas and agglomerates are highly thermally metamorphosed and closely resemble a similar rock assemblage which forms part of the Meall nan Con screen (see below). The Centre 3 cone-sheets, which find geomorphological expression as a series of low, often inconspicuous, terrace features in the Quartz Gabbro and the volcanic rocks, dip at about 50° towards the focus of this centre near Glendrain. They are largely plagiophyric basalts and dolerites with rarer non-porphyritic examples. West of Lochan Sron nan Sionnach, the weathering contrast between the Quartz Gabbro and the Great

Eucrite is demonstrated; the actual contact appears to be complex, with narrow intervening zones of fine-grained gabbro occurring at intervals. There is also evidence for the alteration and crushing of the Quartz Gabbro by the emplacement of the Great Eucrite.

Interpretation

The exposures at Glas Bheinn–Glebe Hill provide an excellent opportunity to demonstrate how the evolution of the Ardnamurchan complex has been elucidated using the intricate field relationships of the various intrusions. Within this relatively small site (*c.* 3 km), representative ring-dyke intrusions belonging to all three centres are in close association, enabling the major phases of development of the complex to be studied. The contacts of the porphyritic dolerite on Glas Bheinn with the surrounding intrusions and cross-cutting cone-sheets shows that, relatively, it must be the earliest. This intrusion is assigned to the first centre of plutonic igneous activity on Ardnamurchan and cuts older Centre 1 agglomerates which appear to belong to the extensive Northern Vents (Richey and Thomas, 1930). The dolerite is similar to the porphyritic dolerite of Ben Hiant and an olivine-bearing variety also occurs to the north of Camphouse.

The Faskadale Quartz Gabbro mass extends in a broad arc from Faskadale (NM 501 708) southwards and westwards to Beinn na Seilg and may well continue further west under a roof of Centre 2 rocks to re-emerge south of Sanna Bay as the Ben Bhuidhe intrusion (Gribble *et al.*, 1976). This mass is the outermost ring intrusion of Centre 3; cone-sheets belonging to Centre 2 are absent and it cuts the earlier Hypersthene Gabbro and Glas Bheinn porphyritic dolerite. Likewise, the Great Eucrite is a younger intrusion of Centre 3, as demonstrated by field relationships described above. The Hypersthene Gabbro, discussed in detail for Beinn na Seilg–Beinn nan Ord (see below), is the earliest ring intrusion of Centre 2 cutting the Porphyritic Dolerite of Glas Bheinn and truncating the Faskadale intrusion. High emplacement temperatures for the ring-dyke intrusions are clearly demonstrated by the sanidinite-facies mineral assemblages developed in adjoining country rocks.

Conclusions

The contact relationships between major ring intrusions and the associated cone-sheet swarms within the site, belonging to each of the three centres of Tertiary plutonic activity on Ardnamurchan, can be used to study the evolution of the complex. The early Glas Bheinn Porphyritic Dolerite of Centre 1 is cut by the Hypersthene Gabbro of Centre 2, which is in turn truncated by the Centre 3 Faskadale Quartz Gabbro and Great Eucrite masses. The high-temperature intrusions have given rise to distinctive thermal metamorphic mineral assemblages in altered, weathered basalt lavas and sandy limestones.

BEINN NA SEILG–BEINN NAN ORD

Highlights

The site contains major arcuate gabbroic and doleritic intrusions of Centre 2, including the Hypersthene Gabbro, part of a layered intrusion, which has severely metamorphosed adjoining country rocks. The later, granophyric Quartz Dolerite of Sgurr nam Meann contains superbly exposed evidence for the former coexistence of basic and acid magmas.

Introduction

This extensive site provides a valuable traverse through the arcuate intrusions of Centre 2 of the Ardnamurchan complex. It is of special importance in demonstrating the contact relationships between the complex and the country rocks surrounding it. The arcuate masses include the outer Hypersthene Gabbro which dominates the geology of the site, the granophyric Quartz Dolerite and quartz gabbros and eucrites. The inner set of cone-sheets associated with Centre 2 is also well represented.

The Beinn na Seilg–Beinn nan Ord area was first investigated in detail by Richey and Thomas (1930), following field surveys between 1920 and 1923. There has been no comprehensive account of the area since, but investigations describing the form, field relations and petrology of the Granophyric Quartz Gabbro have been published by Wells (1954a), and the field characteristics of this intrusion have been excellently described by Skelhorn and Elwell (1966). In addition, the structure and petrology of the

Hypersthene Gabbro has been studied by Wells (1954b) who also investigated its xenoliths (Wells, 1951). A recent study by Day (1989) details much new information about the Hypersthene Gabbro and its contact effects.

Description

At Dubh Chreag (NM 452 633), 2 km west of Ormsaigbeg (Fig. 4.6), the outer contact of the Hypersthene Gabbro is exposed against Lower Lias shales and limestones. This is the outermost major intrusion of Centre 2, the contacts of which dips southwards at angles of 45°–60°. Further to the east, around Lochan Ghleann Locha, the Hypersthene Gabbro intrudes Palaeocene basalt lavas overlying Middle Jurassic (Bajocian) sandstones and limestones; these rocks form a narrow strip parallel to the contact (1.5 km long and with a maximum width of 150 m) and are bounded to the south by a fault which downthrows them against Lower Lias. The sediments are intruded by numerous thin cone-sheets belonging to the outer set of Centre 2 inclined to the north at angles of between 35° and 40°. These are beautifully exposed, standing out in stark contrast against the cream-coloured sandstones and pale limestones of the Lias; they were first noted in the coastal cliffs by MacCulloch (1819) and later by Geikie (1897).

The thermal effects of the Hypersthene Gabbro are widespread, but intense thermal metamorphism occurs only within a few hundred metres of the contact. At Dubh Chreag, the Lias shales have been altered to hard, fissile hypersthene–biotite–magnetite–feldspar hornfels with possible cordierite. The impure limestones are now clinopyroxene–sphene–garnet–anorthite–hornblende hornfels, while the purer, Middle Jurassic limestones are now calc-silicate assemblages of recrystallized calcite, garnet, diopside and tremolite. Within the thermal aureole, the lavas and early cone-sheets also show considerable alteration and have become flinty and more massive. Closer to the contact, they are granular clinopyroxene–olivine–plagioclase–magnetite hornfelses often with orthopyroxene. Thoroughly hornfelsed country-rock xenoliths of igneous origin occur at various points along the contact of the gabbro. Day (1989) has demonstrated that extreme thermal metamorphism produced distinctive rheomorphic melts from the varied country-rock lithologies.

Among the main intrusions of Centre 2, a kilometre to the south of Sonachan, a strip of intensely altered volcanic rocks lies between the Older Gabbro of Lochan an Aodainn and younger quartz gabbros. Although severely metamorphosed to granular hornfelses, the agglomerates can still be matched with those of the 'Northern Vents' of Centre 1, and the basalts have retained their flow structure.

A traverse from the southern margins of the Hypersthene Gabbro towards the focus of Centre 2 at Sonachan crosses several arcuate intrusions identified and named by Richey and Thomas as follows:

(oldest)	Hypersthene Gabbro
	(Granophyric) Quartz Dolerite of Sgurr nam Meann
	Quartz Gabbros of Loch Caorach and Beinn na Seilg
	Eucrite of Beinn nan Ord
	Quartz Gabbro of Garbh-dhail
	Old Gabbro of Lochan an Aodainn
(youngest)	Quartz Gabbro of Aodainn

Representatives of the inner set of Centre 3 cone-sheets, small felsite/granophyre masses near Aodainn (NM 457 658) and near the summit of Beinn na Seilg and the westwards termination of the Faskadale Quartz Gabbro of Centre 3 also occur in the site.

The petrography of the Hypersthene Gabbro is extremely variable. The predominant rock type is a relatively fine-grained hypersthene – olivine gabbro but quartz dolerite and quartz gabbro occur as marginal facies at Dubh Chreag and east of Lochan Ghleann Locha. Masses of troctolitic or allivalitic gabbro grading into peridotite, coarse-grained augite-rich gabbro and gabbro pegmatite are also found in the intrusion; these often have intrusive contacts and quartz–feldspar veins occur towards the outer margins of the mass. One of the most interesting features of the gabbro is the sporadic development of layering produced by a combination of textural variation and variations in the modal mineralogy. Layers rich in pyroxene, magnetite and olivine occur at the base of some of the structures which vary from 10 mm to over 1 m in thickness. A degree of rhythmic layering of feldspar-rich and pyroxene-rich layers is apparent. Intra-layer slump structures and erosional surfaces are also noteworthy features, being comparable with those on Rum (Askival–Hallival). The layering dips inwards and increases from about 5° at the outer margin to over 60° at the inner contacts.

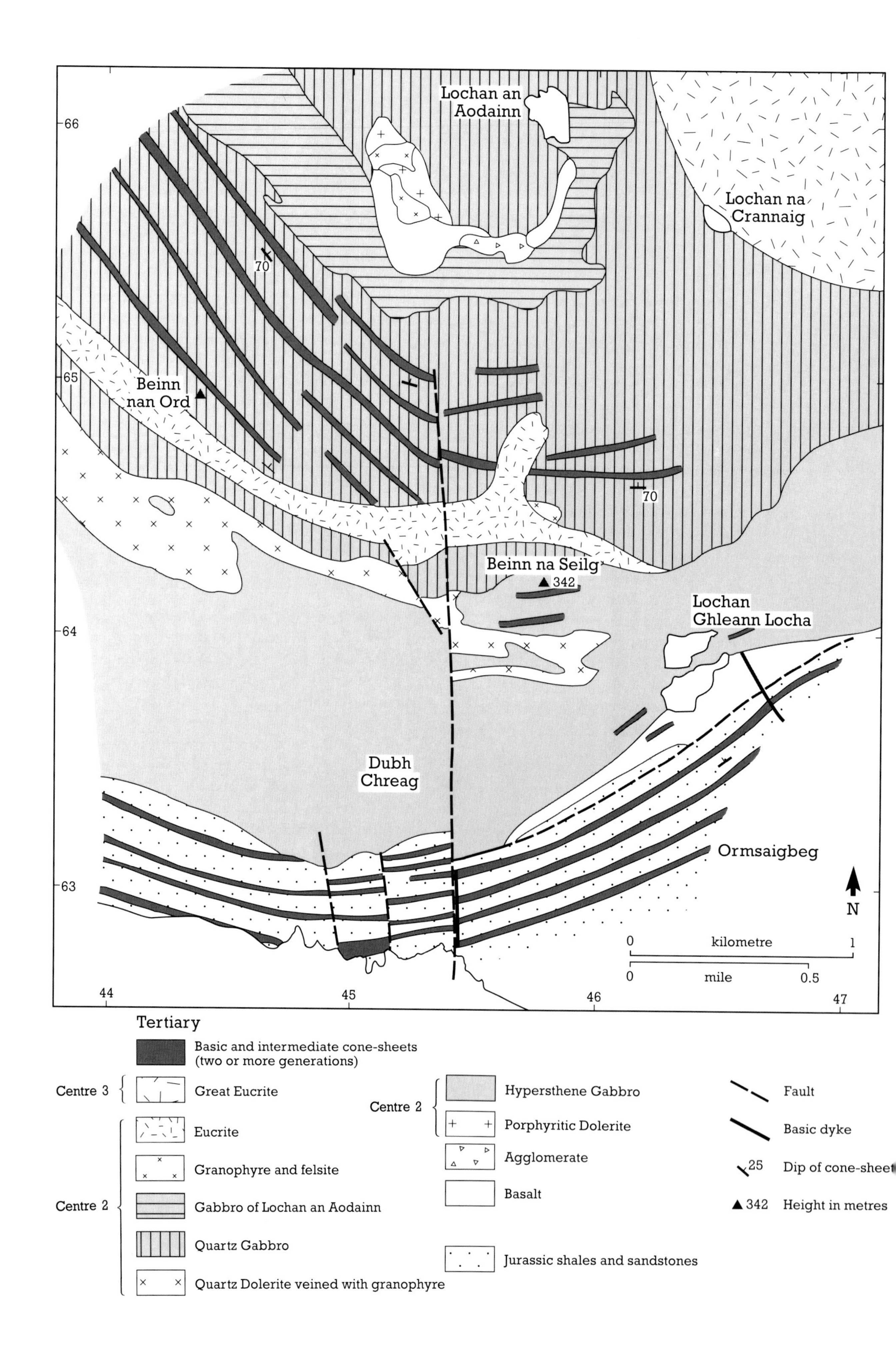

Lochan an Aodainn
Lochan na Crannaig
Beinn nan Ord
Beinn na Seilg
342
Lochan Ghleann Locha
Dubh Chreag
Ormsaigbeg
70
70
66
65
64
63
44
45
46
47
N
0 kilometre 1
0 mile 0.5
Tertiary
Basic and intermediate cone-sheets (two or more generations)
Centre 3
Great Eucrite
Centre 2
Eucrite
Granophyre and felsite
Gabbro of Lochan an Aodainn
Quartz Gabbro
Quartz Dolerite veined with granophyre
Centre 2
Hypersthene Gabbro
Porphyritic Dolerite
Agglomerate
Basalt
Jurassic shales and sandstones
Fault
Basic dyke
25 Dip of cone-sheet
342 Height in metres

The granophyric Quartz Dolerite of Sgurr nan Meann truncates the Hypersthene Gabbro on the north and east. Along most of its inner contact, it adjoins the Quartz Gabbro of Loch Caorach but its southern extremity lies entirely within the Hypersthene Gabbro, and along a short length in the north, its inner contact lies against the Beinn nan Ord Eucrite. The mass is younger than the Hypersthene Gabbro and the contact between the two is complex, with sill-like apophyses penetrating the older rock for distances of up to a kilometre. In contrast, the inner contact is simple and near vertical. The lithology of the Quartz Gabbro intrusion is its most outstanding feature. It comprises an association of feldspar-phyric and aphyric dolerite and gabbro xenoliths, often of great size, which are locally penetrated by an anastomosing plexus of acid net-veining associated with dykes and sheets of felsite/granophyre. A full petrological account of this intrusion has been published by Skelhorn and Elwell (1966); the detailed petrography is therefore not described here. The thicker sheets and dyke-like bodies of granophyre with aphyric dolerite inclusions penetrate the porphyritic dolerite and, in some exposures, the margins of the porphyritic dolerite appear to be chilled and phenocryst free; in others, the phenocrysts are abruptly truncated by the sheets or dykes and the grain size of the dolerite does not decrease towards the contact. Thus, there is conflicting evidence for the relative ages of the two dolerites. However, evidence that the porphyritic dolerite is the earlier is drawn from its inclusion as xenolithic blocks within the aphyric dolerite, with granophyre generally intervening between the two.

The Quartz Gabbros of Loch Coarach (NM 433 656) and Beinn na Seilg were considered by Richey and Thomas (1930) to be parts of a single ring-dyke, although they differ petrologically. The most basic type is an olivine gabbro consisting of olivine, augite, labradorite and accessory ore. On this original assemblage changes have been superimposed, changes caused by 'acid material of late consolidation' (Richey and Thomas, 1930) such as schillerization of feldspars, de-schillerization of pyroxene, development of hypersthene and locally abundant acid mesostasis which has crystallized as alkali feldspar and quartz. To the west and south-west, the Quartz Gabbros are bounded by the Quartz Dolerite (and the Hypersthene Gabbro in the case of the Beinn na Seilg mass). The contact is poorly exposed and is difficult to interpret. These gabbros are not cut by the inner cone-sheets of Centre 2, but they do inject some of the components of the Granophyric Quartz Dolerite which is therefore probably the older intrusion. Along their inner contacts, where the quartz gabbros adjoin the Eucrite of Beinn nan Ord, both rocks are brecciated at exposed junctions. At a few localities sharp contacts show the Quartz Gabbro cutting and chilled against the Eucrite.

The Eucrite of Beinn nan Ord lies to the north and east of the Quartz Gabbros and can be traced from Beinn na Seilg northwards and westwards across Beinn nan Ord to Sanna Bay. It is a moderately coarse-grained rock with abundant olivine, ophitic augite associated with large magnetite crystals and plagioclase (labradorite–bytownite). The olivine is usually associated with biotite and hypersthene where it is unaltered. Variants rich in olivine, pyroxene or plagioclase also occur and acidification and granulitization are widespread. The form of the Eucrite is unusual in possessing two arms projecting inwards to the focus of Centre 2 cutting across the earlier Quartz Gabbro of Garbh-dhail. The rock is resistant to weathering and generally well exposed in conspicuous glaciated crags. Over much of its extent, the rock has been microbrecciated possibly due to explosive shattering by an acid magma (Richey and Thomas, 1930). The outer contact against the Quartz Gabbros of Beinn na Seilg and Loch Caorach has been described above. The inner contact is against an earlier gabbro, the Quartz Gabbro of Garbh-dhail, which is cut by cone-sheets and linear crush lines, both of which are absent in the Eucrite. Both contacts of the Eucrite appear to be steep.

The Quartz Gabbro of Garbh-dhail forms much of the north-eastern part of the Beinn nan Ord and Beinn na Seilg area and, like the other quartz gabbros of Ardnamurchan, displays great variation in composition and texture. More basic and finer-grained varieties prevail in the exterior part of the mass, while the interior is more silicic and internal intrusive contacts suggest a composite nature, an initial injection of basaltic magma being followed by a relatively silicic magma. Flow-banding, dipping at high angles towards the focus of Centre 2, is locally present. The gabbro is chilled against the Hypersthene Gabbro on Beinn na Seilg and against the Old Gabbro of Lochan an Aodainn. Cone-sheets from the inner

Figure 4.6 Geological map of the Beinn na Seilg-Beinn nan Ord site (after Gribble *et al.*, 1976).

set of Centre 2 cut this Quartz Gabbro and both are baked against the Eucrite of Beinn nan Ord.

The Old Gabbro of Lochan an Aodainn outcrops in an arc between Lochan an Aodainn and Achosnish and is considered to be a very early intrusive member of Centre 2. It is now very altered but it was originally a variable fine- to coarse-grained olivine-bearing dolerite with olivine-free, olivine-rich, allivalitic and fine-grained augite-rich varieties. The gabbro has a distinctive dull, matt, dark-grey appearance caused by numerous opaque inclusions in the feldspar and alteration of the mafic minerals. Crushing and shattering are widespread, with segregation and migration of acid material locally producing a rock resembling an augite granophyre. The Old Gabbro is in contact with a small mass of basalt lavas and agglomerates along its north-eastern margin, but elsewhere its margins are entirely determined by later intrusions.

North of the Old Gabbro, the Quartz Gabbro of Aodainn crops out and is heterogeneous both in texture and in the proportion of acid mesotasis. It is generally a moderately fine-grained rock bearing both orthopyroxene and clinopyroxene and a little interstitial alkali feldspar and quartz. Coarser areas appear to be xenolithic towards the porphyritic finer-grained rocks. Internal contacts are both sharp and gradational, and other textural variations indicate the possibility of a composite origin for the intrusion. The Aodainn mass clearly veins and intrudes the Old Gabbro, with chilling and some hybridization occurring along this contact. North-east of Lochan an Aodainn, a strip of dark-grey rock containing phenocrysts of plagioclase, augite and olivine intervenes between these masses; this has been interpreted as a remnant of a still older intrusion. The Quartz Gabbro is truncated to the north by the Great Eucrite of Centre 3 which caused the thermal metamorphism noted in the earlier intrusions.

The Quartz Gabbro of Faskadale, the outermost intrusion of Centre 3, enters the eastern margin of the Beinn nan Ord–Beinn na Seilg area and appears to underlie many of the Centre 2 arcuate intrusions indicating a subsurface, south-westerly extension to Centre 3 (see Glas Bheinn above for a description of the Faskadale intrusion).

Small, roughly circular masses of granophyre or felsite, unrelated to the major intrusions, occur in at least two localities with the site. A small mass of dark-grey, non-porphyritic felsite forms a rocky hill south of Aodainn and, on Beinn na Seilg, a shattered outcrop of pink granophyre is found between the Beinn nan Ord and Garbh-dhail quartz gabbros. Dykes do not commonly cut the Centre 2 intrusions but several basic and a few acid examples are known. A thick (1.0–1.5 m) dyke of greenish, flow-banded spherulitic pitchstone outcrops on the shore at Ormsaigbeg and can be traced northwards for about 2 km, cutting the Hypersthene Gabbro, but terminating short of the quartz gabbros.

Interpretation

The sequence of arcuate intrusions and their intricate contact relationships in this area provide a valuable record of the development of Ardnamurchan, Centre 2. The results obtained by Richey and Thomas (1930), from their survey of the area, attracted international attention and rapidly became of outstanding importance as a demonstration of a 'typical' ring-dyke complex. Parts of the area have been reinvestigated by later workers in the light of newly developed concepts in igneous petrology and, while solving some problems, have encountered more. For most of the area, however, no published later work has superseded that of Richey and Thomas and the majority of the problems that they recognized, but perforce had to leave unsolved, have not received the attention that they deserve. The research potential of the area must rank among the highest in Britain and this greatly augments the international value of the area as a 'type locality' for a ring-dyke complex.

Two of the dominant intrusions in the site, the Hypersthene Gabbro and granophyric Quartz Dolerite of Sgurr nam Meann, have been reinvestigated more recently; the conclusions from some of these studies are discussed below.

The Hypersthene Gabbro is the earliest intrusion of Centre 2 and was initially considered to be a ring-dyke by Richey and Thomas (1930). Subsequent work by Wells (1954a) and Skelhorn and Elwell (1971) has shown that, although the contact between the gabbro and the country rocks is mostly outwardly dipping at moderate angles, in places it is demonstrably steep or flat-lying, or complicated by large xenoliths and stoped blocks of the host rock. Wells (1954a) considered the original overall shape to be an upwardly flaring cone with a domed roof, while Skelhorn and Elwell (1971) suggested a more boss-like form. Wells (1954a) concluded that the form of the intrusion, particularly its considerable width, bears little resemblance to a ring-dyke and was probably forcefully intruded. The layered structures in the Hypersthene Gabbro dip inwards and have a conical shape and, according to Wells, gravity accumulation of crystals in the lower part of the intrusion, which had a conical form, played a significant role. The steepening of the layering at the inner contacts of the intrusion was suggested by Skelhorn and Elwell (1971) to have resulted from the deformation of the roughly circular mass by ring faulting or gradual subsidence of a central block. Palaeomagnetic evidence supplied by Wells and McRae (1969) suggests that if this hypothesis is accepted, then the deformation must have occurred before the rocks had cooled below their Curie temperature. Wells (1954a) also suggested that a quartz gabbro marginal facies to this intrusion crystallized from pre-Hypersthene Gabbro magma which was driven towards the upper surface during the forceful intrusion of the Hypersthene Gabbro. The recent study of this intrusion by Day (1989) contains much new factual information and interpretations.

The Quartz Dolerite was recognized by Richey and Thomas (1930) as a ring-dyke, although they were aware of the extensive apophyses projecting from the main mass. Wells (1954b) described the mass as a 'ring-dyke/sill intrusion'. Skelhorn and Elwell (1966) regarded Wells's hypothesis as one possibility but also presented three others: a sill connected to a ring-dyke or plug which is now replaced by a later intrusion; a ring-dyke with stepped contacts, the present erosion surface coinciding with a shallow step; or a marginal remnant of a ring-dyke cap formed above the block which subsided to allow the intrusion of the ring-dyke. Walker (1975) suggested that the mass may have been formed by successive laterally directed injections of magma into a 'curved flange' fracture caused by the differential subsidence of the rocks adjoining the magma chamber. The form of the mass clearly requires further investigation as was pointed out by Black (in discussion, Skelhorn and Elwell, 1971) who noted in addition that the maps produced by the Survey, Wells (1954b) and Skelhorn and Elwell (1966) differ significantly owing to the use of differing criteria to identify the Quartz Dolerite. Earlier, Black (pers. comm.) demonstrated his view that the so-called 'Quartz Dolerite' was in fact an igneous *mélange* of blocks of gabbro (shown on the maps of other workers), intrusions of various types of dolerite, and felsite emplaced along a partial ring-fracture which also served as a channel for gas fluxing.

The granophyric Quartz Dolerite of Sgurr nam Meann demonstrates clearly the association of acid and basic magmas in the intrusions of Centre 2 and in the Ardnamurchan complex as a whole (Blake *et al.*, 1965). Vogel (1982) has published a detailed petrological study of the acid–basic net-veined intrusion. It was observed that the basic aphyric and porphyritic components occur as pillow-like xenoliths, with cuspate and crenulate margins within the silicic rock and with chilling at many contacts indicative of 'liquid–liquid' relationships. Whole-rock chemistry indicates that magma mixing between basaltic and silicic liquids was a dominant process along with limited crystal fractionation. Vogel (1982) suggested that this net-veined complex presents a rare example of the interaction at a high level between mafic and silicic melts in the Ardnamurchan magma chamber before the silicic magma was lost to surface volcanism. It is a clear example showing that basic and acid magmas coexisted in a single intrusive body.

Conclusions

The Beinn na Seilg–Beinn nan Ord site is a classic locality in the Ardnamurchan complex, showing the succession of arcuate intrusions associated with Centre 2. Although arcuate in overall outcrop, the various major intrusions do not all conform to the classic ring-dyke model and some show features unique within the Tertiary Province. Evidence for the coexistence of acid and basic melts and for magma mixing within the high-level Ardnamurchan magma chamber as well as crystal accumulation and fractionation processes is afforded by many of the intrusions, almost all of which are composite in nature.

ARDNAMURCHAN POINT TO SANNA

Highlights

The site contains some of the clearest examples of net-veining, intrusion breccias and 'liquid–liquid' contacts between basic and acid rocks to be found in the BTVP. The Hypersthene Gabbro exhibits excellent layered structures which frequently simulate structures developed in clastic sediments.

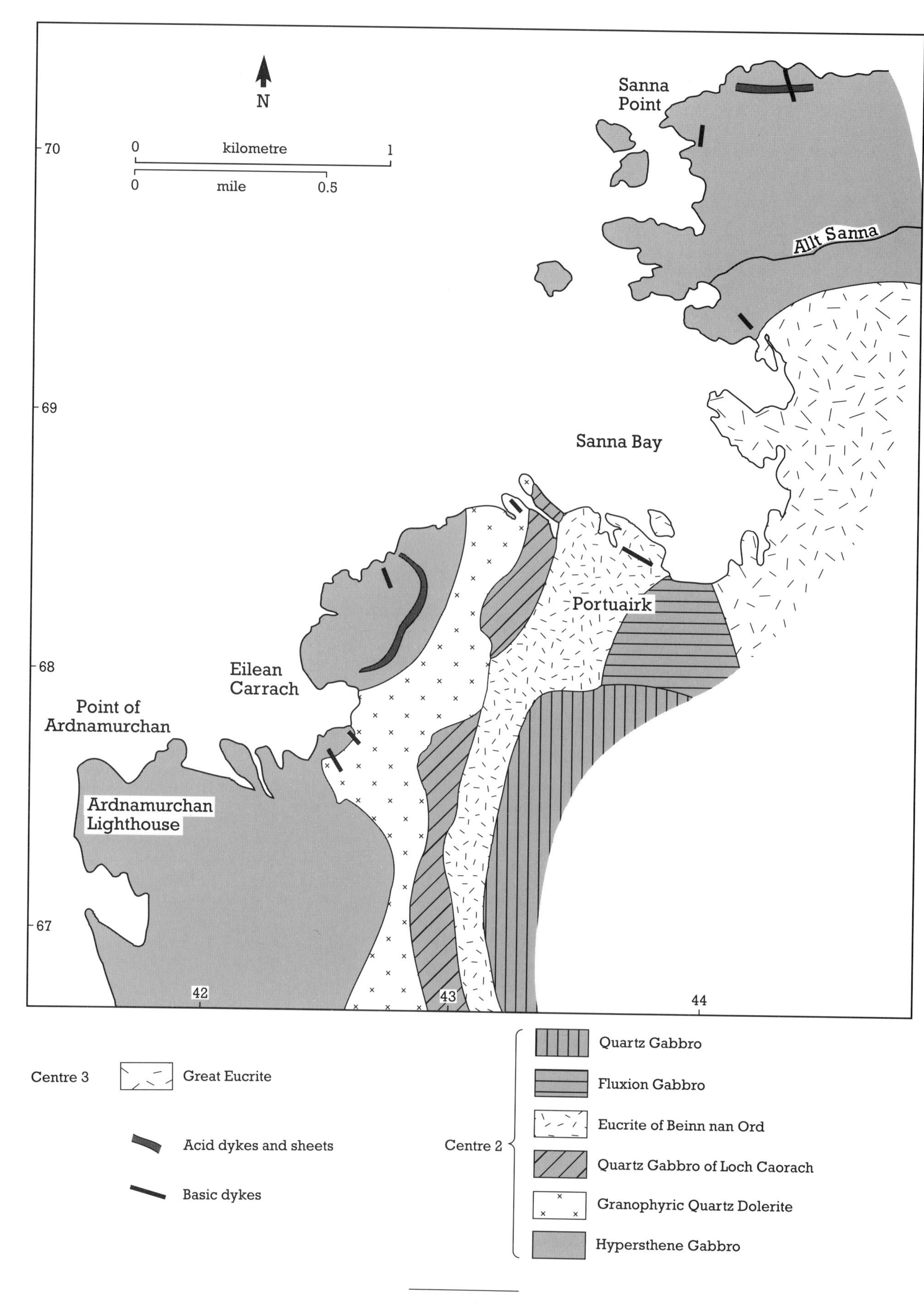

N
0
kilometre
1
0
mile
0.5
70
69
68
67
42
43
44
Sanna Point
Allt Sanna
Sanna Bay
Portuairk
Eilean Carrach
Point of Ardnamurchan
Ardnamurchan Lighthouse
Centre 3
Great Eucrite
Acid dykes and sheets
Basic dykes
Centre 2
Quartz Gabbro
Fluxion Gabbro
Eucrite of Beinn nan Ord
Quartz Gabbro of Loch Caorach
Granophyric Quartz Dolerite
Hypersthene Gabbro

Introduction

The coastal strip from the Point of Ardnamurchan to Sanna Point, including Sgurr nam Meann, provides an excellent traverse across several of the intrusions of Centre 2 and demonstrates their truncation by the Great Eucrite of Centre 3. Many features of petrological interest are easily accessible; these include net-veining, igneous brecciation, igneous layering and the occurrence of pyroxene granulites.

The area has been thoroughly investigated and described by Richey and Thomas (1930); the granophyric Quartz Dolerite of Sgurr nam Meann in particular has received considerable mention by these authors and by Wells (1954b), and was selected for special study in this area by Skelhorn and Elwell (1966).

Description

A traverse from Ardnamurchan Point to Sanna crosses the following intrusions (Fig. 4.7):

1. Hypersthene Gabbro of Ardnamurchan Point
2. Granophyric Quartz Dolerite of Sgùrr nam Meann
3. Quartz Gabbro of Loch Carrach
4. Eucrite of Beinn nan Ord
5. Fluxion Gabbro of Portuairk
6. Great Eucrite of Centre 3

The full sequence is seen between Ardnamurchan Point and the southernmost shores of Sanna Bay where the Great Eucrite is reached. Beyond the outcrop of the latter intrusion, from Sanna to Sanna Point, the Hypersthene Gabbro and granophyric Quartz Dolerite of Sgurr nam Meann (Butchins, 1973) are again exposed, but the Quartz Gabbro of Loch Carrach and Eucrite of Beinn nan Ord are absent, having been truncated by the Great Eucrite. The salient petrographic features and contact relations of all these intrusions, except for the Fluxion Gabbro, are given in the descriptions for Beinn na Seilg–Beinn nan Ord or the Centre 3 areas and are not repeated in this account, where attention is drawn to features not found elsewhere or better displayed in this area.

Figure 4.7 Geological map of the Ardnamurchan Point–Sanna site (after Gribble *et al.*, 1976).

The Hypersthene Gabbro to the west of the lighthouse has a quartz doleritic outer marginal facies and passes eastwards into a coarser, rather silicic quartz gabbro facies. The intrusion is cut by cone-sheets belonging to the outer set of Centre 2 (Richey and Thomas, 1930). On Eilean Carrach, the layering in the gabbro dips to the east at 30° and extensive tabular masses of pyroxene granulite occur parallel to the layers.

The area between the lighthouse pier (NM 423 675) and Eilean Carrach contains superb exposures of the granophyric Quartz Dolerite of Sgùrr nam Meann. In this area, porphyritic dolerite roofs a complex intrusion of aphyric dolerite and granophyre. From place to place, the granophyre varies from being volumetrically dominant to subordinate; small-scale anastomosing veins which cut both dolerites can be traced to sheet-like acid masses which contain a varied assemblage of basic rock types as inclusions (Fig. 4.8). The inclusions frequently exhibit fine-grained, crenulated, chilled edges against the enclosing granophyre. Further to the north, the Quartz Dolerite contains several large, raft-like masses of Hypersthene Gabbro and steeply dipping contacts between the Dolerite and Hypersthene Gabbro are exposed south of Eilean Carrach. Eastwards, the granophyric Quartz Dolerite of Sgurr nam Meann is bordered by a narrow outcrop of the Quartz Gabbro of Loch Caorach, which in turn adjoins an augite-rich variety of the Eucrite of Beinn nan Ord containing felsic veins bearing fine-grained basic xenoliths. The contact relationships of these intrusions have not been convincingly demonstrated.

To the east of the Eucrite of Beinn nan Ord there occurs a mass of laminated gabbro known as the Fluxion Gabbro of Portuairk (Richey and Thomas, 1930). The intrusion is not homogeneous and its variable features were attributed by Richey and Thomas to the modification of basic magma by silicic melt prior to intrusion, probably when the basic rock was in a solid or semi-solid state. It is characterized by feldspar lamination or fluxion texture, which serves to distinguish it from the Beinn nan Ord Eucrite; no clear contacts between these two have been reported. The Fluxion Gabbro is cut by the later Great Eucrite of Centre 3, but again the contact is not readily interpretable because of the lack of contact alteration and the presence of a flow-banding structure in the outer zones of the Great Eucrite.

The Great Eucrite forms the shores of Sanna

Figure 4.8 Granitic net-veining and an intrusion breccia of gabbro and dolerite fragments. Centre 2 ring-dykes, near the lighthouse, western tip of Ardnamurchan. (Photo: C.H. Emeleus.)

Bay, from Portuairk to the north of the Sanna River, and is seen in intrusive contact against the Hypersthene Gabbro to the north. At Sanna Point perfect mineralogical layering dipping to the south at angles of between 10° and 20° towards the Aodainn centre is present in the Hypersthene Gabbro. Occasional anorthosite, peridotite and iron-oxide-rich bands are found and igneous lamination, especially of plagioclase, is locally marked. Rhythmic banding and density stratification are also present in places. From the north of the Sanna River around Sanna Point and for 2.5 km to the east, sill-like bodies of granophyre containing inclusions of aphyric and porphyritic dolerite cut the Hypersthene Gabbro. Butchins (1973) considered those near Sanna Point to be apophyses from the granophyric Quartz Dolerite of Sgurr nam Meann.

Interpretation

Although the features seen in the section from Ardnamurchan Point to Sanna duplicate to some extent those seen in parts of Beinn na Seilg–Beinn nan Ord, many are much better displayed. Exposures of the granophyric Quartz Dolerite of Sgurr nam Meann (Centre 2) provide clear evidence as to its form and contact relationships with the Great Eucrite (Centre 3) and the excellent shore exposures south-east of Eilean Carrach clearly demonstrate the 'mixed magma' character of the granophyric Quartz Dolerite (Skelhorn and Elwell, 1966; Vogel, 1982). This is arguably the best exposure of net-veining, intrusion breccias and chilling of basic magma against acid magma to be found in the BTVP. On Eilean Carrach and elsewhere in the Hypersthene Gabbro, mafic pyroxene-granulite xenoliths are comparable with inclusions found in the mafic plutonic rocks of Skye, Rum and at other localities. Previously considered to be of sedimentary origin (Wells, 1951), an igneous origin for the xenoliths was suggested by Brown (1954) in accordance with the interpretations of similar rocks elsewhere (for example, Rum, Askival–Hallival). As a consequence of the very varied, well exposed geology, this section is of great significance within the Ardnamurchan complex and is used extensively for educational purposes (for example, see Gribble *et al.*, 1976), despite the fact that some of the contact relationships remain poorly understood.

The excellent examples of layering found east of Sanna Bay contain many structures which simulate those developed in clastic sediments. Although extensive reappraisal of the origins of igneous layering have taken place recently (for example, Parsons, 1987), the layering in the Hypersthene Gabbro provides evidence which supports the contention that the structure results from gravity-controlled accumulation of crystals, probably aided by convection currents (Skelhorn and Elwell, 1966; in agreement with Wells, 1954a) and that the gabbro may form part of a large layered intrusion.

Conclusions

The outstanding feature of the Ardnamurchan Point to Sanna section is the exceptionally fine exposure of net-veining and intrusion breccia near Eilean Carrach (Fig. 4.8); this shows clearly that basic and acid magmas coexisted, that basic magma was chilled against acid magma and that both magmas mixed to give a variety of intermediate (hybrid) rocks. This type of relationship is common throughout the BTVP, but the examples exposed here are certainly the most accessible and arguably the best exposed in the Province. The igneous layering in the Hypersthene Gabbro is closely comparable with layering found in gabbros and ultrabasic rocks in Skye and Rum, and lends support to the suggestion that this gabbro is part of a layered intrusion. The manner in which the Hypersthene Gabbro structures are truncated by the Great Eucrite provides evidence for the relatively late emplacement of Centre 3.

CENTRE 3, ARDNAMURCHAN

Highlights

The Great Eucrite in this site forms one of the most perfect annular intrusions in the British Tertiary Volcanic Province; it gives rise to a nearly complete ring of hills which dominates the Ardnamurchan Peninsula. The intrusive sequence in Centre 3 changed with time, from early eucrites and gabbros to late intrusions of intermediate and acid compositions.

Introduction

This site encompasses Centre 3 which represents the final stage in the intrusive evolution of the Ardnamurchan complex. Centre 3 consists debatably of seventeen concentric intrusions of highly variable petrography, ranging from coarse gabbros and dolerites to minor, late, intermediate tonalites and quartz monzonites. The Centre has been conventionally regarded as an almost perfect set of nested ring-dykes and is of international repute as such. The form of the masses is excellently reflected by the topography of the area. However, their contact relationships where exposed are often obscure. The contacts of the Centre 3 intrusions with the earlier rocks of the complex are also demonstrated.

Centre 3 was first recognized and described by Richey and Thomas (1930), who interpreted its intrusions as ring-dykes. More recent work by Smith (1957), Bradshaw (1961), Wills (1970) and Walsh (1971, 1975 and in Gribble *et al.*, 1976) has challenged this interpretation, and new models of the form of the centre and its constituent intrusions have been proposed. The geochemistry and petrogenesis of the intrusions and their relationships has been investigated by Walsh (1975) and Walsh and Henderson (1977).

Description

The rocks assigned to Centre 3 (Fig. 4.9) of the Ardnamurchan complex crop out in a natural amphitheatre spanning 7 km from east to west and 6 km from north to south, the topography accurately reflecting the concentric outcrops of the various intrusions. Richey and Thomas (1930) recognized seventeen intrusions as ring-dykes which generally become progressively younger towards the middle of Centre 3. The following sequence of intrusions was proposed:

(oldest)

A	Quartz Gabbro of Faskadale
B	Fluxion Gabbro of Faskadale
C	Gabbro of Plocaig (outside site)
C1	Gabbro, south-east of Rudha Groulin (outside site)
D	Porphyritic Gabbro of the Meall nan Con screen
E	Great Eucrite
E1	Outer Eucrite

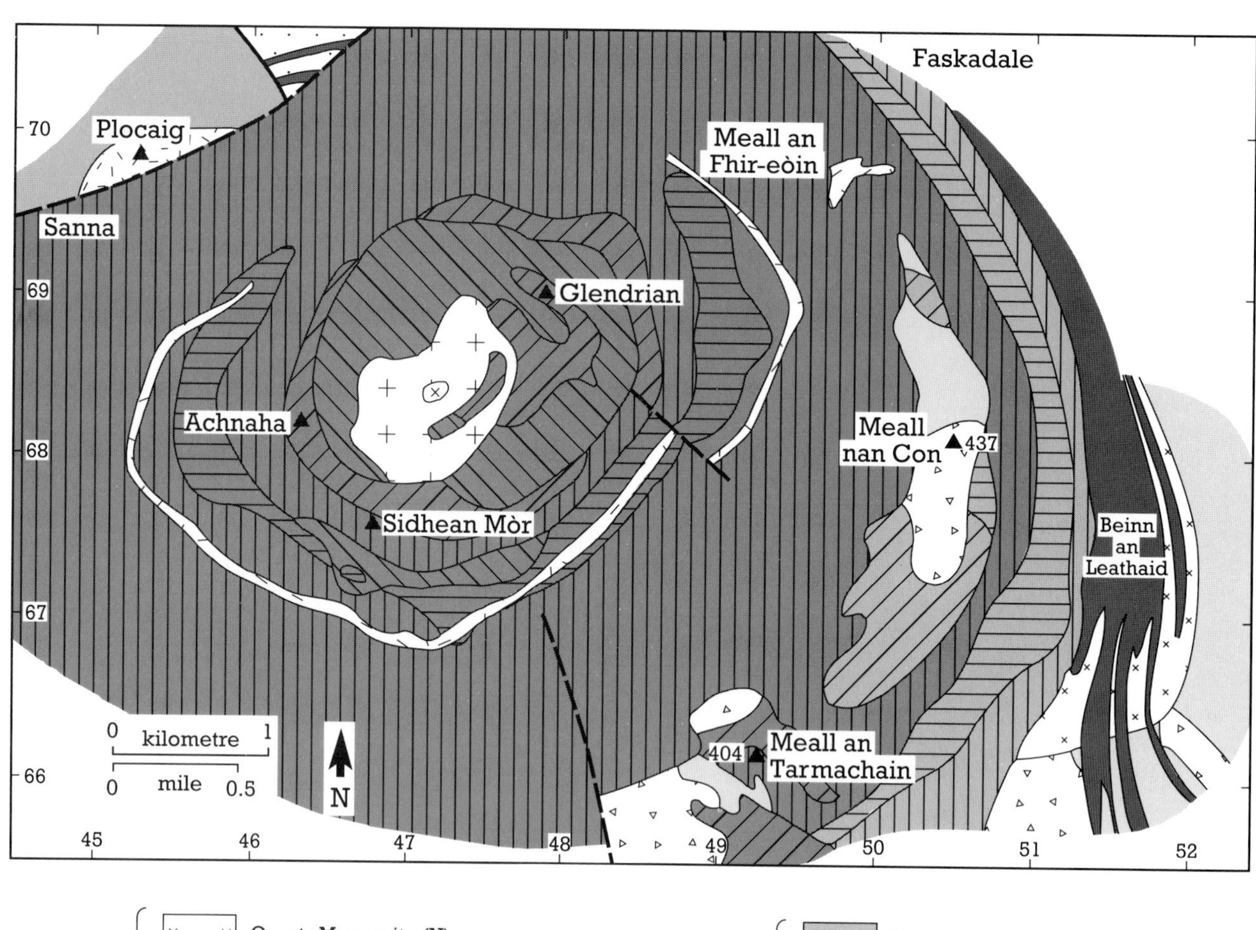

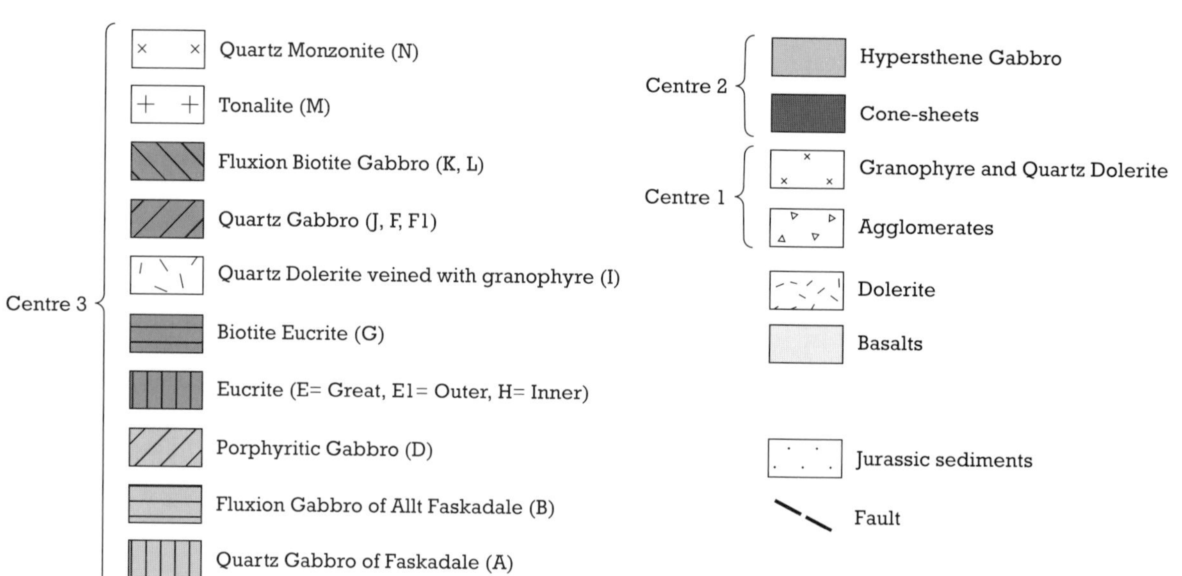

Figure 4.9 Geological map of Centre 3, Ardnamurchan (after Gribble *et al.*, 1976).

Figure 4.10 The natural amphitheatre of Centre 3, Ardnamurchan. The imposing arcuate ridges in the distance are formed by the Great Eucrite. (Photo: A.P. McKirdy.)

F	Quartz Gabbro of Meall an Tarmachain summit
F1	Quartz Gabbro, south side of Meall an Tarmachain
G	Biotite Eucrite
H	Inner Eucrite
I	Quartz Dolerite veined with granophyre
J	Quartz Biotite Gabbro
K	Fluxion Biotite Gabbro of Sithean Mor
L	Fluxion Biotite Gabbro of Glendrian
M	Tonalite
N	Quartz Monzonite

(youngest)

The majority of the rocks are coarse-grained gabbros and dolerites and intermediate rocks comprise only a small fraction of the centre. Few contacts between the individual intrusions are visible and their recognition is based upon minor differences in mineralogy or the presence or absence of structural features rather than clear intrusive relationships.

The intrusions of Centre 3, demonstrated by this site, are described below from the outermost inwards; this order corresponds approximately with the age sequence proposed by Richey and Thomas (1930).

The Quartz Gabbro of Faskadale (A) is the outermost and oldest ring intrusion of Centre 3 and has been described in detail above in Beinn na Seilg–Beinn nan Ord. This mass extends in an unbroken arc from Faskadale Bay, between Beinn an Leathaid and southwards through Meall nan Con and Glas Bheinn to pass beneath the intrusions of Centre 2 on the northern and eastern slopes of Beinn na Seilg. The petrography of the intrusion is very variable and ranges from olivine eucrite to basic granophyre.

On Glas Bheinn (NM 493 648), the truncation of the outer cone-sheets of Centre 2 and Centre 1 country rock by the Faskadale intrusion is very marked, although the actual contact is not visible. Along its contact with the Hypersthene Gabbro of Centre 2 exposures are discontinuous and the relationships of the two rocks are obscured by intervening screens of amygdaloidal basalt and possibly by shattering (Richey and Thomas, 1930). To the south-west of Meall an Tarmachain (NM 493 663), highly metamorphosed remnants of an original volcanic cover form a roof to the Quartz Gabbro, the contact dipping gently to the south-west. On its inner margin, the Quartz Gabbro of Faskadale is in contact with at least three later intrusions of Centre 3 – the Fluxion Gabbro of Faskadale, the Quartz Gabbro of Meall an Tarmachain and the Great Eucrite.

The arcuate Fluxion Gabbro of Faskadale (B) extends from Faskadale along the Allt Faskadale to the south-eastern slopes of Meall an Tarmachain. Its outer margin abuts the Quartz Gabbro of Faskadale and petrographically there is little other than the texture to distinguish the two rocks; at the single exposure of their contact described in the Memoir, the supposedly older Quartz Gabbro shows no thermal metamorphism. The petrological features of the Fluxion Gabbro have been described above.

At Plocaig (NM 453 698), beyond the north-eastern limit of the site near Sanna Bay, a small and apparently lenticular mass of moderately coarse-grained, occasionally feldspar-phyric olivine gabbro (C) lies between the north-western margin of the Great Eucrite and the Hypersthene Gabbro (Centre 2). This intrusion contains xenoliths of granular, hypersthene-bearing hornfels and of the Hypersthene Gabbro. The contacts are obscure but this gabbro is believed to be intermediate in age between these two major intrusions.

The Meall nan Con ridge and summit are formed of highly metamorphosed agglomerates intruded by a small mass of gabbro (D). The gabbro extends in two short, arm-like projections on either flank of the agglomerate. Thermal alteration of the gabbro 'arms' led Richey and Thomas (1930) to conclude that the gabbro is older than the adjoining eucrites; Bradshaw (1961), however, argues that the gabbro might well be part of the Great Eucrite, a view reinforced by the work of Walsh (1971).

The Great Eucrite (E), which here includes the Outer Eucrite (E1), Biotite Eucrite (G) and Inner Eucrite (H) of Richey and Thomas (1930), is the most spectacular unit of the Ardnamurchan complex. It forms an almost continuous, bold, glacially sculptured annular ridge in marked topographic contrast with the subdued landforms developed from the intrusions it encircles. The eucrite is thus the only example in Ardnamurchan of a complete ring-intrusion and, at the present level of erosion, makes up over half of the outcrop of Centre 3. The rock is typically coarse and feldspathic, but like many of the Ardnamurchan intrusions, varies greatly in detail; the proportions of feldspar, augite and olivine vary, often abruptly. Flow-banding is restricted to the inner parts of the intrusion and the outermost zones are rich in tangential, near-vertical, augite-rich pegmatite veins. The Great Eucrite is presumed to be younger than the Faskadale Quartz Gabbro, and Fluxion Gabbro, although the evidence for this is inconclusive because of the poor exposure in the critical areas. According to Richey and Thomas, the Biotite Eucrite lies between the Great and Inner Eucrites, distinguished by the widespread, but by no means ubiquitous, occurrence of biotite. However, Bradshaw (1961) has shown that biotite occurs in all the Ardnamurchan eucrites and there appears to be little convincing evidence to show that the three eucrites are separate intrusions.

The Quartz Gabbro forming the summit at Meall of an Tarmachain (F) is a small intrusion which contains localized olivine-rich and eucritic segregations. The clinopyroxenes in this gabbro are markedly different from all other gabbros, within the complex, being enriched in iron and deficient in calcium (Walsh, 1975), thus supporting its status as an independent intrusion. It is disputably younger than the adjoining eucrite and the Quartz and Fluxion Gabbros of Faskadale. A second mass of quartz gabbro outcrops to the south of the summit (F1); both intrude agglomerates and basalts.

A narrow arcuate intrusion of dolerite veined by granophyre (I) crops out intermittently in small isolated knolls in the low-lying area within the Great Eucrite outcrop. It extends in an extremely narrow, discontinuous horseshoe from Meall an Fhir-eoin (NM 486 698) south, west and northwards to terminate close to the Allt Sanna. It was the first intrusion to be mapped as a ring-dyke on Ardnamurchan by the Survey. The attitude of the contacts with the older Great Eucrite can be closely estimated to dip outwards at about 70°. The dolerite is remarkably homogeneous compared with the coarser gabbros and eucrites, and it contains a relatively higher proportion of biotite and more sodic plagioclase. Localized hornblende and biotite indicate that the mass may be partly a hybrid in origin.

Three masses of quartz-biotite gabbro (J) around Glendrian (NM 479 690), Achnaha (NM 463 682) and Druim Liath (NM 475 683) were presumed by Richey and Thomas (1930) to be parts of a single intrusion which once occupied the central parts of the complex, but was disrupted by later intrusions. Exposure is generally poor here and the contact relationships with the adjoining eucrite are difficult to establish. The gabbros are petrographically heterogeneous and show considerable variations in grain size, from dolerites to pegmatites and in modal mineralogy.

Two intrusions of fluxion gabbro have been mapped within Centre 3; these are known as the Glendrian (L) and Sithean Mor (K) Gabbros respectively. The Glendrian intrusion is a titaniferous, magnetite-rich, basic rock, and it has an almost complete ring-shaped outcrop about Centre 3 and forms a resistant, prominent ridge above the inner, younger intrusions. It is closely associated with a quartz–biotite gabbro which it almost certainly post-dates, xenoliths of the quartz–biotite gabbro being reported, but chilling being absent at the exposed contact. The fluxion texture is produced by the lamination of labradorite crystals dipping inwards towards Centre 3 at 30°-40°. Hybridization by an acid magma before emplacement has been suggested (Richey and Thomas, 1930), this resulting in the local concentration of apatite and abundant interstitial feldspar growths of quartz and alkali feldspar. The Fluxion Gabbro of Sithean Mor (K) crops out as a small, crescent-shaped body which forms a prominent steep-sided ridge surrounded by eucrites. Although termed a fluxion gabbro (Richey and Thomas, 1930), only the northern parts show any alignment of feldspars and the rock grades southwards into a uniform quartz gabbro. The rock appears to be younger than the adjoining Inner Eucrite against which it is chilled and it also bears a few xenoliths of the eucrite.

The Tonalite (M) occupies low-lying, poorly exposed ground in the middle of Centre 3 and its outcrop forms one of the most extensive areas of intrusive intermediate rocks within the BTVP. The mass is roughly ovoid, being elongated north-east to south-west, and has a distinctive appearance with large platy biotites in a leucocratic feldspathic groundmass. The margins of the intrusion against the Fluxion and Quartz gabbros are finer grained, more acidic and leucocratic and have been interpreted as the products of chilling, thus establishing that the Tonalite is the younger. Exposures of the contacts are rare and their form confusing; some dip steeply outwards and others dip steeply inwards. A substantial mass of quartz- and biotite-bearing gabbro lies within the eastern part of the Tonalite.

The later Quartz Monzonite (N) forms a small, poorly exposed oval mass within the Tonalite outcrop and can be distinguished in the field by the presence of more abundant, larger, deep-brown biotite crystals in a finer-grained, feldspathic groundmass. At its contacts with the enveloping Tonalite, the Quartz Monzonite is finer-grained and the contacts appear to dip inwards at 65°. Both rocks contain the disequilibrium assemblage of plagioclase, alkali feldspar, quartz, augite, hornblende, biotite, magnetite, ilmenite, apatite and chlorite and have generally been interpreted as hybrids or the results of crustal contamination and assimilation.

The Centre 3 complex is dissected by many crush lines and minor faults generally trending NS or NNW–SSE, as is clearly shown on aerial photographs (for example, see Stewart, 1965 and cover photograph of Gribble *et al.*, 1976). A set of radial joints and crush lines marked by erosion hollows is also prominent. From Meall nan Con to Beinn an Leathaid the intrusions of Centre 3 truncate those assigned to Centre 1. The earlier intrusions (mainly quartz dolerites and granophyres) cut earlier volcanic rocks and are themselves intruded by dense swarms of cone-sheets. The composite sheet-like body of Beinn an Leathaid belongs to Centre 1. In this intrusion, dolerite at the base is succeeded by a granophyre carrying xenoliths of Moine schist and gneiss; a thin transition zone between the dolerite and granophyre occurs east of the summit. Farther to the east, agglomerates of the Northern Vents are exposed. A few NNW-trending dolerite dykes cut the Centre 3 ring-dykes, for example to the east of Achnaha.

Interpretation

Centre 3, Ardnamurchan was first recognized by Richey and Thomas (1930) who, despite an almost total absence of evidence relating to the contacts between intrusions, interpreted it as a series of nested ring-dykes. In more recent years, Richey's axiom that the contacts dip steeply outwards has been challenged and it has been postulated that the contacts dip gently inwards to produce a saucer- or funnel-shaped intrusion (Smith, 1957; Bradshaw, 1961; Wills, 1970; Walsh, 1971, 1975). Therefore, Centre 3 is important both because of its long acceptance as an almost perfect example of a nested ring-dyke complex and because of the doubts raised by subsequent research as to whether it is a ring-dyke complex at all. The problem at present has not been resolved although geophysical investigations (Bott and Tuson, 1973; Binns *et al.*, 1974) have revealed that the maximum positive gravity anomaly for the Ardnamurchan complex is less than half of that found over the Skye, Rum and Mull complexes, indicating that the Ardnamurchan complex occupies a much smaller volume than the others. Moreover, the gravity maximum is located over Centre 2 and not Centre 3. Although these facts are admittedly inconclusive, they can hardly be said to support the hypothesis that Centre 3 consists of a nested complex of ring-dykes some 3 km in radius, each ring-dyke having been formed by the subsidence of a central block of pre-existing, relatively dense igneous rock.

More recent workers on Ardnamurchan have tended to reduce the number of components in the complex by combining two or more of Richey and Thomas's intrusions into single entities. Thus, Smith (1957) and Bradshaw (1961) suggested that the Great Eucrite (E), the Biotite Eucrite (G) and the Inner Eucrite (H) form a single intrusion and Walsh (*in* Gribble *et al.*, 1976) suggested that the fluxion gabbros B and L are not independent intrusions but parts of the adjacent quartz gabbros (A and J). Stewart (1965) listed the ring-dykes but omitted the gabbro south-east of Rudha Groulin (C1) which was described as a sheet, and combined F and F1 into a single intrusion. The Outer Eucrite (E1) is petrographically identical to the Great Eucrite (E) and, although described as a separate mass on the basis of unsatisfactory evidence (Richey and Thomas, 1930), was not shown as an independent entity on the map which accompanied the Memoir. It is listed by Stewart (1965), but has been tacitly combined with the Great Eucrite in the most recent literature and maps of Walsh (1975) and Gribble *et al.* (1976). If this body of work is accepted *in toto*, the number of discrete intrusions (apart from cone-sheets and dykes) forming Centre 3 is reduced from seventeen to ten.

Richey and Thomas (1930) regarded all the Centre 3 intrusions as having derived from a single magmatic parent. More recently, however, Walsh (1975) has shown that this may be true only of the eucrites, gabbros and dolerites. The pyroxenes in the Tonalite (M) and Quartz Monzonite (N) fail to show the significant iron-enrichment to be expected if they had formed from the same magma as the basic intrusions. On these grounds, Walsh (1975) considered the intermediate rocks to be hybrids of basic magma contaminated by assimilation and partial anatexis of country rock and previously emplaced intrusions, a view also supported by Walsh and Henderson (1977) on the basis of rare-earth-element geochemistry.

It is clear that although Centre 3 has long been regarded as containing some of the most perfect examples of ring-dykes in the BTVP, the status of these intrusions is now far from clear and the centre really demands a concerted and co-ordinated field and laboratory investigation to resolve the outstanding problems of the form of its constituents and their spatial and chemical relationships to one another.

Conclusions

Centre 3 has long been accepted as being a classic example of a ring-dyke complex. New evidence, however, contradicts this interpretation and the Centre is probably a saucer- or funnel-shaped intrusion. The controversy surrounding this Centre gives the site special significance. Many of the originally mapped intrusions have been argued to be part of the same intrusion and it is now suggested that the intrusions number no more than ten. The basic intrusions are probably genetically related to the same parent magma, but the intermediate rocks appear to be the result of contamination of a basic magma with crustal rocks.

Chapter 5

Isle of Mull

INTRODUCTION

The igneous rocks of Mull are arguably the most historically significant in the British Tertiary Volcanic Province. They have been examined extensively for the past two centuries, but it was through the classic Mull Memoir of the Geological Survey of Great Britain (Bailey *et al.*, 1924) that they achieved world-wide attention. Not only was the extremely complex geology of central Mull described and illustrated by the beautiful 'One Inch' geological map (Sheet 44), but important concepts such as ring-dykes, cone-sheets, centres of igneous activity, and magma types and magma series were developed, rendering Mull important in the international context.

Mull is a mountainous island with a long, indented coastline. The coastal exposures provide good sections through the relatively early lavas (Fig. 5.1); however, the intrusions forming the central complex are largely inland where much of the countryside is mantled by peat. This is especially true of the lower ground, where tantalizing, but incomplete, glimpses of the solid geology are provided by stream sections as, for example, in the Allt Molach (see below). Exposure is, however, often good and continuous on the higher ground. The authoritative account of the field geology contained in the Geological Survey's Memoir and maps has provided the background for much subsequent work over the past thirty years. The emphasis of this more recent work has been on the geochemistry, geophysics and isotope geology of the Tertiary igneous rocks; publications have been summarized by Skelhorn (1969) and Emeleus (1983), and by several authors in Sutherland's review of the *Igneous Rocks of the British Isles* (1982). These investigations covered a range of topics, including the deep geology of the Mull volcano, the likely duration of igneous activity, the origins of the granitic and basaltic rocks and the extent to which they have received contributions from crustal as well as mantle sources, and the magnitude and likely effects of extensive hydrothermal systems established during the life of the complex.

The sequence of igneous activity on Mull is summarized in Table 5.1. The general pattern bears similarities with that found on Skye and

Figure 5.1 The flat-lying succession of basalt lavas of the Wilderness area, western Mull, give rise to the trap-type topography. Bearraich site, Mull. (Photo: C.H. Emeleus.)

Table 5.1 The Mull Central Complex: sequence of events (after Skelhorn, 1969, pp. 2–6)

(youngest)
Dykes were intruded throughout the sequence (Loch Bà–Ben More)

Loch Bà Centre (Centre 3; North-West or Late Caldera)

Loch Bà felsite ring-dyke (Allt Molach–Beinn Chaisgidle, Loch Bà–Ben More)
Hybrid masses of Sron nam Boc and Coille na Sroine (Loch Bà–Ben More)
Beinn a' Ghraig Granophyre (Loch Bà–Ben More)
Knock Granophyre (Loch Bà–Ben More)
Late basic cone-sheets (Loch Bà–Ben More)
Early Beinn a' Ghraig Granophyre and felsite (Loch Bà–Ben More)
Glen Cannel complex and some late basic cone-sheets
(Allt Molach–Beinn Chàisgidle, Loch Bà–Ben More)

Beinn Chàisgidle Centre (Centre 2)

Glen More ring-dyke (Loch Sguabain, Cruach Choireadail)
Late basic cone-sheets (Allt Molach–Beinn Chàisgidle), Loch Scridain sheets (intruded towards middle and end of Centre 2 and start of Centre 3)
Ring-dyke intrusions around Beinn Chàisgidle
?Augite diorite masses of An Cruachan and Gaodhail (Loch Bà–Ben More)
Corra-bheinn layered gabbro (Loch Bà–Ben More)
Second suite of early basic cone-sheets
Second suite of early acid cone-sheets
Explosion vents (numerous at margin of the South-East Caldera) (Loch Bà–Ben More)

Glen More Centre (Centre 1; including the Early or South-East Caldera)

Ben Buie layered gabbro
Loch Uisg granophyre-gabbro
First suite of early basic cone-sheets (Loch Bà–Ben More)
Early acid and intermediate cone-sheets (Loch Bà–Ben More)
Acid explosion vents containing porphyritic rhyolite material (Loch Bà–Ben More)
Glas Bheinn and Derrynaculen granophyres (Loch Spelve–Auchnacraig)

Updoming and folding in south-east Mull as a result of rising diapir (Loch Spelve–Auchnacraig).

Lava eruption on to eroded surface of Mesozoic and older rocks. Latest flows overlap in time with formation of the South-East Caldera where pillow lavas are found. (Lavas: Bearraich, Ardtun, Carsaig Bay, Loch Bà–Ben More. Pillow lavas: Loch Sguabain, Cruach Choireadail)
(oldest)

Ardnamurchan, but it differs in detail. Activity commenced with the eruption of basaltic lavas fed from linear NW–SE fissures now represented by the extensive Mull dyke swarm. A thickness of up to 2 km of lava is preserved and studies of zeolite zones (Walker, 1971) indicate that the original succession may have exceeded 2.2 km in thickness. Eruption of the younger lavas overlapped the establishment of a central complex, as flows, including pillow lavas, were erupted into caldera lakes associated with central subsidence and ring-dyke formation in and around the Early or South-East Caldera (Centre 1; Table 5.1).

Three distinct centres of igneous activity have been recognized within the Mull Central Complex (Table 5.1); in each, both acid and basic intrusions are present and granitic rocks may predominate at outcrop, particularly in Centre 3. However, the centres coincide with a major gravity high and it is clear that dense, gabbroic or peridotitic rocks underlie all the centres (Bott and Tuson, 1973); thus, basaltic or picritic magmas have been the driving force in the Mull central complex as in the other centres of the BTVP. The gravity surveys of Mull also show that the areally extensive granitic rocks are relatively thin, probably less than 2 km in thickness (Bott and Tuson, 1973, see also Bott and Tantrigoda, 1987). Other geophysical investigations include detailed studies of the magnetostratigraphy by Mussett *et al.* (1980) and Dagley *et al.* (1987), which suggest that the igneous activity spanned several magnetic reversals. When combined with radiometric age determinations, a sequence of reversed–normal–reversed polarities is apparent, covering about 3.5 Ma in the middle Palaeocene (between about 60 and 57 Ma; Mussett *et al.*, 1988; Table 1.1).

The arcuate, centrally-focused character of the intrusive igneous activity is strikingly demonstrated by the ring-dykes and cone-sheets of the Mull complex, particularly those of Centres 2 and 3. This feature, together with the especially clear examples of cone-sheet swarms and the perfection of individual ring-dykes such as the Loch Bà Felsite, makes Mull a classic area for these forms of intrusion, as it is also for the clear demonstration of a shifting focus of igneous activity as the complex developed.

Granite magmas were available during the activity of all three centres on Mull. Geochemical and isotopic studies (Walsh *et al.*, 1979) provide evidence that the early granites of Centre 1 contain substantial contributions of partial-melt products from Lewisian gneiss, in addition to magma derived from fractional crystallization of basaltic magmas. The granites of the later Centres 2 and 3, however, were principally derived by fractionation of basaitic magma, with only minor contributions from crustal sources. Presumably the initial rise of basaltic magmas into the crust melted out most of the available low-melting-point constituents from the gneisses, contributing to the Centre 1 granites, but leaving little available for the acid rocks in Centres 2 and 3 when subsequent batches of basalt magma rose through the, by then, extremely refractory restites. In order to make deductions about the sources of the granitic rocks it is necessary to assume that no major event has affected their compositions since their emplacement. It has been demonstrated that virtually all of the Mull central complex rocks were affected by massive circulation of heated meteoric waters (Taylor and Forester, 1971; Forester and Taylor, 1976), which clearly might have been expected to alter the rock compositions. However, the alteration appears to have been limited and it has been shown that careful sampling can provide material which has not undergone reorganization of either elemental or isotopic compositions (Walsh *et al.*, 1979).

The stratigraphy of the Palaeocene lavas of Mull is not known in the same detail as the Skye lava succession. Little has been published specifically on this topic since the appearance of the Mull Memoir (Bailey *et al.*, 1924). In this, the authors distinguished a major Plateau Group (Table 5.2) of olivine basalts overlain by another thick group of generally non-porphyritic basaltic lavas which comprised the Central Group – flows largely restricted to outcrops within the central complex or immediately adjoining areas (Bailey *et al.*, 1924, plate III and table X). Within the upper part of the Plateau Group, high on Ben More, the distinctive sequence of pale-coloured olivine basalts, conspicuously feldspar-phyric basalt (Big-Feldspar basalts) and a thick mugearite flow were mapped. A further subgroup, which closely resembled basalts of the Central Group, was found at the base of the Plateau Group around Loch Scridain and on Staffa. This was termed the Staffa Type. The majority of flows in the Plateau Group were silica-poor, olivine-phyric basalts and these voluminous basalts were held to be the products of a Plateau Magma-Type. The finer-grained, generally olivine-poor and more silica-rich Central Group lavas were distinguished as the Non-Porphyritic Central Magma-Type basalts. These and other magma types were thought to form a genetically linked group of basalts and more fractionated rocks and to constitute a Normal Mull Magma Series (Bailey *et al.*, 1924, Chapter 1). The magma types and magma series were defined on chemistry, mineralogy and petrography and therefore included both lava flows and intrusive rocks. The terminology and correlations of the Mull lavas are summarized in Table 5.2; the Mull Tertiary igneous sites are shown in Fig. 5.2.

Although little regional mapping appears to have been carried out on the Mull lavas since the publication of the Memoir, a large and valuable amount of geochemical and petrological data has built up, particularly in the last fifteen years. Using fresh samples collected outside the zone of intense pneumatolytic alteration around the central complex, Beckinsale *et al.* (1978) were able to subdivide the lavas (mainly of the Plateau Group) into three groups on geochemical criteria. The approximate positions of these groups are indicated in relation to the Memoir stratigraphy in Table 5.2. The olivine basalts of Group 1 were thought to have come from partial melting of

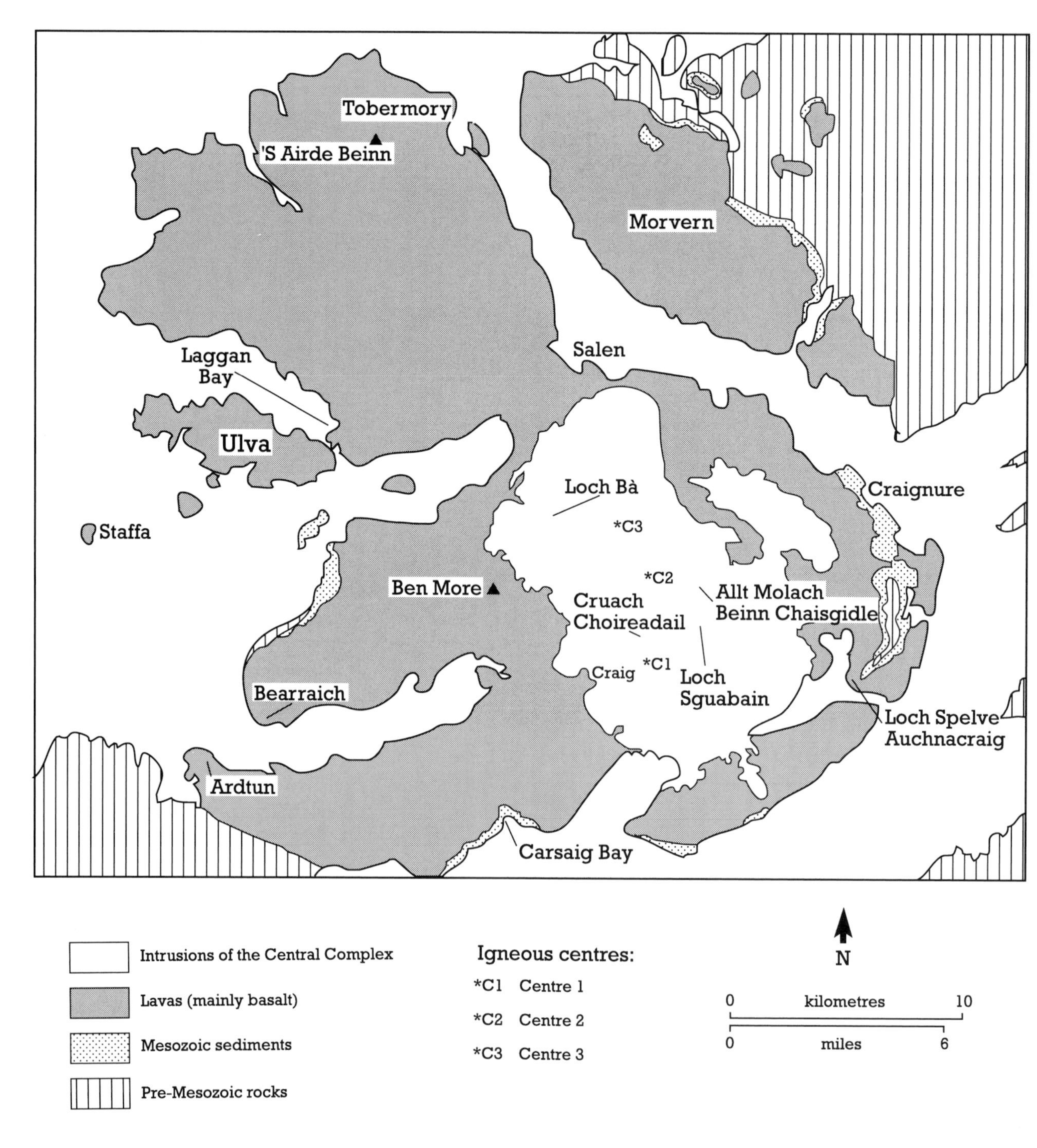

Figure 5.2 Map of the Isle of Mull, showing localities mentioned in the text.

garnet-lherzolite mantle. The Group 2 lavas, equivalent to the Staffa Type, appeared to come from a higher-level mantle source made of plagioclase lherzolite and the Group 3 flows were thought to represent mixtures of Group 1 and Group 2 magmas (or sources).

Essentially the same parts of the succession, and sometimes the same flows and localities, were examined by Morrison (1978), Thompson *et al.* (1982), Morrison *et al.* (1985) and Thompson *et al.* (1986), who came to rather different conclusions about the sources and origins of the flows. Particular attention was paid to the Staffa Type flows (Group 2 of Beckinsale *et*

Table 5.2 Classification and correlation of the Mull lavas

Mull Memoir (Bailey *et al.*, 1924)	Beckinsale *et al.* (1978)	Morrison (1978) Thompson *et al.* (1982) Morrison *et al.* (1985) Thompson *et al.* (1986)
Central Group (= NPCMT)	Not dealt with in detail	Some samples analysed, all zeolitized or hydrothermally altered.
(Includes pillow lavas in central complex)		
Plateau Group (majority = PMT)	Group 1 olivine basalts (mainly sampled in north-west Mull)	Mull Plateau Group (MPG) Note that many are transitional between alkali basalt and tholeiite, and compare closely with Skye Main Lava Series. Some lower crust contamination.
Pale Group of Ben More (= PMT)	and	
(with interlayered mugearite and Big-Feldspar Basalt)	Group 3 olivine basalts (mainly sampled around Lochaline, Morven)	
(Staffa Type at base = NPCMT)	Group 2 of south-west Mull	Staffa Magma Type (SMT) Variably enriched in lower and upper crustal contaminants.

(NPCMT = Non-Porphyritic Central Magma Type) later = tholeiitic basalt
(PMT = Plateau Magma Type) later = alkali olivine basalt but many flows are in fact transitional between alkali basalt and tholeiite
Total thickness of Mull lavas estimated about 2000 m (Bailey *et al.*, 1924)

al., 1978), where it was found that virtually each flow examined had significant chemical and/or isotopic features; that is, each was unique. This variability was attributed to differing degrees of contamination of the basaltic magmas as each followed its own unique course towards the surface. Picritic magmas were thought to have been formed by partial melting of a spinel lherzolite source, the magma ponded at the Moho and evolved to olivine-basalt magma by olivine fractionation. The fractionated batches of basalt magma pursued their own paths upwards, some were caught in density traps within the crust where they produced silicic rheomorphic magmas from the melting of the adjoining crustal rocks. Where this occurred, the basaltic magma assimilated all or some of the partial melts and became distinctively fingerprinted, taking on, in very diluted form, chemical and isotopic characteristics of that particular country rock. A few flows passed through the crust without reaction or mixing and lack the distinctive contaminated features of the majority. The actual amount of contamination undergone by the flows is estimated to have been between 5–10%, insufficient to mask features attributable to the mantle sources and equilibration at the base of the crust. However, the possible contaminants – granulite- or amphibolite-facies Lewisian gneisses and Moine rocks – have distinctive lead, neodymium and strontium isotopes and concentrations of elements such as potassium, rubidium, titanium, phosphorus and the rare earths which it is possible to analyse with extreme accuracy using modern techniques (cf. Chapter 1), thus enabling deciphering of these fingerprints, and allowing possible models for Palaeocene subcrustal plumbing to be constructed. Some possibilities are shown diagrammatically in Fig. 5.3. All the lavas except F underwent variable degrees of fractional crystallization at Moho depths; F rose directly to the surface without either appreciable fractionation or contamination. All the remainder, except E, were held at the boundary between Lewisian granulite-facies gneisses and overlying amphibolite-facies gneisses, where mixing with small amounts of melt from the granulites occurred. Subsequently, A ponded at the Lewisian amphibolite gneiss/Moine schists boundary and acquired an 'amphibolite' fingerprint from small

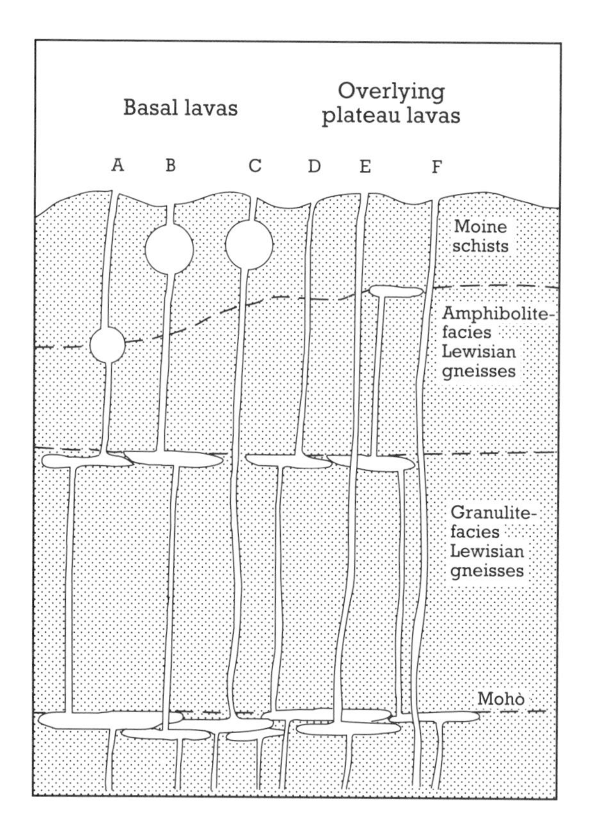

Figure 5.3 Sketch of the magmatic plumbing beneath south-west Mull during extrusion of the Palaeocene basaltic lavas (after Morrison *et al.*, 1985, fig. 4). See text for explanation.

amounts of crustal melt, whereas B was held within the Moine schists and mixed with small amounts of melt generated from the schists. C was also contaminated by Moine-derived melt but did not receive a contribution from either Lewisian gneiss source.

It is apparent that there are currently two very different views on the origin of the diversity of compositions among the Mull (and other) lavas. In one, the variability is attributed to partial melting occurring at different levels in the mantle, with the possibility of some mixing of the melt products; the mantle is thought to have had long-term vertical heterogeneity prior to the Cenozoic (Beckinsale *et al.*, 1978). A radically different viewpoint is that the magmas originated from a spinel lherzolite mantle source, formed when basaltic magmas on their way to the surface were ponded and their heat triggered localized small amounts of partial melting in the adjoining crustal rocks (Thompson *et al.*, 1982; Morrison *et al.*, 1985; Thompson *et al.*, 1986). If such events were on a sufficiently large-scale, granite-producing anatectic melts could form. The absence of rhyolite flows in the Mull lava succession therefore implies that large volumes of silicic melt were not generated at any given time during this early stage in Tertiary magmatism in Mull. However, conditions obviously changed as the central complex became established and its dense, hot, mafic root grew within the crust below Mull (cf. Bott and Tuson, 1973).

The emplacement of British Tertiary Volcanic Province central complexes was frequently accommodated by folding, faulting and other structural adjustments. On Mull some of the most spectacular examples of these features are provided by the annular folding around the early centre which is particularly well displayed in the Loch Don area (see Loch Spelve below). The significance of early folding has been examined by Walker (1975) who suggested that the deformations may be gravity structures developed in association with central updoming caused by the early diapiric emplacement of the granite magmas.

With a few notable exceptions, the researches into the Tertiary igneous geology of Mull since the publication of the Mull Memoir have been either largely geochemical or geophysical; few studies have involved intensive field-work. This is clearly a tribute to the high quality of the Survey's original field mapping and interpretation. However, the great strides made over the past 65 years in such areas as structural geology and physical volcanology, indicate that there must be considerable scope for further field-based research into many aspects of the Tertiary igneous geology of Mull. The size and complexity of the Mull centres prompts the suggestion that this should be undertaken on the basis of team-work.

BEARRAICH

Highlights

The site contains an excellent succession of basalt flows at the base of the Mull lavas, with good development of columnar jointing, pegmatitic veins and segregations, and associated lignite beds. The lavas are seen to envelop 'MacCulloch's Tree'.

Introduction

The south-western part of the Ardmeanach Peninsula at the entrance to Loch Scridain shows a representative and relatively well-exposed, continuous section through the lower lavas of the Plateau Group (Bailey *et al.*, 1924) of Mull and Morvern (Fig. 5.1) The majority of the lavas are alkaline to transitional olivine basalts belonging to the Mull Plateau Group (see Table 5.2 for lava correlations, etc.), but the flows at the base of the succession form part of the Staffa Magma Type (cf. Thompson *et al.*, 1986). The basal flow envelops 'MacCulloch's Tree' (Fig. 5.5).

The most comprehensive investigation, to date, on the lavas of Mull was published in the classic Mull Memoir by Bailey *et al.* (1924); it remains the only account with comprehensive field data. The geochemistry of the lavas was subsequently reassessed by Tilley and Muir (1962) and Beckinsale *et al.* (1978) proposed a new subdivision on the basis of preliminary trace-element and isotope data. Further detailed geochemical investigations by Morrison (1979), Morrison *et al.* (1985) and Thompson *et al.* (1986) suggest that the lavas were significantly contaminated by crustal partial melts during their passage to the surface (see also Introduction to this chapter). The distribution of zeolite minerals within the lavas of Mull, including Ardmeanach, has been studied by Walker (1971).

Description

From Tavool House (NM 437 273; Fig. 5.4) to the Wilderness (NM 405 290), the coastal cliffs provide continual exposure through lava flows which demonstrate spectacular examples of two-tier columnar jointing and complex, auto-intrusive admixtures of massive columnar and scoriaceous basalt. The majority of flows are fine-grained, aphyric basalts which have a very compact appearance. These are the Staffa Magma Type lavas of Thompson *et al.* (1986) which show transitional, tholeiitic–alkaline, olivine-basalt affinities; they were considered to be pyroxene-rich variants of the mildly alkaline Plateau basalts (Tilley and Muir, 1962). Many of the flows higher in the succession are olivine basalts belong to the Mull Plateau Group (Table 5.2), while others are basaltic hawaiites (Beckinsale *et al.*, 1978).

Thin tuff and lignite beds occur towards the base of the lava succession which are contemporaneous with the fluvio-lacustrine sediments at Ardtun. At this horizon, John MacCulloch discovered the well-known large tree fossils, the most famous of these being 'MacCulloch's Tree' at Rubha na h-Uamha (NM 402 278), (Fig. 5.5), an upright coniferous trunk some 12 m high engulfed by lava (MacCulloch, 1819).

Above the basal columnar flows of Staffa Magma Type basalt lies the main succession of Mull Plateau Group basalts. This is a thick sequence of medium- to coarse-grained olivine basalt lava flows, some of which contain glomeroporphyritic aggregates of olivine and plagioclase, but others are feldspar-phyric. Chemically, the lavas have mildly alkaline to transitional alkaline–tholeiitic olivine-basalt affinities. Between 25 and 30 flows can be identified from the base of the suite to the summit of Bearraich (432 m), each generally less than 15 m thick and forming the characteristic stepped topography of the region. Of particular note is the presence of red, scoriaceous flow tops, indicating subaerial extrusion and contemporaneous weathering.

Pegmatite segregation veins of augite, feldspar and analcite are observed in a few of the lavas of the site. The lava flow in the cliffs below Dun Bhuirg (NM 422 262) 2 km west of Tavool House on Loch Scridain contains a good example of this phenomenon. The lava, an ophitic, olivine basalt bearing violet-coloured augite and abundant fresh olivine (Plateau Magma Type) is traversed by numerous small veins of analcite and contains cavities filled by zeolites associated with analcite (see Walker, 1971 for zeolite distribution in Mull). Coarse-textured veins of euhedral pyroxene and strongly zoned plagioclase in a matrix of analcite and feldspar also occur. The alkaline nature of their matrix is shown by the presence of aegirine augite, both in the zoned margins of titanaugite crystals and forming individual crystals; segregation of this alkaline residual material could generate rocks of phonolitic, nepheline, syenite composition (cf. the Shiant Isles).

Subsidiary interests of the site include fragments of Mesozoic and basal Tertiary sediments in the small landslip at Aird na h-Iolaire (NM 403 288). Minor Tertiary intrusions include sills of the xenolithic Scridain Suite (Bailey *et al.*, 1924) and pitchstone dykes.

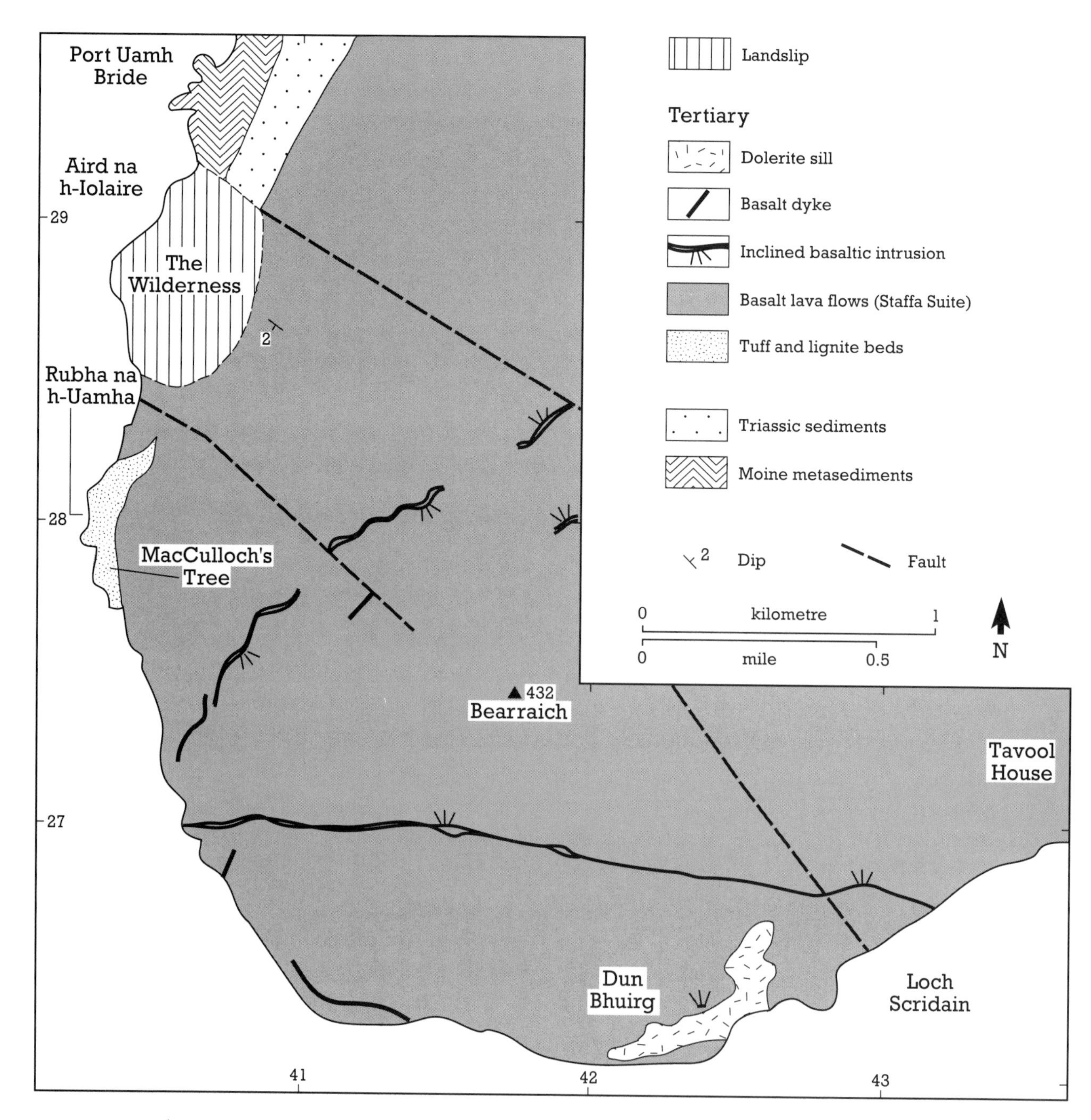

Figure 5.4 Geological map of the Bearraich site (adapted from the British Geological Survey 'One Inch' map, Sheet 43, Iona).

Interpretation

The site provides a virtually uninterrupted section through the base of the Plateau lava succession on Mull, recording the onset of Palaeocene igneous activity. The occurrence of red-weathered tops to lava flows, lignite beds and the nearby Ardtun leaf beds attest to a subaerial, warm temperate environment of lava effusion. Thick red boles above the weathered flow tops in the vicinity of the site (for example, on Aird Kilfinichen, NM 494 278) may indicate possible lava extrusion in bodies of shallow water, although conclusive evidence is lacking. Walker (1971), in a study of the regional distribution of zeolites in the Ardmeanach lavas, has shown them to have suffered burial sufficient to place them in the laumonite and mesolite zones. The systematic

Figure 5.5 'MacCulloch's Tree' on Rubha na h-Uamha (NM 402 278), an upright coniferous trunk 12 m high engulfed by lava of Staffa Magma Type. Bearraich site, Mull. (Photo: C.J. MacFadyen.)

distribution of zeolite zones reflects the depth of burial of the lavas and, on Mull, these zones are discordant to the stratification of the flows. Walker (1971; see also Craig, 1983, Fig. 13.5) compared the Mull zonation with the more complete picture obtained from flows in Iceland, and estimated that the Mull lava succession was originally over 2200 m in thickness.

The lowermost lavas exposed on this site, including the flow enveloping 'MacCulloch's Tree', belong to the Staffa Magma Type of Bailey *et al.* (1924) and Thompson *et al.* (1986) (= Group 2, Beckinsale *et al.*, 1978). These strikingly columnar flows have provided some of the evidence used by Morrison *et al.* (1985) and Thompson *et al.* (1986) to postulate contamination by crustal partial melts – see Introduction to this chapter. At the present time, the areal distribution of these flows appears to be largely restricted to the base of the lavas either side of the entrance to Loch Scridain and to Staffa; however, there is a lack of modern stratigraphic information about the distribution of these and the other types of lavas recognized on Mull and Morvern, and their spatial distribution remains one of the outstanding problems of Tertiary igneous geology in the Hebrides, requiring a combination of modern geochemical work and careful field investigations.

'MacCulloch's Tree' is one of a number seen enveloped in the lavas within the site. It was assigned to the genus *Cupressinoxylon* by Gardner (1887). The conditions of eruption leading to preservation of this and the other trees have not been discussed in detail, but the enveloping lava must have behaved in a very fluid manner; had a typical, fragmenting flow-front advanced across the area, it is difficult to envisage that the trees remained upright. Further evidence about the environment at the start of lava eruption comes from Ardtun.

Conclusions

The site contains an important, continuous section through the lower part of the Mull lava succession. Geochemical investigation of the lowest flows shows that these were contaminated by crustal material on their path to the surface and this knowledge has been used to construct models of likely magmatic plumbing systems beneath the Mull lavas. The lowermost lavas are tholeiitic–transitional basalts, the main body of the overlying thick succession consists of alkali to transitional basalts. They were erupted subaerially and the lowermost flow engulfed growing trees.

ARDTUN

Highlights

The lacustrine sedimentary deposits between the Palaeocene lavas contain leaves of temperate plant species. The sediments and plants make up the renowned Ardtun Leaf Beds which are the prime interest of this site and are of international importance. The underlying lava contains pillows and was erupted into a shallow lake.

Introduction

The coastal cliffs and gullies within this site provide internationally important exposures of sedimentary deposits within the basal lavas of the Plateau Group in south-west Mull. These sediments are fluvial sands and gravels, which contain the renowned Ardtun Leaf Beds of important palaeobotanical value, are the prime interest of the site.

The sediments were originally discovered by a local man from Bunessan but were first fully investigated by the Duke of Argyll (1851). The discovery was of major importance in the study of the volcanic rocks of Mull since the sediments contained a rich terrestrial fossil flora and therefore allowed the associated lavas to be relatively dated. A full history of research, including a comprehensive list of the extracted plants, is given by Seward and Holttum in the Mull Memoir (Bailey *et al.*, 1924). In addition, a full description of the site is provided in a field excursion guide to the Tertiary volcanic rocks of Mull by Skelhorn (1969). Radiometric age studies on the lavas above and below the leaf beds have been carried out by Evans (1969), Mussett *et al.* (1973) and Mussett (1986).

Description

Although exposure is almost continuous along the coastal cliffs, the most accessible sections are located within small gullies leading down the cliffs (Figs 5.6 and 5.7). A sedimentary succession, which varies in thickness between 4 m and 15 m, lies upon the upper slaggy amygdaloidal zone of a thick columnar lava flow which exhibit well-developed, twisting columnar joints in the cliffs and sea stacks. Above the sediments, a second major columnar lava flow is exposed. Both lavas are olivine basalts of the Staffa Magma Type (Thompson *et al.*, 1986), or Group 2 of Beckinsale *et al.* (1978), (Table 5.2). The best section through the sediments is to be found in the ravine at Slochd an Uruisge (NM 377 248) and is described in detail by Skelhorn (1969). The sediments are predominantly flint-bearing conglomerates and grits, but contain three finer-grained horizons of silty sandstone and clay. The latter are known as the Ardtun Leaf Beds (Top, Middle and Bottom), containing an abundant leaf flora, including remains attributed to *Platanus* (plane), *Corylus* (hazel), *Quercus* (oak) and *Ginkgo* (maidenhair tree) (Skelhorn, 1969). The specimens are exceptionally well preserved and

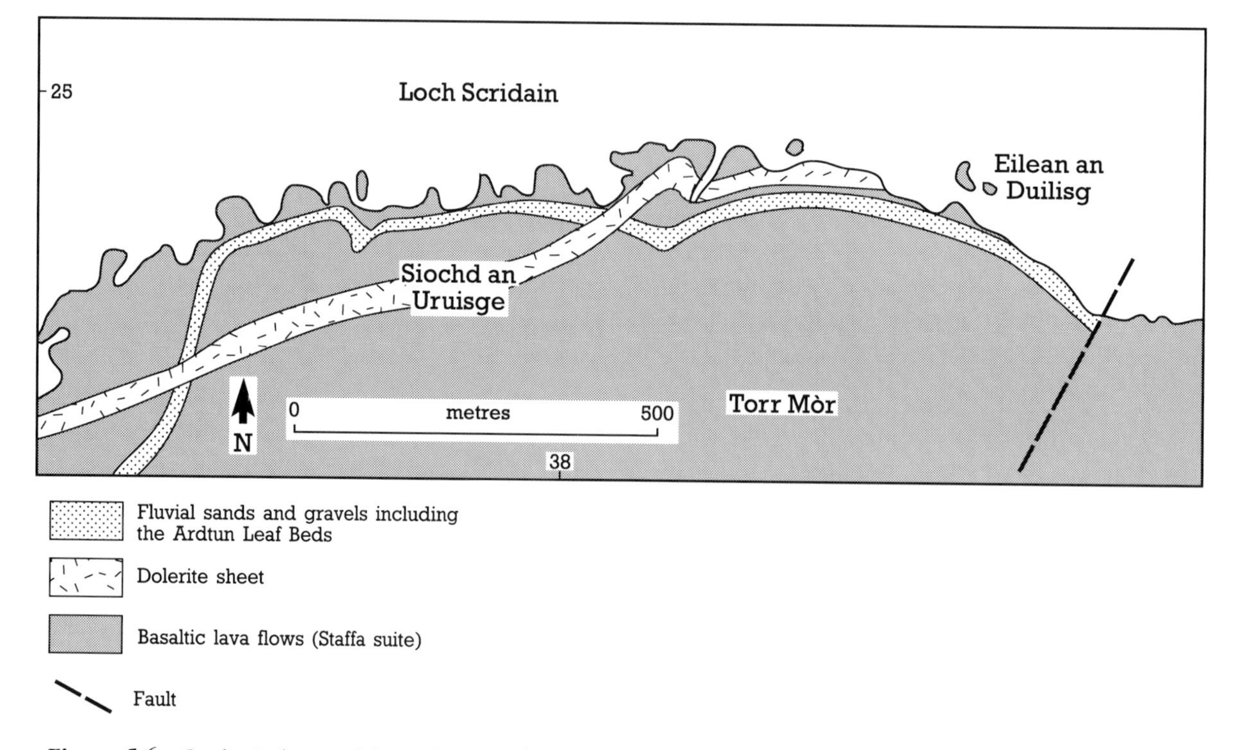

Figure 5.6 Geological map of the Ardtun site (adapted from the British Geological Survey 'One Inch' map, Sheet 43, Iona).

Figure 5.7 The best section through the Ardtun Leaf Beds at Slochd, an Uruisge (NM 377 248). Ardtun site, Mull. (Photo: C.J. MacFadyen.)

some are thirty or more centimetres in length, although they are extremely fragile. In addition, a sparse molluscan fauna has been recorded (Cooper, 1979) and fragments of a terrestrial beetle *Carabites scoticus*, a possible weevil and a ceropid insect have also been identified (Gardner, 1887; Cockerill, *in* Bailey *et al.*, 1924). The lowest leaf bed is underlain by a thin coal seam which passes down into 0.15–0.20-m-thick, whitish, concretionary root clay on top of the lower basalt flow (Bailey *et al.*, 1924).

The coarser-grained sediments are rich in angular quartz grains and derived flints and silicified chalk pebbles are abundant in the conglomerate. Pebbles of porphyritic igneous rocks are also present which may be of Palaeocene age and derived from nearby flows, or of Lower Old Red Sandstone age (Bailey *et al.*, 1924); conclusive evidence for their source is lacking.

The gullies to the east of the main locality (NM 377 248) expose a dolerite sill transgressive towards the lavas and the intercalated sediments. Against the sediments, the sill has a well-developed tachylitic margin which has partly devitrified. The flow beneath the sediments contains pillows which may be either whole or fragmented, probably formed when the lava flowed into a small, temporary lake (Skelhorn, 1969).

A little to the south-east of Eilean an Duilisg (NM 386 249), the sedimentary horizon is downthrown below sea level by a N–S-trending fault.

Interpretation

These deposits provide valuable evidence for extended periods of sedimentation during the period of lava effusion in south-west Mull. The sediments and associated flora indicate deposition in a shallow temporary lake surrounded by marshland and which intermittently received sediment and plant debris from surrounding areas. Lava occasionally flowed into this wet environment, resulting in the development of pillow structures such as those immediately beneath the sediments. The flora in the sediments shows that a temperate climate prevailed. These conditions were established on a land surface created by the first large Palaeocene lava fields in this region. Coarse, clastic sediments were probably derived from the surrounding volcanic landscape and possibly from the tuffs containing pebbles of Cretaceous flint exposed near Malcolm's Point, Carsaig (Skelhorn, 1969). The finer-grained Leaf Beds may represent deposits formed during periods of stagnation when little sediment was washed into the lake and a limited freshwater fauna flourished. The plant material in the Leaf Beds has been correlated with other temperate northern floras and dated as early Eocene (Gardner, 1887; Seward and Holtum, *in* Bailey *et al.*, 1924), although their age is now generally accepted as being Palaeocene (but see Simpson, 1961). In addition, radiometric dates for the lavas above and below the Leaf Beds reveal an age of around 58 Ma (Evans, 1969; Mussett *et al.*, 1973, 1980). A geochemical discussion of these lavas is presented in the 'Introduction' to this chapter.

Conclusions

The rich, well-preserved flora includes temperate-climate tree species such as plane, hazel, oak and ginkgo. They were preserved during relatively quiet periods of sedimentation when muds and silts accumulated; at other times, when more vigorous sedimentation gave sands and gravels, the delicate plant debris was probably destroyed. The coarser-grained debris was largely derived from the surrounding volcanic landscape but the presence of fragments of Cretaceous flint and silicified Chalk may indicate that older rocks may also have been exposed to erosion. The sediments were laid down in shallow lakes, or from rivers flowing across the lavas; the basaltic flow underlying the plant-bearing sediments was probably erupted.

LOCH SGUABAIN

Highlights

The site contains excellent examples of pillow lavas formed when basalts were erupted into the early, South-East Caldera lake. The pillow lavas are veined by granite and thermally metamorphosed where they are intruded by the Glen More ring-dyke.

Introduction

The site encompasses the north-western slopes of Bheinn Fhada above Loch Sguabain where a representative section through lavas belonging to the Central Group within the early South-East Caldera is exposed (Table 5.1). Exceptionally well-developed pillow lavas dominate the interest of the site and other features include acid veining and thermal metamorphism of the lavas at the contact with the quartz-gabbro/granophyre Glen More ring-dyke. The lavas within this site have received no detailed investigation since Bailey *et al.* (1924) carried out the regional survey. However, the geochemical work of Beckinsale *et al.* (1978) and Morrison (1978) on the Mull lavas in general has applications to the tholeiitic Central Group lavas of this site.

Description

The lava flows exposed in the craggy outcrops above Loch Sguabain (Fig. 5.8) are sparsely feldspar-phyric and aphyric tholeiitic to transitional basalts forming part of the outer, earlier zone of lavas in the South-East Caldera (Bailey *et al.*, 1924). The basalts here lie within the zone of pneumatolysis associated with the Mull central complex; the effects of this hydrothermal alteration include the albitization of feldspar, decomposition of olivine and secondary epidote and quartz in veins and vesicles.

Excellent pillow structures are located at the summit of the first ridge on Beinn Fhada (Fig. 5.9), south-east of the loch (NM 630 304). The individual pillows here average 0.6–0.7 m in diameter and have clearly identifiable chilled margins, within which lie concentric zones rich in vesicles. Altered, fine-grained, hyaloclastite tuffs surround the pillows, and thin-bedded tuffs are occasionally present in the lava sequence.

The South-East Caldera lavas have been intruded by what appears to be an extension of the Glen More ring-dyke composed of quartz gabbro which passes upwards into granophyre. South-east of Loch Sguabain, granophyric net-veined breccia becomes progressively more evident at higher levels in the intrusion, with the uppermost portions showing a preponderance of granophyre to gabbro or coarse dolerite. Granophyric veins are also common in the lavas near the contact. The lavas have been thermally altered by the intrusion and a thin zone of granular hornfels is

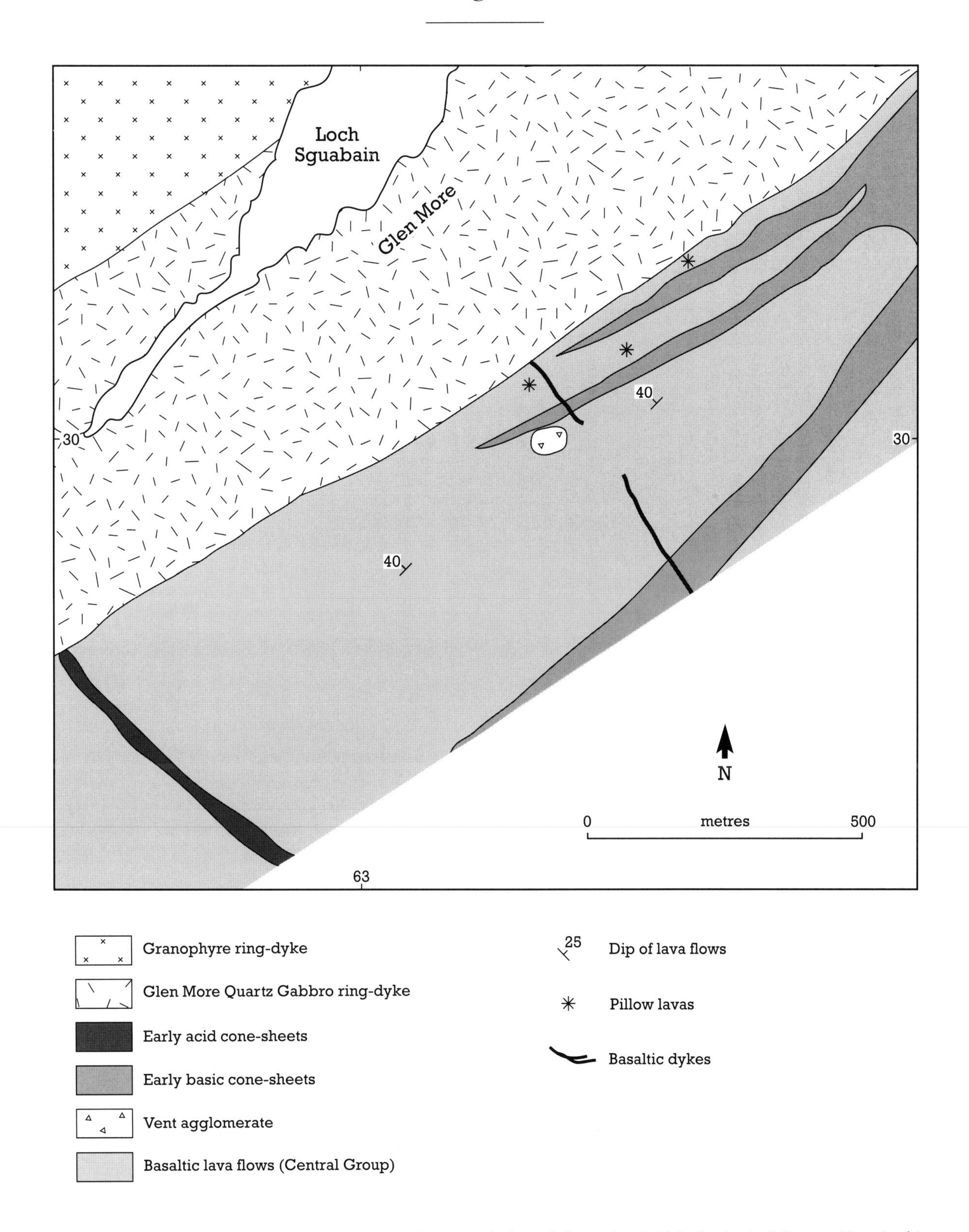

Figure 5.8 Geological map of the Loch Sguabain site (adapted from the British Geological Survey 'One Inch' map, Sheet 44, Mull).

Figure 5.9 Basaltic pillow lavas, formed in a caldera lake. Loch Sguabain site, Mull. (Photo: A.P. McKirdy.)

developed at the contacts. In addition, early basic cone-sheets, which dip north-westwards at angles approaching 45°, intrude the lavas within the site.

Interpretation

Pillow lavas occur in the basal parts of several of the lava fields of the BTVP (for example Mull, Ardtun; Rum, Fionchra) but they are not recorded within central complexes except on Mull. The widespread occurrence of pillows in the basalts exposed in Glen More, and particularly on this site, and the manner in which they appear to lie within an arcuate zone (Bailey *et al.*, 1924, fig. 18), has been interpreted as being due to accumulation within a caldera which was, from time to time, occupied by a lake. The early-established, arcuate fracture defining the caldera guided later intrusions belonging to the Glen More Centre (Centre 1, Table 5.1). This occurrence is also of importance since it records overlap between the last-preserved stages of major basalt lava accumulation on Mull and the establishment of strongly centralized igneous activity. The occurrence is of general importance since it provides a clear example of a caldera within the BTVP and can thus be used to help elucidate the structure of areas where calderas may be present but are much less well preserved (for example Skye, Kilchrist; Rum, Cnapan Breaca).

Granophyric veins cutting the lavas next to the dolerite and gabbro of the Glen More ring-dyke were thought to have been derived from the intrusion (Bailey *et al.*, 1924) which elsewhere shows the intimate association of basic and acidic rocks (Cruach Choireadail).

Conclusions

A caldera collapse structure formed at an early stage in the history of the Mull central complex. It was filled by the latest tholeiitic lavas of the regional lava succession, which developed distinctive pillow structures as they came into contact with the waters of a caldera lake. The arcuate fracture defining the caldera continued to be active after lava accumulation ceased and guided subsequent ring-dyke intrusions in the Glen More Centre. The lower, gabbroic and doleritic part of the Glen More ring-dyke intrudes and bakes the pillow lavas and is probably also the source of numerous white, granitic veins cutting the lavas near the contact.

LAGGAN BAY

Highlights

A columnar-jointed basalt flow is overlain by a well-exposed ash band and a mugearite lava flow. The mugearite flow and the ash are closely associated with mugearite which forms a volcanic plug, attesting to a phase of localized vent activity.

Introduction

The coastal and inland exposures about 1.5 km north-east of Ulva Ferry at the head of Loch Tuath provide a demonstration of columnar basaltic and mugearitic lava flows at the base of the Mull lava pile. The mugearites are associated with a volcanic plug, now infilled by ash and cut by a mugearite vent intrusion. As with most of the Mull sites, there have been no detailed studies specific to the lavas exposed at Laggan Bay, although brief descriptions of the area are contained in the Mull Memoir (Bailey *et al.*, 1924). Mugearites collected from the site have, however, been incorporated into the geochemical reconnaissance study of the Mull lavas (Beckinsale *et al.*, 1978).

Description

Columnar, transitional olivine basalts, including pyroxene-rich variants and basaltic hawaiites, occur at the base of the lava succession at Laggan Bay (Fig. 5.10). At Ulva Ferry (NM 445 400) a basaltic hawaiite is exposed in the coastal cliffs and is overlain by an ash which, in places, is altered to a pinkish-grey bole. The ash forms the lower parts of the raised sea cliffs and, although poorly exposed, can be traced northwards and westwards around Laggan Bay. Near Na Torranan (NM 452 415) it attains a thickness of 10 m in places. The ash is overlain by a fissile, mugearite lava flow and the whole sequence is intruded by a mugearite plug, 800 m in diameter. The plug displays vertical flow-banding and forms the headland of Na Torranan and the flat ground to the east. Northward, at Camas an Lagain (NM 448 418), the irregular scoriaceous and rubbly base to the mugearite is exposed above the ash which is much reduced in thickness here.

A section measured in the Allt an Eas Fors (NM 445 423) is as follows:

		thickness
10.	olivine basalt	3 m
9.	intermittently exposed bole	0.1 m
8.	massive, sparsely feldspar-phyric basalt with a vesicular, rubbly base	8–10 m
7.	red bole with banded clay-rich upper part	0.4 m
6.	scoriaceous lava with reddened top	5 m
	(gap in exposures in stream bed 8 m)	
5.	irregular, platy-jointed transgressive basic sheet	2–5 m
4.	pinkish, fissile ash with massive indurated bands	1.5 m
3.	columnar mugearite with base and pipe amygdales	12 m
2.	main ash band	3–5 m
1.	cliff sequence – undetermined	10–12 m

South-east of the waterfall (NM 445 422), the mugearite lavas are exposed lower in the cliff sequence and the upper flow bears a close resemblance to that of the Na Torranan vent intrusion.

Interpretation

The site provides fine exposures of alkaline to transitional lavas which include basalts, basaltic hawaiites and more evolved mugearites which Bailey *et al.* (1924) considered to be close to the base of the Plateau Group lava succession. However, the geology of Ulva suggests that the mugearite could be as much as ten flows above the base; Staffa Magma-Type basalts have not been recorded from Ulva or this site (BGS Sheet 43 (Iona)). The sequence is part of the basalt–hawaiite–mugearite–benmoreite Group I trend of Beckinsale *et al.* (1978). Evolved lavas such as mugearites are relatively rare in this part of the lava succession and the field evidence here suggests that, although the basaltic lavas were erupted from fissures, the more evolved lavas were erupted from central vents (Beckinsale *et al.*, 1978). The volcanic plug within the site

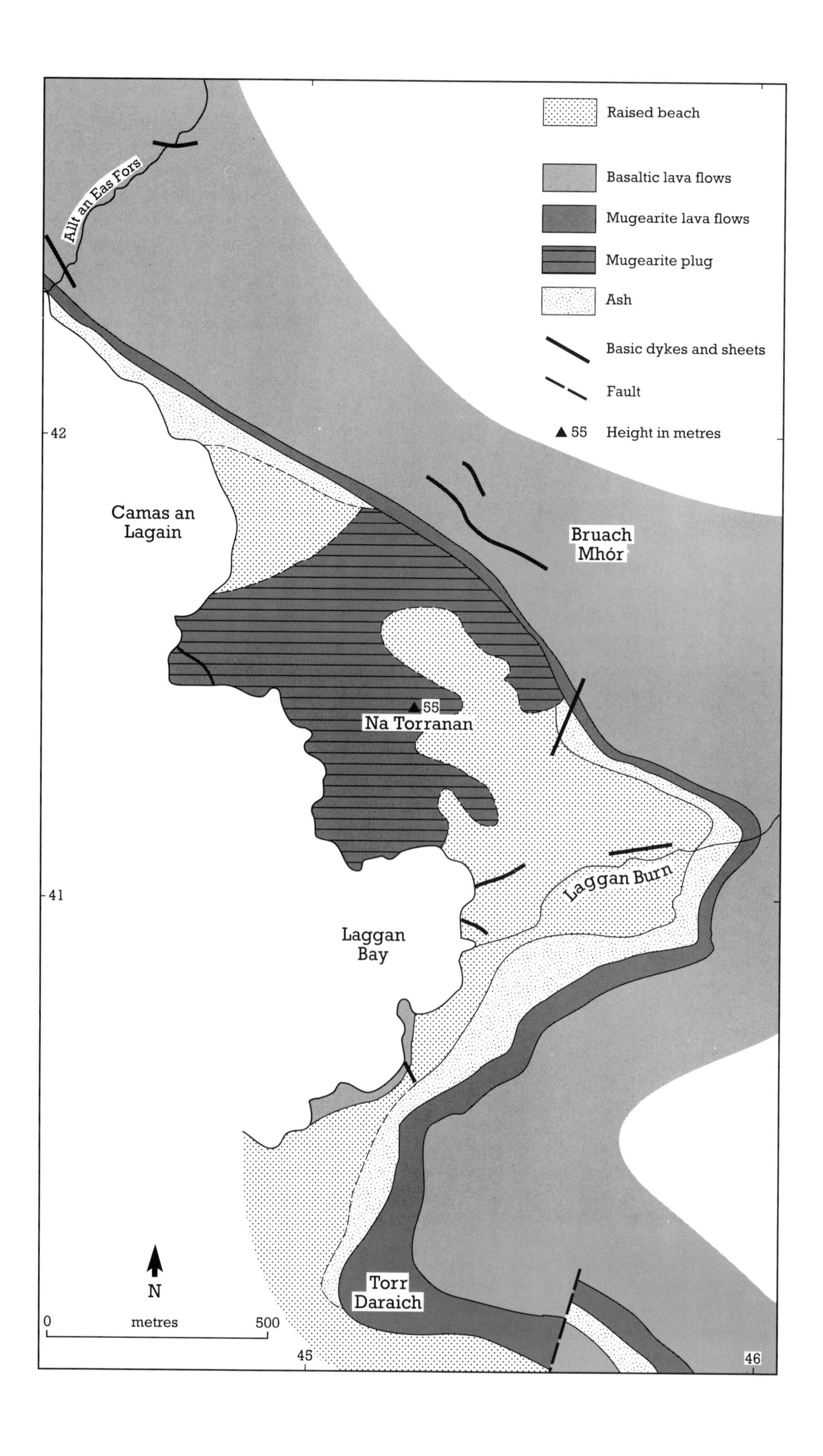

Raised beach
Basaltic lava flows
Mugearite lava flows
Mugearite plug
Ash
Basic dykes and sheets
Fault
▲ 55 Height in metres
Allt an Eas Fors
Camas an Lagain
Bruach Mhór
55
Na Torranan
Laggan Burn
Laggan Bay
Torr Daraich
N
0 metres 500
42
41
45
46

provides such evidence; the vent intrusion could have been the source for the associated mugearite lava and the ash the result of earlier explosive activity which established the vent (Bailey and Anderson, 1925). A significant time interval must have intervened between the eruption of the ash and the mugearite lava, as the upper surface of the ash is weathered to a bole in many places. It appears that the plugs on Mull lie in linear arrays which were probably fissure controlled. The evolved nature of the mugearites suggests that small magma chambers, in which crystal fractionation occurred, were located beneath the plugs along the fissure length. In addition, some basaltic lavas were undoubtedly erupted from vents such as that of 'S Airde Beinn. Similar vents are lacking or rare in the other Hebridean lava fields, although they are common in the Palaeocene lava fields of Antrim where there is a clear association with the regional dyke swarm (Walker, 1959).

Conclusions

The early basaltic and rarer mugearite lavas, ash and the mugearite volcanic plug within this site provide evidence that:

1. the first lavas erupted on Mull were largely basaltic;
2. the eruptive style and products were notably different at different times and more-evolved lavas and ash were erupted from explosive central vents located along the regional fissure trend.

Shallow-level magma reservoirs, in which crystal fractionation occurred, probably existed beneath the vents.

'S AIRDE BEINN

Highlights

A large, elongate dolerite plug intrudes the lava plateau along a regional structural lineament. The plug has produced a marked thermal aureole in the surrounding lavas; the aureole has been divided into three distinct zones on the basis of their mineralogy. High-temperature alteration of amygdales gave rise to larnite, rankinite and other uncommon calc-silicate minerals.

Figure 5.10 Geological map of the Laggan Bay site (adapted from the British Geological Survey 'One Inch' map, Sheet 43, Iona).

Introduction

'S Airde Beinn is a small but conspicuous rock-girt hill with a central depression occupied by a lochan (Fig. 5.11). The hill is carved from a doleritic plug which rises through the flat-lying, trap-featured lavas of the Mull Plateau. Such plugs are of restricted occurrence in the Tertiary Igneous Province and on Mull are best developed in the north of the island. It is not only the largest but also the best-known plug on Mull, having been described by Judd (1874), Geikie (1897), Bailey *et al.* (1924) and Richey and Thomas (1930). The thermal effects of the intrusion on the adjacent lavas and their amygdales have been detailed by Cann (1965); the mineralogy of the aureole is a key feature of the site.

Description

The 'S Airde Beinn plug (NM 470 540) rises as much as 60 m above the surrounding countryside and it measures 850 by 440 m. It is elongated in a NNW direction parallel to the Tertiary regional structural trend and dyke swarm. The rock is a coarse-grained dolerite composed of olivine, titaniferous augite, labradorite and magnetite and it is mineralogically identical to the lavas which it intrudes. Bailey *et al.* (1924) suggested affinities to some of the pillow lavas of the Mull complex, but Beckinsale *et al.* (1978) showed the gabbro to be quartz- and hypersthene-normative assigning it to Group 2 (Table 5.2). In thin section, the olivine is either fresh or partly altered to iddingsite and forms large irregular crystals. Large zoned feldspars are enclosed ophitically by clinopyroxene and a second generation of acicular augite associated closely with magnetite occurs interstitially in a chloritized residuum. The walls of the intrusion are close to vertical but there is no obvious abrupt contact between the plug and the adjacent lavas; the lavas appear to grade locally into the dolerite. The thermal effect of the intrusion upon the basalt country rock explains this apparent gradational contact. There is an increase in the granularity of the basalts as

Figure 5.11 A view of 'S Airde Beinn from the south. 'S Airde Beinn site, Mull. (Photo: C.J. MacFadyen.)

the intrusion is approached, accompanied by a change in amygdale compositions which normally contain minerals such as thompsonite, natrolite, analcite, heulandite, stilbite and gyrolite. The changes in basalt and amygdale composition in the thermal aureole around the plug are discussed in detail below.

Interpretation

The 'S Airde Beinn plug is one of several plugs which intrude the lavas of Mull and Morvern, most of which have NNW–SSE elongations, approximately parallel to the regional dyke swarm. It also lies on a major fault zone traceable in a similar direction from Druim Fada (NM 465 555) to the shores of Loch Frisa (NM 475 510). The numerous plugs of north Antrim are also elongate parallel to the regional dyke swarm, where the dykes and plugs are seen to merge. Like 'S Airde Beinn, many of the Antrim plugs are surrounded by pronounced thermal aureoles (cf. Tilley and Harwood, 1931; Preston, 1963); they are considered to have been long-lived feeders for lavas high in the Antrim Palaeocene lava field. A similar explanation is most likely for the Mull and Morvern plugs, which must have been coeval with the dykes and intruded as part of the same phase of magmatism, connected with Palaeocene crustal extension.

The principal value of the site is in the presence of the conspicuous metamorphic aureole associated with the plug. The apparent gradation from the coarse dolerite into the basalt at the contact was considered by Bailey *et al.* (1924) to indicate that the ascending magma which formed the plug melted and mingled with the lava wall-rock. Cann (1965), however, argued that only the most highly metamorphosed lavas have reacted with the magma. He distinguished three zones of progressive thermal metamorphism. The first is characterized initially by an increase in the amount of interstitial chlorophaeite, by the degree of alteration of pyroxene and olivine and by olivine being replaced by hypersthene and iron ore to produce a fine-grained hypersthene–augite–plagioclase–iron ore granulite. The second stage is characterized by the reappearance of olivine and an increase in granularity. Where a reaction with the magma has occurred, the third stage is reached, and olivine becomes the dominant ferromagnesian mineral.

Cann (1965) has also identified three classes of

amygdale minerals on the basis of their behaviour on metamorphism. The first class consists of amygdale assemblages originally dominated by zeolite minerals such as thompsonite, natrolite and analcite with rare heulandite and stilbite. These show a consistent sequence of metamorphic change directly related to the stage of metamorphism attained by the surrounding basalts; many have been converted to plagioclase late in the first stage of metamorphism. Gyrolite is the dominant original amygdale mineral in the second class and this passes first into reyerite and then to wollastonite during metamorphism. A rim of aegirine-augite surrounds the wollastonite on its first appearance, caused by a reaction between the wollastonite and the basaltic magma. Amygdales originally filled with calcite constitute the third class. These have been altered to anhydrous calc-silicate assemblages of larnite, rankinite and wollastonite which form concentric monomineralic zones decreasing in Ca/Si outwards. The hornfelsed basalt around the amygdales has had its composition altered by the loss of Si and, at a late stage, of Mg and Al and has gained principally Ca. In places, melilite has replaced the amygdale walls and, near the amygdales, the basalt is unusually rich in augite. Metamorphosed 'amygdales' consisting largely of ferromagnesian minerals (hypersthene, olivine, hornblende) are also present and are attributed by Cann (1965) to the infilling of vesicles or voids in partly formed amygdales during the metamorphism.

Conclusions

The 'S Airde Beinn plug caused distinctive, high-temperature alteration of the surrounding basalt lavas. Three zones of thermal alteration have been recognized in the basalts and their amygdales, on the basis of mineralogy and petrography. The formation of the calc-silicate minerals larnite and rankinite provides particularly compelling evidence for high temperatures in the aureole and there has also been reaction between the basalt lavas and the marginal dolerite at the edge of the plug. This plug, in common with others in Mull and Morvern, probably acted as a long-lived feeder for lava flows since removed by erosion. It was intruded at the same time as dykes cutting the lavas.

CARSAIG BAY

Highlights

The site is notable for:

1. the presence of the thickest development of basal Palaeocene sediments on Mull, which contain debris derived from weathering of Cretaceous and earlier rocks under desert conditions;
2. a complete section through a composite (basic–acid) sill containing cognate and accidental xenoliths. Some of the latter have been partially fused and their high-temperature minerals include mullite.

Introduction

Carsaig Bay is a site of multiple interest, extending between Rubh' a' Chromain and Carsaig House, including the off-shore island of Gamhnch Mhor and the cliff at An Dunan. The interest of the site can be divided into the following categories:

1. Palaeocene sedimentary rocks, which attain their maximum development in the site and comprise sandstones and mudstones with numerous chalk clasts from the underlying Cretaceous strata.
2. The Rubh' a' Chromain composite xenolithic sill: this sill belongs to the Scridain Suite of sills which exhibit felsitic, basaltic and hybrid portions rich in cognate and accidental xenoliths.
3. The Gamhnach Mhor Syenite.

There has been no recent investigations into the geology of this site and the Mull Memoir is again the main source of information. A recent study concerned with magmatic processes which operated during the emplacement of a Scridain Suite sill to the west of the site has been made by Kille *et al.* (1986). This investigation, which has some applications to the Rubh' a' Chromain sill, demonstrated localized turbulent flow of the magma during emplacement and selective assimilation of wall rocks. The geology of the site is depicted in Fig. 5.12.

Description

Palaeocene sediments are well represented in the gully above Aird Ghlas (NM 534 212) and at

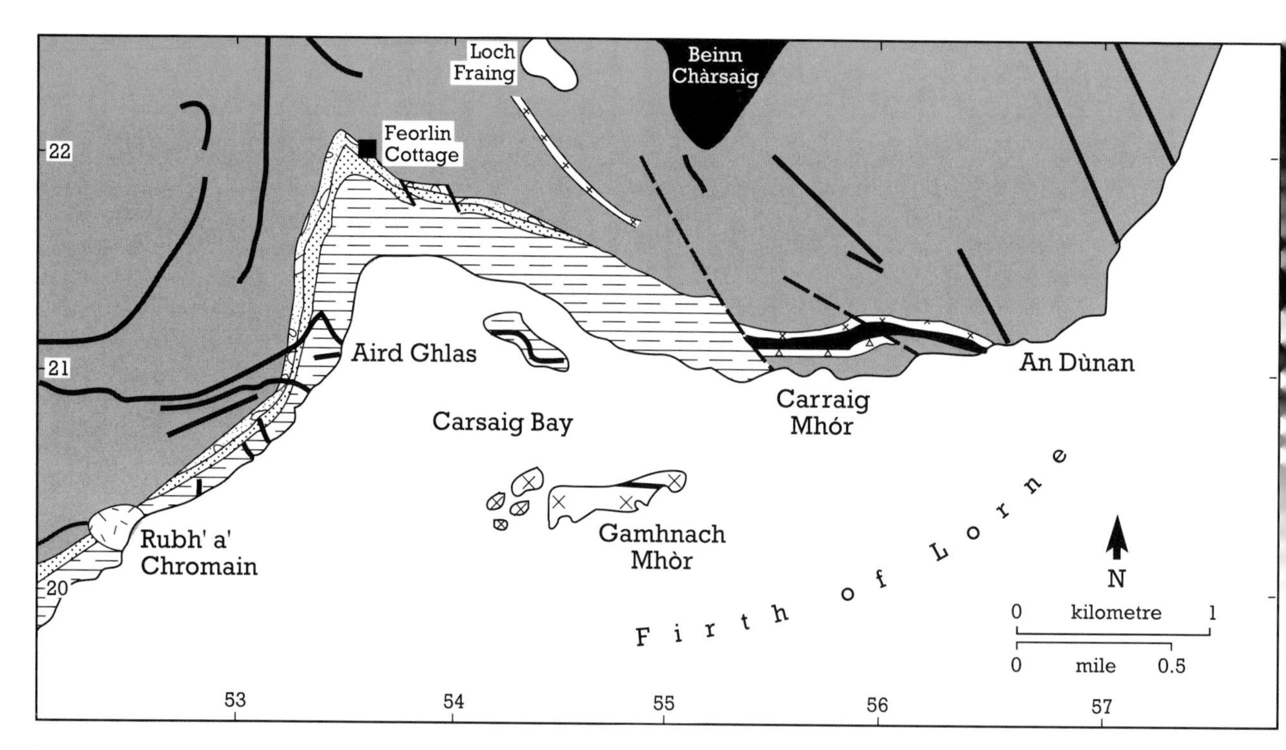

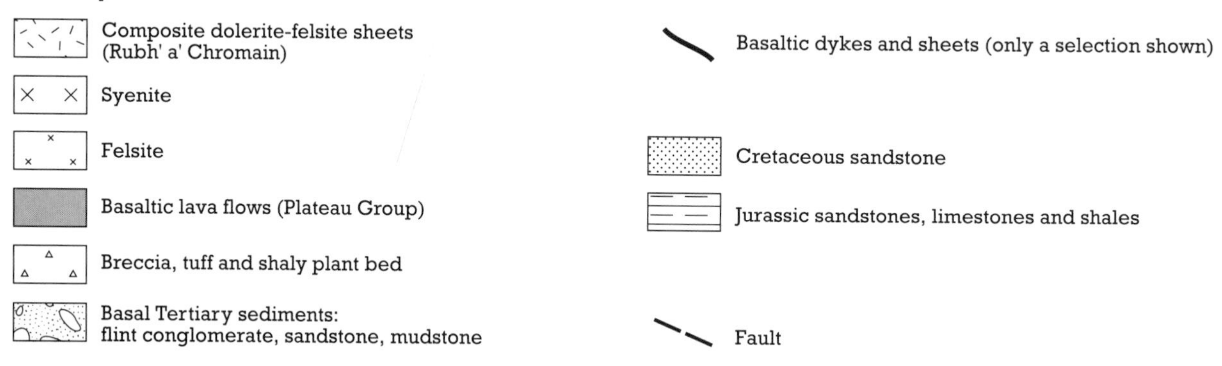

Figure 5.12 Geological map of the Carsaig Bay site (adapted from the British Geological Survey 'One Inch' map, Sheet 44, Mull).

Feorlin Cottage (NM 535 221), where they attain their maximum thickness of 1–2 m. At Aird Ghlas, Bailey *et al.* (1924) have recorded the following sequence:

6. Basalt lava
5. Soft, shaly sandstone with flint fragments — 0.30 m
4. Greenish sandstone with flints and silicified chalk clasts — 0.60 m
3. Purple mudstone — 0.30 m
2. Greenish, partly glauconitic sandstone with chalk and flint fragments — 1.60 m

Gap

1. Cenomanian sedimentary rocks (for details see Lee and Bailey, 1925, pp. 118–20)

The purple mudstone, which commonly occurs beneath the basal lava flow in Mull and Morvern, is interbedded with other sediments at this locality. Chalk fragments in the sandstones are hard and siliceous; their size and angularity suggests that they were derived locally, the silicification probably took place after the deposition of the sandstone (Bailey *et al.*, 1924). Abundant Cretaceous foraminifera have been identified within the chalk clasts (Jones, *in* Judd,

1878). The sandstone which underlies the mudstone consists of beautifully rounded, presumably wind-worn, quartz grains cemented by a dark, cherty matrix which also surrounds the chalk fragments. The latter are less abundant to the east, distinct horizons rich in chalk clasts being absent at Eas Mheannain (NM 538 221), where the fragments are irregularly scattered within the underlying thick, red-stained, pebbly sandstone.

The sequence within the stream section at Feorlin Cottage (NM 535 221) differs from that at Aird Ghlas and is as follows:

5.	Basalt lavas	
4.	Purple mudstone	1.0 m
3.	Chalk fragments in a black sandy matrix	0.20 m
2.	Massive, white pebbly sandstone with some iron staining and containing a lenticular bed of white flints towards the top	6.0 m
1.	Glauconitic sandstone (Cenomanian) passing down into concretionary limestones	

Still farther east in the site, no chalk is encountered within the sandstone, which grades rapidly downwards into a more glauconitic sandstone containing crushed bivalve debris thought to have been derived from the Cenomanian; the rock may or may not be of Palaeocene age. Sediments comprising 3–7 m of pale-green to buff-coloured breccia containing clasts of fine-grained vesicular basalt in a black, calcitic matrix, occur in the cliffs between Carraig Mhor (NM 555 211) and An Dunan (NM 563 212), within the lava succession. Plant remains have been located in shaly deposits a little below this horizon.

The first basalt lavas, lying immediately above the sediments, are markedly columnar – a feature characteristic of basaltic flows found elsewhere near the base of the Mull sequence. However, it is not known if these flows belong to the typically columnar Staffa Magma Type flows, examples of which occur at Ardtun. The basal lavas reach their maximum development of around 350 m, with an average flow thickness of 15 m in the Carsaig area (Bailey *et al.*, 1924).

The Rubh' a' Chromain Sill, exposed at the western edge of the site, is related to the Scridain Suite of sills (Bailey *et al.*, 1924). It is a striking and clearly exposed example of a composite intrusion having a sheet-like form and dipping to the north-west. The centre of the sheet contains felsite distinctly separated from basaltic margins which are normally chilled against the country rock (Fig. 5.13). The special interest of this composite intrusion lies in the presence of a varied suite of cognate and accidental xenoliths up to a metre in diameter; these are described below.

In detail, the intrusion cuts an earlier, irregular trachytic (bostonite) intrusion with a fine-grained, purplish-grey, aphyric appearance. Like the sill, the bostonite is xenolithic; a xenolith of black fossiliferous limestone has been tentatively assigned to the early Lias age, possibly related to the Broadford Beds (Lee and Bailey, 1925).

The lower contact of the Rubh' a' Chromain Sill abuts the Lias which appears to have been locally fused, resulting in the formation of tridymite. Kille *et al.* (1986) have described widespread fusion of pelites where doleritic Scridain sheets cut Moine metasediments to the west of the site at Traigh Bhan na Sgurra (NM 424 185). The following section through the composite intrusion at Rubh' a' Chromain is based on the work of the Survey:

	Top
Bostonite, at upper contact	
Chilled tholeiitic basalt	0.08 m
Tholeiite with cognate gabbroic xenoliths	1.20–1.80 m
Quartz dolerite with densely packed aluminous xenoliths	0.60–1.50 m
Porphyritic felsite with scattered xenoliths of sandstone at the margins and sandstone and shale xenoliths more centrally	6–9 m
Banded, possibly hybridized felsite	0.50 m
Quartz dolerite with numerous cognate xenoliths and a few accidental types	0.60–1.50 m
Chilled tholeiitic basalt	0.10 m
Lias sandstones	
	Bottom

The quartz dolerite is poor in olivine with brown, elongated, often curved crystals of augite. The interior felsite has a sharp contact with the quartz dolerite and contains sparse crystals of labradorite in a felsitic groundmass. Some chilled patches of rhyolite or acid pitchstone (Bailey *et al.*, 1924) occur in the centre of the sill; they exhibit skeletal and devitrification textures under the microscope.

The cognate xenoliths are generally dark-coloured, coarsely crystalline dolerites which

Figure 5.13 The Rubh' a 'Chromain composite sill exposed at the western edge of the Carsaig Bay site, Mull. (Photo: C.J. MacFadyen.)

appear as glomeroporphyritic patches of bytownite, hypersthene and rare greenish augite in a fine-grained variolitic matrix. A few olivine pseudomorphs are also present. Accidental xenoliths are much more variable and, although not all of the following types occur at Rubh' a' Chromain, they characterize the Scridain Suite as a whole (see Bailey *et al.*, 1924 and Kille *et al.*, 1986 for other localities where similar xenoliths are described within the Scridain sheets).

1. Micaceous gneiss
   Granulite
   Quartzite
2. Granite
   Pegmatite
3. Sandstone
   Shale
   Carbonaceous rock
   (bituminous shale/coal)
4. Basalt lava

A significant feature of the Rubh' a' Chromain Sill is the distinctive, high-temperature mineralogy of the thermally metamorphosed aluminous xenoliths. They appear to have partially fused and have a reaction rim containing anorthite, pink sillimanite, green hercynite/pleonaste spinel, corundum and cordierite. Of these, the corundum or sapphires occur isolated as small, blue, tabular crystals. The feldspars often take on a rosy hue owing to the inclusion of needles of rare pink sillimanite. The inclusions are generally completely recrystallized to granular hornfels. Shaly and sandy xenoliths are frequently fused to buchite. A fuller description of these xenoliths is given by Thomas (1922); the sillimanite was later identified as mullite ($3Al_2O_3.2SiO_2$), a common mineral in bricks (Bailey *et al.*, 1924, p. 268).

At the entrance to Carsaig Bay, the low island and surrounding reefs of Gamhnach Mhor are formed from an alkaline syenite. The body is assumed to be a sill although the contacts are not visible; it was tentatively given a Tertiary age by Bailey *et al.* (1924). The rock is yellowish to greyish-brown with a rough, irregular fracture and composed mostly of sodic orthoclase, with subordinate aegirine and pale-green aegirine-augite. Accessory minerals include a strongly blue-green, alkali amphibole, magnetite and apatite. Phenocrysts are absent, as is nepheline in hand specimen. The intrusion is traversed by numerous segregation veins, basic dykes and a tholeiitic sheet bearing cognate xenoliths belonging to the Scridain group of sills.

Interpretation

The Palaeocene sediments at Carsaig Bay are of special significance as it is here that they attain their maximum development. The sandstones, containing wind-worn quartz grains and rounded chalk fragments, were probably deposited in a desert environment, which persisted from late Cretaceous times when major uplift of the Chalk occurred and Palaeocene sedimentation commenced (Bailey *et al.*, 1924). The persistent purple mudstones at the base of the lavas and above the desert sandstones occur throughout Mull and have probably formed by the lateritic decomposition of basaltic ash in a warm and moist climate (Bailey *et al.*, 1924). The ash may have been the product of an initial volcanic explosion on Mull followed by a long repose period prior to the eruption of the first lavas.

The Scridain sheet swarm is an important and conspicuous feature of Tertiary igneous activity on Mull. The numerous sheet-like masses have intruded the lava plateau, Mesozoic sediments and Precambrian basement rocks of the Ross of Mull and Gribun peninsulas. Bailey *et al.* (1924) considered the sheets to belong to a single complex containing a wide range of rock types from basalt through to rhyolite. The sheet exposed within the site is, however, a fairly uncommon example of a Scridain sill by virtue of its composite nature (a few other examples are quoted by Bailey *et al.*, 1924, p. 287). It provides evidence for the coexistence of acid and basic magmas during Tertiary volcanism on Mull, which is in line with evidence from numerous other localities quoted in this volume. The absence of a chilled contact between the tholeiite and the felsite in the sill implies that felsitic magma was injected into the centre of the sill before the tholeiite had cooled; both were probably liquid together, as shown by the presence of banded, possibly hybridized felsite at the contact.

There have been no published, detailed, petrological investigations into the origin of the different magma types in composite sills or in the suite as a whole. Bailey *et al.* (1924) suggested that the wide range of magma types represented by the Scridain Suite were the result of crystal fractionation in a magma reservoir. This reservoir probably existed beneath the lava pile and the evidence provided by the accidental xenoliths suggests that it disrupted Precambrian and Mesozoic country rocks. The xenoliths were regarded by Bailey *et al.* as having been broken off the walls of the magma chamber and accumulated towards the base, hence the more basic parts of the sheets are richer in xenoliths. Alternatively, it is suggested that fusion of the country rock could create an acidic and intermediate melt by mixing. Ample evidence that assimilation of country rock by the basic magma is a real process comes from partly digested xenoliths, fusion of sandstone at sill margins and evidence for fusion of metasediments at Traigh Bhan na Sgurra (Kille *et al.*, 1986). However, no firm conclusion can be reached on this issue until considerable further field-work and petrological investigations have been undertaken.

The occurrence and alteration of the cognate xenoliths in this composite sill are very important features of this site since they:

1. provide information about subsurface rocks in this part of Mull; and
2. prove that the magmas of the Scridain sills were hot on emplacement, possibly over 1100°C (cf. Kille *et al.*, 1986).

Fusion of sediment at the lower contact indicates that these temperatures were sustained, thus the turbulent flow envisaged during sill emplacement elsewhere probably also applies at Rubh' a'Chromain (cf. Kille *et al.*, 1986).

Bailey *et al.* (1924) suggested that the intrusive Scridain Suite was probably contemporaneous with the emplacement of late basic cone-sheets associated with the central complex. The syenite is therefore of an earlier age; apart from this, little is known about the rock, which is a distinctly unusual type in the BTVP.

Conclusions

The site is of exceptional value as it contains several unique features of Tertiary igneous activity on Mull. Basal Tertiary sediments are well developed in Carsaig Bay and record a continuation of an arid, desert environment from the late Cretaceous into the Palaeocene. Warm, temperate conditions then prevailed when an initial explosive volcanic event marked the onset of Tertiary volcanism and spread ash over much of the region to form a basal mudstone. The Rubh' a' Chromain sill provides a clear example of a composite intrusion which belongs to the extensive Scridain Suite, emplaced during a single event along with the late basic cone-sheets associated with the central complex (Table 5.1). It provides further evidence from the BTVP for coexisting acid and basic magmas; the acid portion derived either from crystal fractionation or crustal assimilation in a shallow-level magma reservoir. Abundant accidental xenoliths appear to have been derived from Precambrian, Mesozoic and Palaeocene rocks, possibly from the walls to the magma chamber. The Gamhnach Mhor Syenite is an earlier intrusion of probable Cenozoic age and is unusual by reason of its thoroughly peralkaline character.

LOCH SPELVE–AUCHNACRAIG

Highlights

This is a classic locality of international importance for demonstrating concentric folding associated with the emplacement of an igneous central complex. Exposure of Dalradian and Moine rocks in fold cores proves the nature of the Precambrian basement beneath Mull and show that the Great Glen Fault was offset by emplacement of the central complex.

Introduction

The rocks exposed on the western shores of Loch Spelve and on the peninsula lying between its northern arm and the Firth of Lorne lie within a site of multiple interest. Principally, the exposures provide a traverse through clearly defined circumferential folds around the earliest centre (Centre 1) of the Mull central complex, which are best developed in the south and east of the island. In addition, the site contains a unique succession of Triassic strata, which is the thickest sequence in the Hebrides, together with sections through Devonian lavas, Lower and Middle Jurassic (Lias, Inferior Oolite, Great Estuarine) and Cenomanian beds unconformably succeeded by Palaeocene sedimentary rocks. Basement Precambrian rocks are also exposed as inliers. Lavas dominate the Tertiary (or other) igneous rocks present in the site but there are also exposures of the early Glas Bheinn Granophyre within the western margin of these, and basaltic dykes, especially in coastal exposures (Fig. 5.14).

The folds were first described in the Mull Memoir by Bailey *et al.* (1924), who chose the area covered by this site to be the most favourable for the demonstration of the structures. Subsequent discussions relating to the nature of the folding have been published by Cheeney (1962), Rast *et al.* (1968) and Skelhorn (1969). Most recently, Walker (1975) has related the structures of this area to the early evolution of the central complex.

Description

Bailey *et al.* (1924) have distinguished the following structural elements within the area around Auchnacraig:

1. Marginal tilt (outermost zone). A general centripetal dip of up to 50° towards the caldera of Centre 1 is easily recognizable along the coast between Grass Point (NM 748 308) and Rubha na Faoilinn (NM 729 274) where Mesozoic rocks underlie early Cenozoic sediments and lavas.
2. Duart Bay Syncline. The axis of this fold passes just east of Auchnacraig and traverses the eastern slopes of Garbh Dhoire. The Tertiary lavas, which exceed more than 300 m in thickness, have been affected by this fold; Bailey *et al.* (1924) recorded the highest lava as being Big-Feldspar basalt (Table 5.2), while Cheeney (1962) recognized lavas of the Central Group (Table 5.2) a short distance west of Gortenanrue (NM 728 279). The syncline is approximately symmetrical with maximum dips of around 45°.
3. Loch Don Anticline. The crest of this anticline extends through Loch a'Ghleannain and Meall Reamhar (NM 726 301). It is offset by over 100 m by a WNW-trending, sinistral, strike-slip fault which passes through Port Donain (NM 740 290). Immediately south of the fault the anticline bifurcates. In the north of the site, the core of the structure exposes Dalradian schists and limestones which are succeeded by Devonian lavas similar to lavas of the Lorne Plateau, Mesozoic sediments and Palaeocene lavas. The fold plunges southwards so that Dalradian and Devonian rocks pass below ground to the south of the Port Donain Fault. The Mesozoic rocks of the eastern branch of the fold do not extend beyond Carn Ban (NM 722 289) but reach the coast near the ruined chapel (NM 709 284) in the western branch. The anticline is

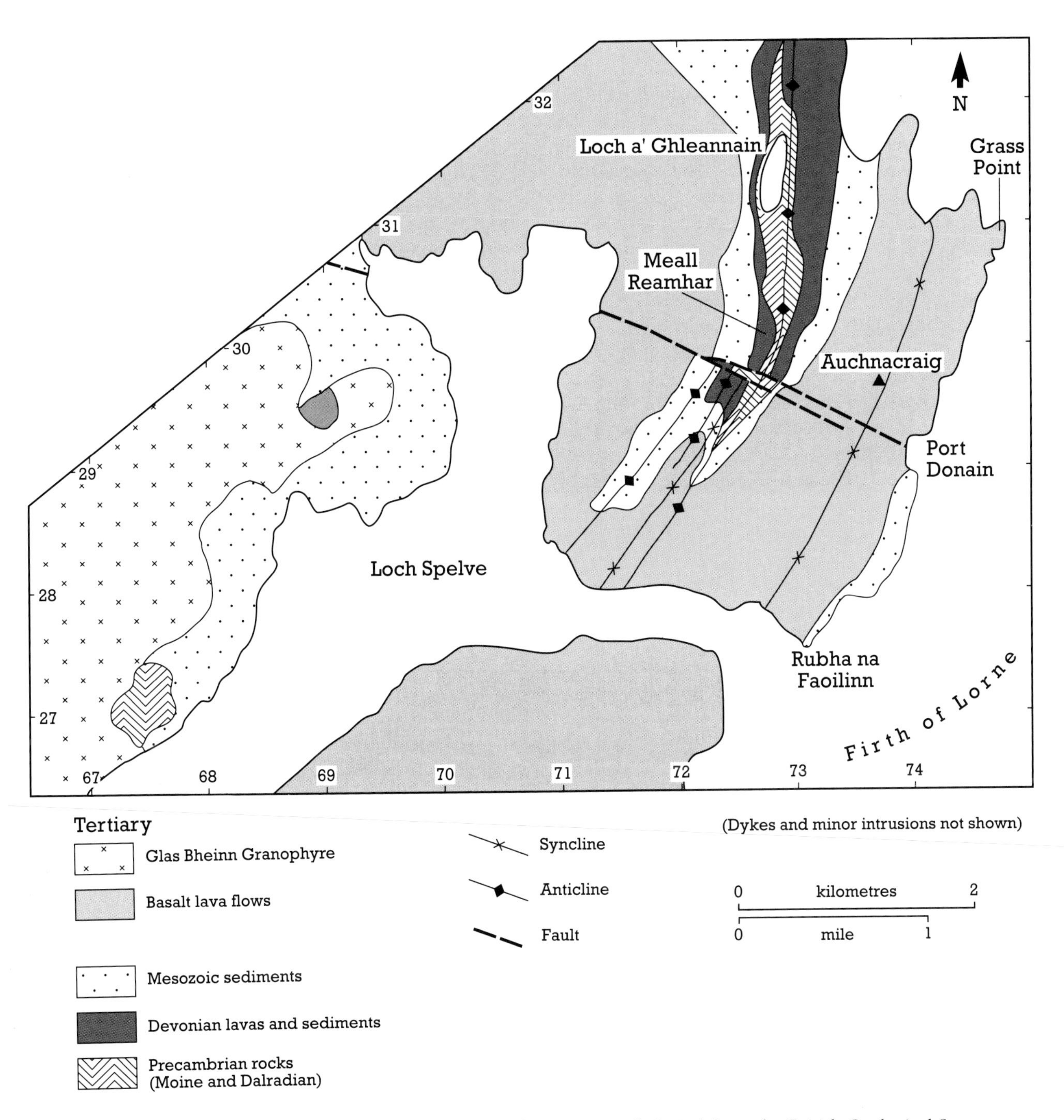

Figure 5.14 Geological map of the Loch Spelve–Auchnacraig site (adapted from the British Geological Survey 'One Inch' map, Sheet 44, Mull).

highly compressed, the eastern limb is, in places, vertical or overturned, while the western limb normally dips at between 30° and 60°, but occasional attitudes of up to 80° are recorded. The intensity of the folding has led to the brecciation of the Devonian and Tertiary igneous rocks (Bailey *et al.*, 1924). The former are generally crushed while the latter show only local crushing, a difference attributed to post-Devonian but pre-Tertiary movements along the Great Glen Fault, which is coincident with the southern margin of the Mull complex. Northwards, the Loch Don Anticline weakens and passes into a faulted monocline and then into a series of *en échelon* folds of the Craignure Anticline

(Bailey *et al.*, 1924).

4. Coire Mor Syncline. This syncline is best developed to the north beyond the site boundary where it is expressed by the distribution of Big-Feldspar basalt flows. However, it is clearly demonstrated around Rubha na Cille (NM 706 283) where its eastern limb adjoins the Loch Don Anticline marked by Mesozoic and older strata along the crest. The western limb of the syncline is recognized in the exposures of vertical, crushed Trias and Lias beneath Tertiary lavas at NM 706 286. The fold is again very compressed with steep and overturned limbs.
5. Loch Spelve Anticline (innermost zone). The core of this fold contains Moine schists to the south of Balure (NM 675 268) and Mesozoic sediments at Rubha na Faing (NM 707 297) on the north-eastern shore of Loch Spelve. The fold is tight and markedly asymmetrical; its inner margin coincides with the boundary of the South-East, or Early Caldera (Bailey *et al.*, 1924; Table 5.1).

Interpretation

The annular fold belt surrounding the Mull central complex is the most clearly defined in the BTVP. The cores of the folds expose Precambrian rocks including both Moine and Dalradian schists, proving:

1. that the area lies across the line of the Great Glen Fault; and
2. that this major pre-Tertiary structure is notably offset south-eastwards by about 5 km away from its normal trend (Walker, 1975).

In addition to revealing the deeper basement beneath this part of Mull, the folds also show that the area is underlain by Devonian lavas similar to those of Lorne and by Triassic and Jurassic rocks.

Formation of the arcuate folds was tentatively regarded by Bailey *et al.* (1924) as one of the earliest events in the Mull central complex. Their origin was ascribed to the intrusion of the early granophyres of Glas Bheinn and Derrynaculen; thus, forcible intrusion of acid magmas was clearly envisaged at this early stage in research on Mull. Cheeney (1962) observed an angular discordance of about 60° between Mesozoic sediments and overlying lava flows near Auchnacraig, leading him to recognize two periods of folding, the first post-late Cretaceous, but before eruption of the earliest Palaeocene lavas, the second after the main series of lavas had been erupted. Further evidence for an early phase of folding was claimed by Rast *et al.* (1968), who considered that the folding commenced as early as Triassic times; however, this view has received little support in subsequent discussion. A further contribution came from Skelhorn (1969) whose (unpublished) observations suggested that the folding may be related to the first centre (Centre 1 or Glen More Centre; Table 5.1). The folding pre-dates all the intrusions associated with this centre, but folding has deformed some of the (regional) NW–SE-trending dykes. There is clearly a consensus that folding occurred at an early stage in the development of the Mull central complex, but further, detailed field investigations are required to determine how many phases of folding occurred, exactly when they took place and whether the folding was an ongoing process possibly preceding and overlapping activity at Centre 1.

The cause of the folding was attributed by Walker (1975) to initial updoming of central Mull by an acid diapir, probably initiated before eruption of the first lavas since these rest on Moine gneisses and Triassic rocks in the area of folding, whereas they are underlain by Cretaceous or Jurassic elsewhere. Walker envisaged the folds as gravity structures, producing as much as 500 m of upper crustal shortening. He speculated that the folding could have resulted in unroofing of the acid diapir, giving rise to the breccias classed as vent agglomerates, which are full of granophyre fragments, and volcanic fragmental rocks replete with rhyolitic debris now found in the Coire Mhor syncline (part of the annular fold system 4 to 8 km NNW of the site) and 1 km south of the site on the southern side of Loch Spelve.

The tectonic effects of the emplacement of the Mull central complex are the finest developed in the BTVP and are regarded as textbook examples of international importance. They are, however, still poorly understood in detail and would merit a thorough reinvestigation.

Conclusions

Clearly defined concentric folds associated with the early emplacement of Centre 1 on Mull are the finest expression of this type of deformation associated with igneous activity in the British

Tertiary Volcanic Province (BTVP). Initial structural disturbances, which probably commenced before lava effusion, updomed the area while later gravity folding around the dome involved lavas and early Centre 1 acid volcanics. The folds are responsible for the preservation of Moine and Dalradian inliers, which define the course of the Great Glen Fault, possible Devonian lavas and successions through Mesozoic strata.

CRUACH CHOIREADAIL

Highlights

There is a superb, continuous section from gabbro at the base of the Glen More ring-dyke through intermediate rocks to granophyre at the top of the intrusion. The ring-dyke cuts basaltic pillow lavas which formed in a caldera lake, and is itself cut by numerous late basic cone-sheets.

Introduction

The site contains a virtually continuous section through the Glen More ring-dyke which clearly shows a gradual upwards transition from gabbro at the base to granophyre at the top (Fig. 5.15). The ring-dyke is the last major intrusion associated with the Beinn Chaisgidle Centre (Centre 2, Table 5.1). Two dyke-like arms which protrude upwards from the ring-dyke at Cruach Choireadail are connected by a horizontal sheet. Also exposed within the site are the arcuate Coire 'an t-Sailein quartz gabbro body, the Ben Buie olivine-gabbro intrusion, gabbroic plugs, some volcanic rocks and late basic cone-sheets.

The Mull Memoir (Bailey *et al.*, 1924) contains a detailed petrological description of the Glen More ring-dyke. The authors advanced a crystal fractionation model to account for the petrological variations. However, Holmes (1936) and Fenner (1937) challenged this view, while Koomans and Kuenen (1938) favoured the fractionation model. Skelhorn (1969) has produced a field guide to the Cruach Choireadail region and discusses the models for the petrogenesis of the Glen More ring-dyke.

Description

Between the Coladoir River and the summit of Cruach Choireadail (NM 595 305), a 450 m vertical section through one of the dyke-like

Figure 5.15 Cruach Choireadail, viewed from the Coladoir River, exposing gabbro/granophyre of the Glen More ring-dyke. Cruach Choireadail site, Mull. (Photo: C.J. MacFadyen.)

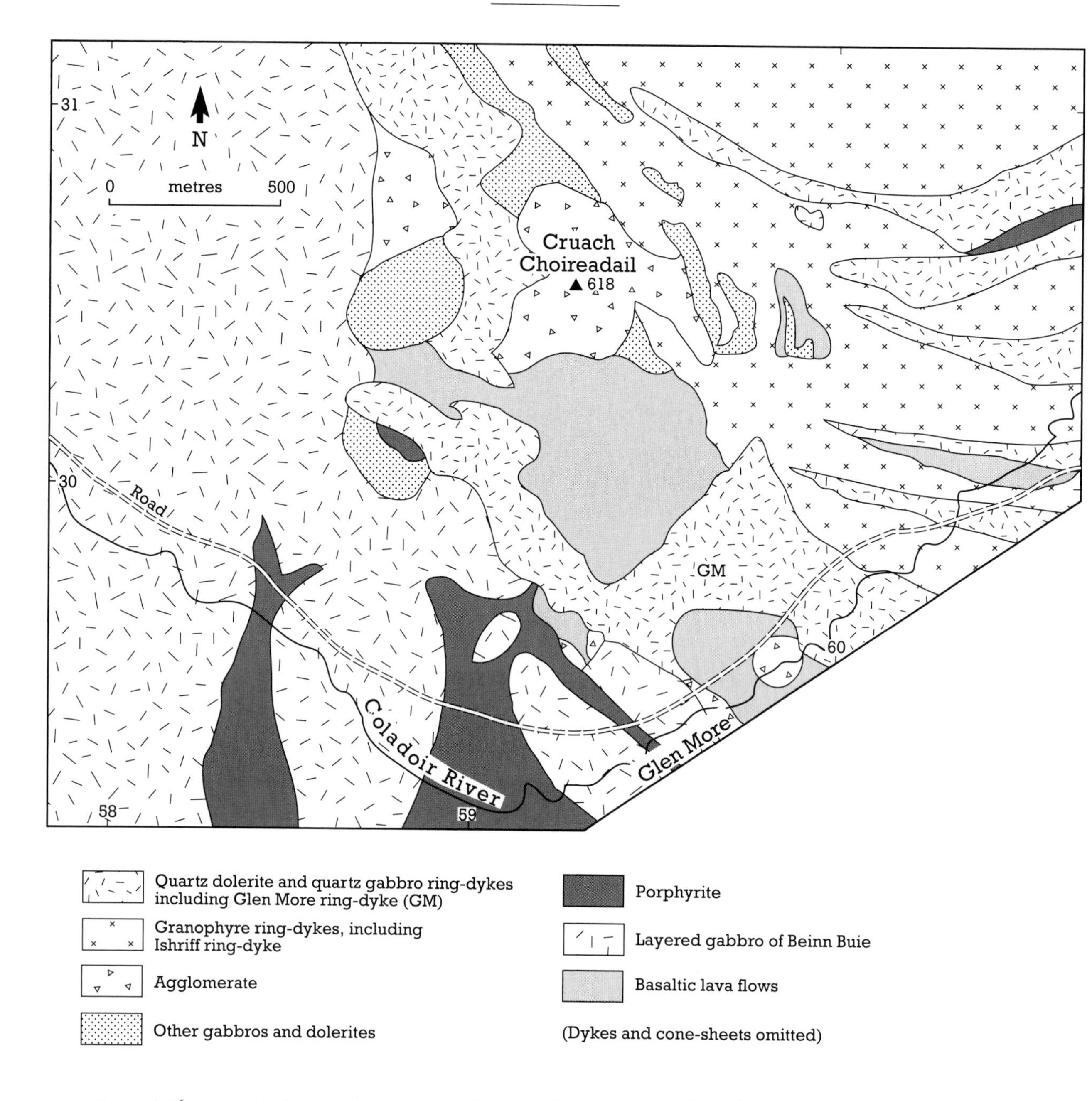

Figure 5.16 Geological map of the Cruach Choireadail site (adapted from the British Geological Survey 'One Inch' map, Sheet 44, Mull).

apophyses of the Glen More ring-dyke is exposed (Fig. 5.16). The lowermost rocks are the most basic, containing about 44% silica (Koomans and Kuenen, 1938), and are moderately coarse-grained, speckled quartz gabbros/dolerites containing labradorite (40%), augite (40%), ilmenite + magnetite (12%) and minor amounts of quartz + alkali-feldspar intergrowths. Pseudomorphs of chlorite after olivine also occur but form less than 5% of the modal mineralogy (Koomans and Kuenen, 1938). The distribution of feldspar is sometimes uneven, resulting in leucocratic segregations and horizontal segregation veins of an acidic residuum up to a metre in thickness. These are common in the basic portion of the intrusion. There is a gradual and remarkably regular decrease in grain size upwards through the intrusion as the rocks generally become paler and more leucocratic. The proportion of acid mesostasis likewise increases upwards, olivine completely disappears, plagioclase becomes less calcic and the habit of augite changes from ophitic to acicular. The quartz gabbro passes upwards into a pinkish-grey, augite diorite with a hybridized appearance and finally into pinkish leucocratic 'granophyre'. The summit 'granophyre' is a fine-grained, quartz-microporphyritic, spherulitic, sodic felsite essentially composed of albite, orthoclase, quartz and alkali-feldspar intergrowths, and quartz and fibrous green hornblende replacement after acicular augite.

In the Coladoir River, the Glen More ring-dyke is separated from the earlier Ishriff Granophyre ring-dyke by a narrow screen of basalt and dolerite. These thermally metamorphosed rocks are probably early members of the late, basic, cone-sheet complex (Table 5.1). This screen is traceable for some distance up the slopes of Cruach Choireadail. The numerous NW-trending sheets traversing the lavas are truncated by the Glen More ring-dyke and form the main wall rocks; however, the ring-dyke is itself cut by late basic cone-sheets (Bailey *et al.*, 1924) in the north-west, showing that it has been bracketed by the late basic cone-sheet emplacement episode. Examples of the lavas associated with the South-East Caldera (Centre 1) are located to the south of Cruach Choireadail and the small lochans and are notably packed with large feldspar phenocrysts in many instances. Occasional pillow structures are observed in these altered basalts indicating that the lava sometimes flowed into standing water within shallow lakes. The pillows vary in size from a few centimetres to over one metre in length and have chilled margins but seem to be devoid of concentrically arranged vesicles. Well-developed pillow lavas along the Beinn Fhada plateau above Loch Sguabain in upper Glen More, are described in the Loch Sguabain report (see above).

Between the two dyke-like parts of the Glen More intrusion, a small gabbroic plug is exposed, cut by late basic cone-sheets. Identical plugs are found around Cruach Choireadail having chemical affinities with the caldera lavas and they may be regarded as potential feeders for them. Other features of interest within the Cruach Choireadail site include the agglomerates or breccias towards the summit and an exceptionally large xenolith of the Ben Buie layered gabbro (exposed at the western end of the site) at a height of about 250 m. This gabbro is the youngest intrusion of the earlier Glen More Centre (Centre 1; Table 5.1).

Interpretation

The continuous vertical section through the Glen More ring-dyke, showing an unbroken transition from basic rocks at low levels in the intrusion to acid rocks at high levels, is one of the most clearly demonstrated examples of such compositional variation in the BTVP. The close association of basic and acid rocks, and by implication magmas, in the same intrusion is a common feature of intrusions in the Province. The fine exposures in this site provide an excellent opportunity to study the problem.

The gradual change in rock type at different levels in the intrusion was attributed by Bailey *et al.* (1924) to *in situ* differentiation involving crystal settling towards the base and migration of residual, progressively more evolved magma into the upper parts of the ring-dyke. Objections have been raised to this mechanism. Skelhorn (1969) pointed out that the gabbros at the base lack rhythmic layering, igneous lamination and other features characteristic of cumulate rocks. Further evidence against a model involving fractionation of basaltic magma by crystal settling is the lack of marginal border rocks representing the parental basaltic magma (cf. Wager and Deer, 1939) and the observation that granophyre is in direct contact with the country rocks, against which it is chilled. Another possible mechanism involves the simultaneous intrusion of acid and basic magmas, with mixing of the two contrasted magmas to give intermediate rock types which certainly have the textural attributes of hybrid igneous rocks. The magma mixing and hybridization model was advanced by Holmes (1936) and Fenner (1937), but Koomans and Kuenen (1938) argued against it on the grounds that:

1. there was an absence of biotite, hornblende and orthopyroxene which they claimed were characteristic of a 'hybrid series';
2. the predominantly curved trends on the chemical variation diagrams (as opposed to straight-line relationships produced by magma mixing); and
3. the lack of intrusive breccias or sharp contacts between the various rock types.

They suggested that gravitative separation and settling of augite and iron oxides (+ plagioclase) resulted in the formation of the granophyre, and that subsequent 'pneumatolytic emanations from the lower reaches of the column' were responsible for the textural features of the intermediate rocks.

Clearly, the close association of basic and acid rocks, and the apparent height control on their occurrence in the intrusion, have important petrogenetic implications. However, it is obvious from a cursory glance at the literature that a thorough petrological investigation of the Glen More ring-dyke is long overdue. This would have to take into account modern work on magma mixing and zoned magma chambers; until this is done the site will not attain the international status which it almost certainly merits.

Conclusions

The Glen More ring-dyke shows continuous variation from gabbroic rocks in its lower levels through intermediate rocks to granophyre in the highest parts of the intrusion. This variation may have been caused by:

1. settling of early-formed crystals (augite, magnetite, calcic plagioclase, plus some olivine) towards the base of the intrusion, allowing the remaining magma, enriched in silica, potassium and sodium, to crystallize as granophyre at the top of the ring-dyke; or, alternatively,
2. basaltic magma may have been injected into the base of the ring-dyke at the same time as granitic magma entered the top; the two magmas then mingled to give the rocks of intermediate composition now found between the gabbro and granophyre.

Pillow lavas in the site formed when basaltic magma flowed into a caldera occupied by a lake (see also Loch Sguabain).

ALLT MOLACH – BEINN CHÀISGIDLE

Highlights

Continuous stream sections expose acid and basic ring-dykes and numerous cross-cutting cone-sheets associated with Mull Centre 2. Extensive acid veining of basic rocks occurs where these intrude acid rocks. Screens of country rock (earlier volcanic and igneous rocks) are present between the ring-dykes. Acid intrusions of Mull Centre 3, including the classic Loch Bà ring-dyke, truncate the earlier Centre 2 intrusions.

Introduction

The geological significance of the Allt Molach–Beinn Chàisgidle site lies in the exposure of various members of the second intrusive centre on Mull – the Beinn Chàisgidle Centre (Centre 2) (Table 5.1). The intrusions are predominantly ring-dyke structures of felsite and quartz dolerite/gabbro compositions and the screens between them are formed by earlier volcanics erupted into the South-East Caldera. The Loch Bà felsitic ring-dyke, together with granophyre bodies belonging to the Glen Cannel complex, truncate the ring intrusions and are associated with the third, and last, Loch Bà Centre (Centre 3).

The classic Mull Memoir (Bailey *et al.*, 1924) remains the main reference source for this site, although a field guide to the area has been published by Skelhorn (1969), which contains a more detailed petrological description of the intrusions than is presented here.

Description

The site is conveniently divided into three geographical sections (Fig. 5.17):

1. The stream section and valley sides of the Allt Molach from the head of Loch Sguabain at Ishriff to Sgulan Beag (NM 610 320).
2. The bleak upland plateau between Sgulan Beag and Beinn Chàisgidle.
3. Beinn Chàisgidle and northwards to Glen Cannel.

(a) The Allt Molach–Sgulan Beag

The section along the Allt Molach contains 14–15 separate, subvertical intrusions of felsite, granophyre and quartz dolerite/gabbro. These are cut by a suite of cone-sheets inclined towards either Centre 2 or Centre 3. Composite dykes are also present in this section. Bailey *et al.* (1924, p. 308) have recorded the following sequence of ring-dyke intrusions striking NE–SW across the stream from Loch Sguabain, the detail of which is not shown on the accompanying site map (but see Skelhorn, 1969, fig. 3): (No time sequence is implied here; sequence commences at Loch Sguabain.)

1. Glen More ring-dyke – gabbro merging to granophyre
2. Ishriff ring-dyke – granophyre

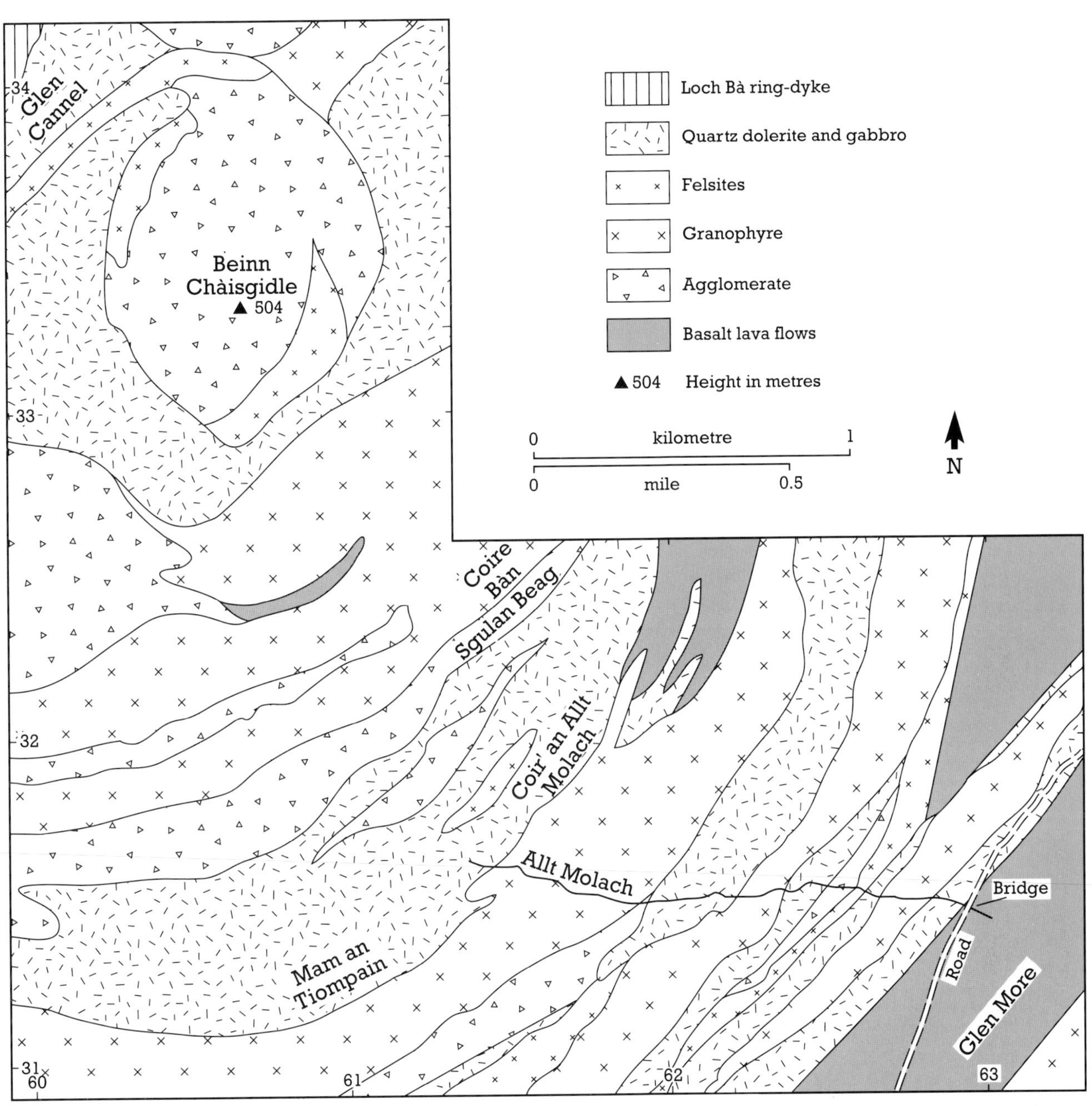

Figure 5.17 Geological map of the Allt Molach–Beinn Chàisgidle site (adapted from the British Geological Survey 'One Inch' map, Sheet 44, Mull).

3. Quartz gabbro cut by quartz + alkali feldspar segregation veins
4. Granophyre–feldspar-phyric felsite
5. Acidic quartz gabbro merging to granophyre
6. Xenolithic feldspar-phyric felsite
7. Quartz gabbro
8a. Granophyre
8b. Felsite
9. Quartz gabbro passing uphill into granophyre
10. Quartz dolerite possibly merging uphill into granophyre
11. Non-porphyritic granophyre
12. Feldspar-phyric granophyre
13. Vesicular quartz dolerite

Acidic net-veining is common in the basic intrusions which are often chilled against the acidic ones. Late basic cone-sheets cut most of the ring-dykes and also show net-veining by rheomorphic acidic material.

Discontinuous arcuate patches of basic and

acid rock intervene between the various ring-dykes and have been interpreted by Bailey *et al.* (1924) as screens of: earlier porphyritic basaltic lava between intrusions 2/3; non-porphyritic basaltic lava between intrusions 3/4, 5/6, 4/6, 6/7; fine-grained dolerite possibly metamorphosed cone-sheets between intrusions 6/7; agglomerate between intrusions 6/7; cone-sheets, agglomerate and associated brecciated early dolerites between intrusions 10/11, 9/11; and tuff and undifferentiated lava between intrusions 12/13.

Various dykes, some of them composite (tholeiite–felsite–tholeiite), cut many of the ring-dykes but, in general, exposure is not sufficient to conclude that they cut all.

(b) Sgulan Beag–Beinn Chàisgidle

The ring-dykes of the Allt Molach area pass uphill into a series of acid and basic intrusions cut by many cone-sheets and separated from one another by screens of brecciated basic sheets, agglomerates and basaltic and rhyolitic lavas which form prominent, upstanding N- and NW-dipping ridges. Exposure is again poor in many places, especially on the grassy slopes of Coire Ban (NM 613 328) to the east, and Mam an Tiompain (NM 604 315) to the west.

(c) Beinn Chàisgidle–Glen Cannel

The summit and northern slopes of Beinn Chàisgidle are formed by agglomerates, quartz dolerite and a profusion of cone-sheets, all of which are truncated abruptly by a number of intrusive bodies associated with the later Loch Bà centre (Centre 3). Parts of the Glen Cannel peralkaline granophyre and the Loch Bà felsite ring-dyke are poorly exposed on Beinn Chàisgidle and in the valley. The granophyre forms low, smooth, striated knolls within the arcuate caldera fault intruded by the Loch Bà ring-dyke. It has a uniform appearance, weathering to a pale-pink colour and is moderately coarse-grained, consisting of perthitic orthoclase, quartz, aegirine-augite, magnetite and sphene. Occasionally the granophyre may appear spherulitic or more obviously granophyric as it is immediately south of the old burial ground at Gortenbuie (NM 599 345), just north of the site.

The Loch Bà felsite ring-dyke is arguably the most spectacular of its kind in the Province and was the first complete steep-sided ring-dyke to be described in the world. It is described in detail in the Loch Bà report (see below). Although not continuously exposed, this intrusion is well represented in the Allt a Choire Bhain (east of the site) and intermittently on the northern slopes of Beinn Chàisgidle and towards Breapadail (NM 586 328) outside the site. Stream sections flowing northwards from Beinn Chàisgidle expose the ring-dyke and also expose evidence in the rocks on either side of the ring-dyke (such as crushing) for tectonic movements along the ring fault now occupied by the felsite. Considerable subsidence within this ring fault, of at least 150 m, has been estimated by Lewis (1968).

Interpretation

The site contains representative sections through the intrusions associated with Centre 2 and clearly demonstrates the cross-cutting effects of the final intrusive centre (Centre 3), which developed when the focus of activity shifted to the north-west. The ring-dyke intrusions of the Centre 2 have been conventionally regarded as a suite of separate intrusions. However, Skelhorn (1969) has challenged this view and has suggested that gabbros 3, 5 and 7 are possibly members of the same intrusion split up by a series of later acidic ring-dykes. This could also apply to many of the other intrusions. The time sequence and structure of the intrusions and the status of the intervening screens are also open to reinterpretation. The sequence of Bailey *et al.* (1924) was based upon the fact that acidic veins connecting with the acid intrusions cut the basic ring-dykes and cone-sheets alike. However, as many of these intrusions are chilled against granophyre, the net veining may well be the result of the localized fusion of adjacent acid rock by the basic intrusions which generated rheomorphic magmas resulting in the acid back-veining of the basic rocks. Many of the late basic cone-sheets are also back-veined and chilled against the intrusions, although others apparently truncate the veining, indicating several distinct episodes of cone-sheet emplacement.

Skelhorn (1969) has thrown some doubt on the interpretation of some of the granophyres as ring-dyke structures; for example, intrusion 4 has subvertical contacts which dip at 60°–90° towards Centre 2 and should perhaps be considered as a cone-sheet. As exposure is rather poor in some parts of the site, this problem is not fully resolvable but the relationships between topography and outcrop do suggest steep contacts.

Finally, the Palaeocene history of the site can be summarized as follows (after Skelhorn, 1969 and Bailey *et al.*, 1924):

Intrusions	Centre
Dykes Loch Bà ring-dyke and ring fault Late basic cone-sheets Glen Cannel complex – granophyre	Centre 3
Dykes Glen More Intrusion Late basic cone-sheets Various ring intrusions (2–14) (around Beinn Chàisgidle) Early basic cone-sheets Early acid cone-sheets	Centre 2
Explosion vents associated with the South-East Caldera First lavas and vents	Centre 1

Conclusions

This site provides valuable and informative sections through numerous intrusions of dominantly basic to acidic compositions associated with Centre 2. It shows the complexity of the centre, the evolution of which has, as yet, to be fully resolved. Both acid and basic magmas were emplaced as ring-dykes, together with several generations of basic and acid cone-sheets. Additionally, earlier volcanic rocks erupted into the South-East Caldera (Centre 1) are preserved as altered screens between the ring-dykes. Acidic bodies belonging to the final intrusive centre on Mull, the Loch Bà Centre (Centre 3), truncate Centre 2 intrusions and mark a shift in the focus of activity to the north-west. Clear examples of rheomorphic acid net-veining occur where basic intrusions cut and chill against lower-melting-point acidic rocks.

LOCH BÀ–BEN MORE

Highlights

The Loch Bà Felsite ring-dyke is the international type example of a ring-dyke intruding arcuate faults within which central subsidence has taken place. The ring-dyke and the hybrid intrusions provide evidence that acid and basic magmas coexisted on Mull and the compositional variation shown by porphyritic glassy inclusions in the ring-dyke suggests that the Mull Centre 3 was underlain by a zoned magma chamber. There is a coastal section across the Mull dyke-swarm axis where 12% crustal dilation has been demonstrated. These features make this a site of international importance.

Introduction

The area between Loch Bà, Ben More and Loch na Keal is a site of multiple interest and of international petrological importance. The Plateau Group lavas attain their maximum thickness here and representative flows belonging to the Central Group lavas are present in various states of metamorphism. Major intrusions belonging to the Glen More and Beinn Chàisgidle centres (Centres 1 and 2 respectively) are demonstrated, together with the full suite of intrusive rocks associated with the Loch Bà Centre (Centre 3). The site has international status as a type locality for ring-dyke structures, the type example being the Loch Bà felsite ring-dyke (Fig. 5.18).

As is the case with many of the other sites on Mull, the Memoir remains the most comprehensive account of the geology of the Loch Bà–Glen More area (Bailey *et al.*, 1924). More recent field investigations in the site have been carried out by Skelhorn (1969) who has revised some of the earlier interpretations. The Loch Bà felsite ring-dyke was discovered by the Geological Survey and recognized as a prime example of an arcuate ring-dyke closely associated with ring-faulting. Bott and Tuson (1973) have studied the subsurface extent of the central complex by gravity surveys and showed it to be underlain by a large basic/ultrabasic cylindrical mass. Beckinsale *et al.* (1978) and Morrison (1978) studied the geochemistry of the Mull Plateau and Central Series lavas on a regional scale, the results of which can be broadly applied to this site and are summarized in the 'Introduction' to this chapter. Sparks (1988) has given a detailed account of the petrology and geochemistry of the Loch Bà ring-dyke.

Description

Features of the lavas within the site are described first; these are followed by an account of central complex intrusions.

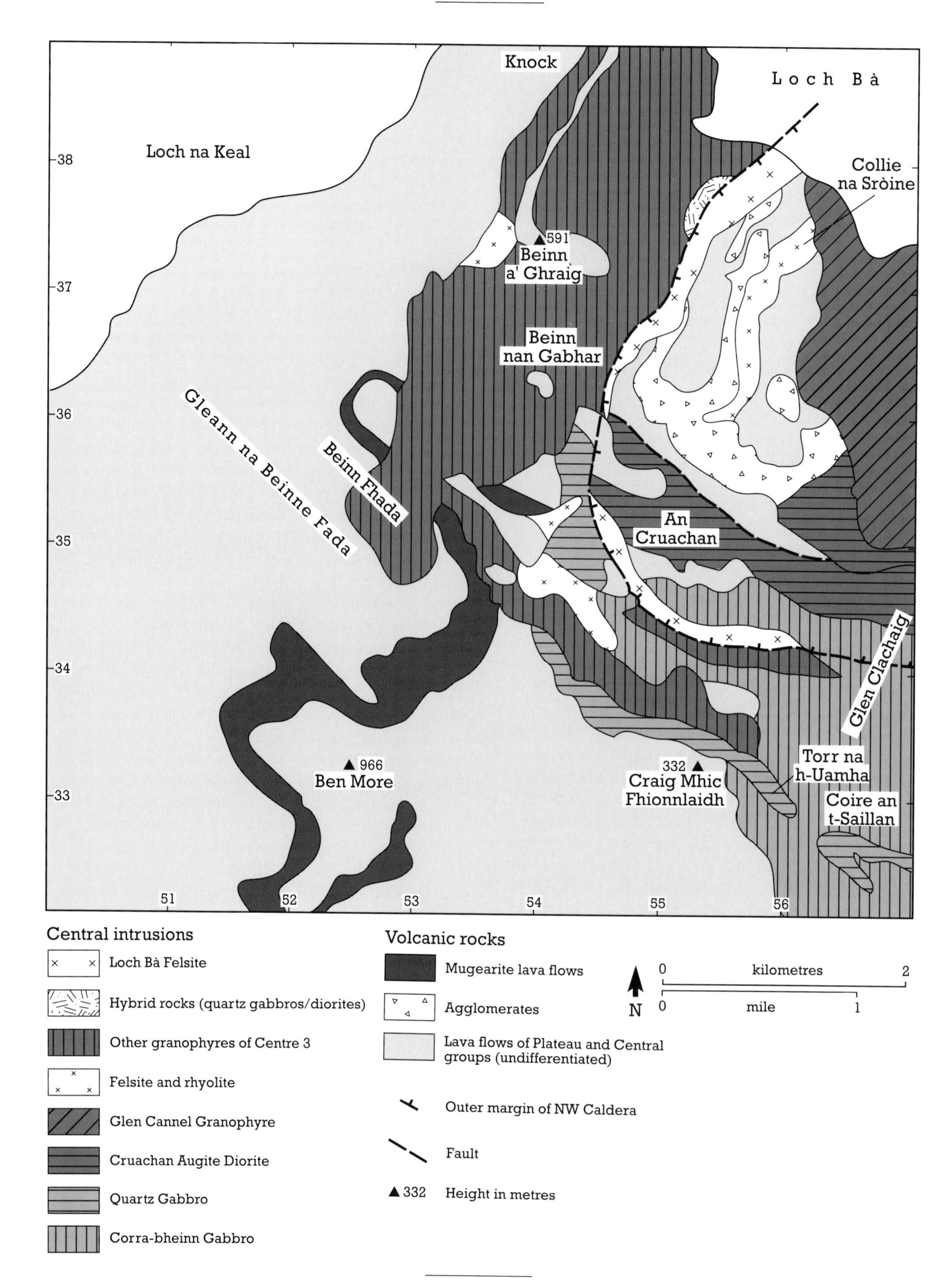

Knock
Loch Bà
Loch na Keal
Collie na Sròine
591
Beinn a' Ghraig
Beinn nan Gabhar
Gleann na Beinne Fada
Beinn Fhada
An Cruachan
Glen Clachaig
966
Ben More
332
Craig Mhic Fhionnlaidh
Torr na h-Uamha
Coire an t-Saillan
38
37
36
35
34
33
51
52
53
54
55
56
Central intrusions
Loch Bà Felsite
Hybrid rocks (quartz gabbros/diorites)
Other granophyres of Centre 3
Felsite and rhyolite
Glen Cannel Granophyre
Cruachan Augite Diorite
Quartz Gabbro
Corra-bheinn Gabbro
Volcanic rocks
Mugearite lava flows
Agglomerates
Lava flows of Plateau and Central groups (undifferentiated)
Outer margin of NW Caldera
Fault
332 Height in metres
N
0 kilometres 2
0 mile 1

(a) The lavas

The Plateau Group lavas (Table 5.2) attain their maximum thickness of over 900 m on Ben More (NM 526 331). They lie within the limit of pneumatolysis around the central complex and the alteration is most severe in the east of the lava outcrop, where the flows adjoin the major intrusions. As a consequence, basalt flows in this area no longer carry fresh olivine and generally contain epidote in amygdales and veins. Trap featuring, reflecting the different weathering properties of the resistant flow centres and the crumbling, amygdaloidal tops, is still discernible and a few recognizable boles are present, attesting to subaerial weathering between successive eruptions.

The lowest division of the Plateau Group, the Survey's Staffa Type basalts, do not crop out in the site as the gentle easterly dip of the lavas has taken the earlier flows below sea-level. The lowest flows preserved are mildly alkaline to transitional basalts which are either aphyric or contain olivine pseudomorphs. These are succeeded by pale-weathering, alkali-olivine basalts belonging to the Survey's Pale Group of Ben More which contains a prominent horizon of mugearite and feldspar-phyric basalt near the base (Table 5.2). The Pale Group forms the upper part of Ben More and isolated outcrops cause the north-western terminations of the ridges of Beinn Fhada and Bheinn a' Ghraig (NM 542 373).

Many of the flows are amygdaloidal and the amygdale minerals are of considerable interest, both in their own right and also in studies concerned with the hydrothermal effects associated with the emplacement of the central complex. A full description of these minerals is contained in the Memoir and zeolite zones in the lavas have been delineated by Walker (1971). Numerous basaltic dykes and sheets, and some of felsite and craignurite compositions, cut the Plateau Group lavas.

The basaltic lavas of the overlying Central Group are exposed outside the central complex and as isolated masses among the intrusions of the complex. South of Creag Mhic Fhionnlaidh (NM 554 333) these lavas are separated from the Plateau Group to the west by a fault and are terminated to the north and east by intrusions associated with Centre 2. Large, detached masses of Central Group lavas occur on the slopes west of Glen Clachaig, within the North-West or Late Caldera of Centre 3 (Table 5.1). Smaller masses are preserved as screens between the intrusions of Centre 2. The lavas to the south of Creag Mhic Fionnlaidh, outside the caldera, are cut by early acid and basic cone-sheets whereas the flows preserved within the caldera are virtually free from cone-sheets. A few basic dykes also intrude the lavas. The Plateau Group and Central Group lavas to the west of the central complex are pierced by volcanic vents which, in turn, are cut by early basic cone-sheets, the Corra-bheinn gabbro and granophyre veins. The largest vent (at NM 558 324) is about 700 m in length. The agglomerate in this and other smaller vents contains fragments of Moine gneiss (Fig. 5.19). Gneiss fragments occur in vents outside the calderas, but are not found in the numerous agglomerates within these structures (Bailey *et al.*, 1924, fig. 29) where subsidence has presumably brought the gneisses and schists below the level at which the magmas were liable to rapid vesiculation and explosion.

Towards the later, major intrusions of the central complex, all lavas show progressive effects of thermal metamorphism. These changes are particularly well demonstrated by alteration in the low-temperature amygdale mineral assemblages and successively higher-temperature phases are found as the later intrusions are approached. The lavas also show signs of thermal metamorphism and have been hornfelsed up to pyroxene-granulite facies near to the intrusions.

(b) The Central Complex

Three major intrusions of somewhat uncertain age occupy a 2-km wide, NW-trending tract of ground extending from the south-eastern extremity of the site to the southern slopes of Beinn Fhada and Beinn nan Gobhar (NM 538 317). These bodies separate the lavas from the intrusions of Centre 3 and form the margin of the Central Complex (Fig. 5.18).

The Corra-bheinn Gabbro is one of the intrusions in the extreme south-east of the site, forming the eastern slopes of Sleibhte-coire. Its western and southern contacts are intrusive against Central Group lavas, vent agglomerate and (outside the site) the Derrynaculen Granophyre. To the north-east it is split up by late, basic cone-sheets and probable continuations of the Glen

Figure 5.18 Geological map of the Loch Bà–Ben More site (adapted from the British Geological Survey 'One Inch' map, Sheet 44, Mull).

Figure 5.19 Vent agglomerate containing fragments of Moine gneiss (NM 558 324). Loch Bà–Ben More site, Mull. (Photo: C.J. MacFadyen.)

More ring-dyke. The Corra-bheinn Gabbro is probably an early intrusion of Centre 2 (Skelhorn, 1969; but see Bailey *et al.*, 1924). The gabbro exhibits rhythmic layering, igneous lamination and other structures characteristic of cumulate rocks. Layers dip to the north-east between 15° and 75° (but averaging 40°) towards Centre 2. Skelhorn (1969) has identified nine major rhythmic units in the gabbro to the south-west of Coir' an t-Sailein (NM 566 326) and recorded horizons of tabular pyroxene-granulite xenoliths, some of which disrupt the layering which is also banked up against some xenoliths (Skelhorn, 1969).

An interrupted chain of composite quartz-gabbro masses, changing internally to granophyre towards their tops, extends from Coir' a' Mhàim (NM 580 318) to the south of the site, through Coir' an t-Salein and Torr na h-Uamha (NM 560 332) and terminates on Beinn Fhada. These are probably continuations of the Glen More ring-dyke of Centre 2 (Bailey *et al.*, 1924) and cut the Corra-bheinn Gabbro and the lavas north of Creag Mhic Fhionnlaidh.

The Cruachan Augite Diorite intrusion occurs within and outside the Loch Bà felsite. It is cut by numerous late, basic cone-sheets and, although its outcrop is restricted to narrow screens between the sheets, it is mapped as an entity. The composition of the intrusion becomes progressively more acidic upwards and the highest parts are granophyre or felsite rather than diorite. The mass is assigned to Centre 1 and is probably the equivalent of the Gaodhail Augite Diorite (Bailey *et al.*, 1924), although Skelhorn (1969) has suggested that it could be the earliest member of Centre 3.

The main interest of this site derives from the exposure of the western quadrant of Centre 3, the Loch Bà Centre. A complete and unique suite of intrusions and associated rocks is demonstrated and Skelhorn (1969) has revised the sequence of events for the evolution of this centre as follows:

6. The Loch Bà Felsite ring-dyke
5. The hybrid masses of Sron nam Boc and Coille na Sroine
4. The Beinn a' Ghraig Granophyre
3. The Knock Granophyre
2. The early Beinn a' Ghraig Granophyre and Felsite
1. The Glen Cannel Complex: This consists of an early quartz dolerite plug, vents, intrusive rhyolites, and a late felsite and granophyre.

A mass of pyroxene granophyre dominates the Glen Cannel complex and this forms a low dome elongated along a north-west to south-east axis. It thermally alters a felsite mass (also part of the complex) on the south side of Glen Clachaig. The granophyre has yielded a Palaeocene age (58 ± 3 Ma, Beckinsale, 1974). To the west and south of Glen Clachaig, two quartz-dolerite plugs belonging to the complex have been metamorphosed by the granophyre. Other members of the complex occupy ground between Allt Beithe and Coille na Sroine and include agglomerate-filled vents into which a rhyolite dome has been intruded, along with tuffisite dykes containing fragments of Moine schist (which the agglomerates hereabouts do not, see above). These small masses have intruded Central Group lavas.

The early granophyre and felsite of Beinn a' Ghraig is exposed on the ridge immediately east of the summit. It is chilled on both its east and west contacts against Plateau Group lavas at the base of the Pale Group of Ben More, and a screen of lavas occurs between the Beinn a' Graig and Knock Granophyres at its eastern tip.

The Knock Granophyre has a dyke-like outcrop 100–300 m wide, extending north-eastwards for 3 km from the western edge of the Beinn a' Ghraig ridge. The granophyre characteristically develops shearing and cataclastic textures which have been attributed to the emplacement of the Beinn a' Ghraig Granophyre. These two granophyre intrusions are separated by a particularly clear example of a screen: this is composed of highly metamorphosed basaltic lavas and cone-sheets, it extends for about 3 km, has a vertical range of over 500 m and a width of between 3 m and 100 m. The screen rocks have been recrystallized to fine-grained granulites consisting of plagioclase, augite, hornblende and magnetite and the adjoining granophyres have been contaminated with basic material derived from the screen. Screens are abundantly developed in the Mull Central Complex and elsewhere in the BTVP: this example is one of the clearest since the topography of the site, coupled with reasonable exposure, shows the three-dimensional form to best advantage, and the rocks forming the screen have been thoroughly altered by the adjoining igneous intrusions.

A small mass of hybrid rocks consisting of acid-veined quartz gabbros and diorites lies between the Beinn a' Ghraig Granophyre and the Loch Bà Felsite south of Coille na Sroine (at about NM 553 376). The hybrids are chilled against the Beinn a' Ghraig Granophyre which has been partially melted and back-veins the chilled, marginal hybrid rocks (Rast, 1968).

The site contains the western quadrant of the Loch Bà Felsite ring-dyke (Fig. 5.20), including the historic exposures where its association with a fault and its annular outcrop were first described. The ring-dyke is well exposed and both its outer and inner contacts can be seen to dip outwards at angles of 70°–80°. Between Loch Bà and Beinn nan Gabhar the felsite separates the Beinn a' Ghraig Granophyre from various members of the Glen Cannel Complex and Central Group lavas. On the western slopes of Beinn nan Gabhar, shortly after 'entering' the intrusions of Centre 2 cut by a profusion of late basic cone-sheets, the felsite fails. However, the fault into which it was intruded can be traced southwards until, after a gap of several hundred metres, the felsite reappears and can be traced across the slopes of An Cruachan and the River Clachaig. On the south side of the river its outcrop narrows and disappears once more just beyond the site boundary.

Compositionally, the ring-dyke varies from an apparently flow-banded rhyolite to felsite and typically contains phenocrysts of alkali feldspar and mafic minerals. A particularly striking feature is the common occurrence of dark, fine-grained, elongate wispy xenoliths. The basic character of the xenoliths was first reported by Blake *et al.* (1965); however, Sparks (1988) has shown that these dark, aphanitic rocks vary continuously in composition from basaltic andesite to andesite, dacite and rhyolite. The phenocrysts in the enclosing rhyolite include plagioclase (An_{32} to An_{24}), sanidine, hedenbergite, fayalite, magnetite, ilmenite, apatite and zircon. The phenocrysts are often aggregated; however, it is unusual to find hedenbergite and fayalite in the same cluster. The dark wispy inclusions are usually aphyric; where phenocrysts occur they include plagioclase (An_{65} to An_{30}), a continuous range of clinopyroxenes from augite to pure hedenbergite, pigeonite, magnetite, ilmenite and rare apatite (Sparks, 1988, p. 446). The significance of this range of phenocrysts, and particularly the compositional spectrum of glassy inclusions, is discussed subsequently.

The site includes an excellent section through the axis of the Mull dyke swarm on the southern shores of Loch na Keal, to the west of Eilean Feoir (NM 531 389). The majority of the dykes are basaltic and many are multiple, comprising as

Figure 5.20 Columnar jointing in the Loch Bà Felsite ring-dyke (NM 552 371). Loch Bà–Ben More site, Mull. (Photo: C.J. MacFadyen.)

many as four or five intrusions, which may show cross-cutting relationships. Some 142 dykes have been recorded, with an aggregate thickness of 249 m along a 2-km traverse of the Loch na Keal shore (Bailey *et al.*, 1924). The average dyke thickness is 1.8 m and a total crustal dilation of 12.4% has been calculated. The dykes show the same effects of alteration as the lavas within the zone of pneumatolysis, although often to a lesser extent.

Interpretation

The Loch Bà–Ben More site supports some of the most spectacular geology and scenery on Mull, and is arguably the most important locality as it demonstrates representative rocks from every major Tertiary igneous event on the island. Centre 3 is fully represented and marks a final shift of intrusive activity to the north-west as a possible result of the elongation of the magma chamber in this direction, parallel to the axis of the regional dyke swarm (Skelhorn, 1969). According to Skelhorn, the Centre 2 Glen More ring-dyke, represented in this site by the quartz-gabbro masses, is the earliest intrusion to reflect this elongation. The arcuate, fragmented ring-dykes of Centres 1 to 3 have intricate intrusive relationships from which the history of the complex is determined. They also exhibit a large compositional range from acid to basic rock types. Acidic intrusions, despite their wide aerial extent, represent only a small proportion of the total igneous mass of the Mull central complex, as shown by the gravimetric work of McQuillin and Tuson (1963) and Bott and Tuson (1973). A gravity survey of the Glen Cannel granophyre revealed that it is no more than 1220 m thick and must have a sheet-like form. Walker (1975), however, suggested that the granophyre mass once formed the upper part of a large body of acid magma which had migrated into a curved, flange fracture and thus caused the north-west transfer of the igneous centre to the Loch Bà area, and that further portions of this acid magma migrated to form the younger Knock, Beinn a' Ghraig and Loch Bà intrusions.

The site provides evidence which supports the view that acid and basic magmas probably coexisted in a compositionally zoned magma chamber beneath Mull, and that this mixed magma was injected into fractures associated with ring-faulting and central block subsidence. The Coille na Sroine hybrids provide good evidence for magma mixing, as well as the rheomorphic generation of acid melts. Within the magnificently exposed Loch Bà felsite ring-dyke, there is also excellent evidence for coexisting magmas of contrasted compositions: the dark, glassy inclusions in the felsite (which make up about 15% of the rock) often have lobate, cauliflower-shaped margins and are frequently wisp-like and convoluted. Marshall and Sparks (1984) considered that the inclusions were liquid when they were incorporated into the felsite magma and they explained the heterogeneous, mixed-magma intrusion by envisaging incomplete mechanical mixing between different components in a vertically zoned magma chamber, triggered off by subsidence of the central block. Subsequently, Sparks (1988) made a detailed examination of the rhyolite and inclusion compositions, together with the compositions of phenocrysts in both. The really striking feature to emerge is the compositional range of the glassy inclusions and their phenocrysts (which appear linked by crystal–liquid equilibria). The mafic glasses vary continuously in composition from basaltic andersite through to dacite and rhyolite suggesting that the Loch Bà centre was underlain by a zoned magma chamber, capped by rhyolite magma. Crystals were precipitating from all levels of this chamber, which covers the compositional range of the Middle and Upper zones of the Skaergaard Intrusion in East Greenland (Wager and Deer, 1939; Wager and Brown, 1968; McBirney, 1975), when this structurally ordered sequence was catastrophically disrupted to form the heterogeneous Loch Bà ring-dyke. Sparks' (1988) discoveries and interpretation have obvious implications for the subsurface geology of Mull – the gravity high (Bott and Tuson, 1973; Bott and Tantrigoda, 1987) indicates subsurface, dense, gabbroic rocks which may include a Skaergaard-like body. They also shed further light on the possible mode of crystallization of layered basic intrusions: the evidence from Loch Bà suggests that the compositionally contrasted layers could have built up essentially simultaneously, rather than sequentially as envisaged in the classic models of layering (for example, Wager and Brown, 1968). Sparks (1988) regards the felsite as an intrusive, banded, rhyolitic, welded tuff.

The Loch Bà site provides sound evidence for the frequent suggestion that emplacement of the Mull Central Complex (and others) has involved subsidence of a central block, or blocks, bounded by arcuate ring-faults. Richey (1932) calculated that the downthrow on the block circumscribed by the Loch Bà ring-dyke and associated fault was about 1000 m. Lewis (1968), however, pointed out that much of the displacement could have occurred on two NW-trending faults which transect the block and suggested that the maximum downthrow on the ring-fault, now occupied by felsite, was only about 150 m.

Conclusions

The thick succession of basalt lavas on Ben More and surroundings has been extensively altered by the later intrusions of the central complex, with the result that original minerals (for instance, olivine) were replaced and new ones (for example epidote, chlorite) formed by circulating heated waters and gases. The site lies astride the Mull dyke swarm which locally produced crustal dilation of over 12%; the dykes mostly pre-dated the central complex and thus also are generally altered. A variety of basic, granitic and hybrid intrusions belongs to the youngest centre, Centre 3 of Mull. They provide compelling evidence for the mingling of basic and acid magmas. The Loch Bà felsite ring-dyke is the type example of a ring-dyke. It intruded along a ring-fault bounding a central block which may have subsided by as much as 1 km in aggregate. The ring-dyke intrusion provides spectacular evidence for the explosive disintegration of a crystallizing, zoned magma chamber in which rhyolite liquid passed by continuous compositional variation downwards through intermediate compositions to basaltic liquids. These discoveries are of major scientific importance as they provide a glimpse of the details of the dense rocks known from gravity surveys to underlie Mull; they are of more general importance since they suggest ways in which layered igneous intrusions may crystallize.

Chapter 6

Isle of Arran

INTRODUCTION

Arran has long been known as an area of complex and diverse geology and outstanding natural beauty (Fig 6.1). It is the most accessible of the Scottish Cenozoic igneous centres and is used extensively as a training ground for geologists. Interest in the geology of Arran goes back over more than 200 years to the writings of Pennant (1774), Hutton (1795), MacCulloch (1819), Boué (1820) and others, but the first detailed survey of the island, other than that by Ramsay (1841), was not carried out until the start of the present century by Gunn (1903) and this was subsequently completed by Tyrrell (1928, reprinted 1987). The sites selected as illustrating Arran's varied igneous geology are indicated on Fig. 6.2.

The Tertiary igneous interest is concerned mainly with the varied intrusive rocks; remnants of a former lava cover now only exist as subsided blocks within the Central Igneous Complex. The largest body of intrusive rocks is formed by the granites of the scenically spectacular Northern Mountains (Fig. 6.1). Two distinct biotite granites; one a coarse-grained rock forming the outermost parts and the other a younger, finer-grained rock within the first, are present (see Harker *in* Gunn, 1903; Tyrrell, 1928; Flett, 1942). These two units comprise the Northern Granite. Their significance to the British Tertiary Volcanic Province (BTVP) as a whole lies in the fact that they demonstrably pre-date the Central Ring Complex and, additionally, that emplacement of the outer granite caused updoming and faulting of the enveloping Palaeozoic sediments and Dalradian schists (Bailey, 1926; Woodcock and Underhill, 1987; England, 1988, 1990). Clear examples of the distortion of the country-rock envelope during granite emplacement are often difficult to demonstrate in the Province although predicted by Walker (1975); special significance thus attaches to the north Arran granites. The Glen Catacol site has been selected to represent both granites, their relationships to one another, to the country rocks and to later minor intrusions.

The Central Igneous Complex, between Brodick and Blackwaterfoot, is well known as an area of highly variable geology (Necker de Saussure, 1840; Ramsay, 1841; Gunn, 1900, 1903; Tyrrell, 1928). Its true complexity was realized by King

Figure 6.1 The Northern Granite Mountains of north Arran. Cir Mhor, Arran. (Photo: C.H. Emeleus.)

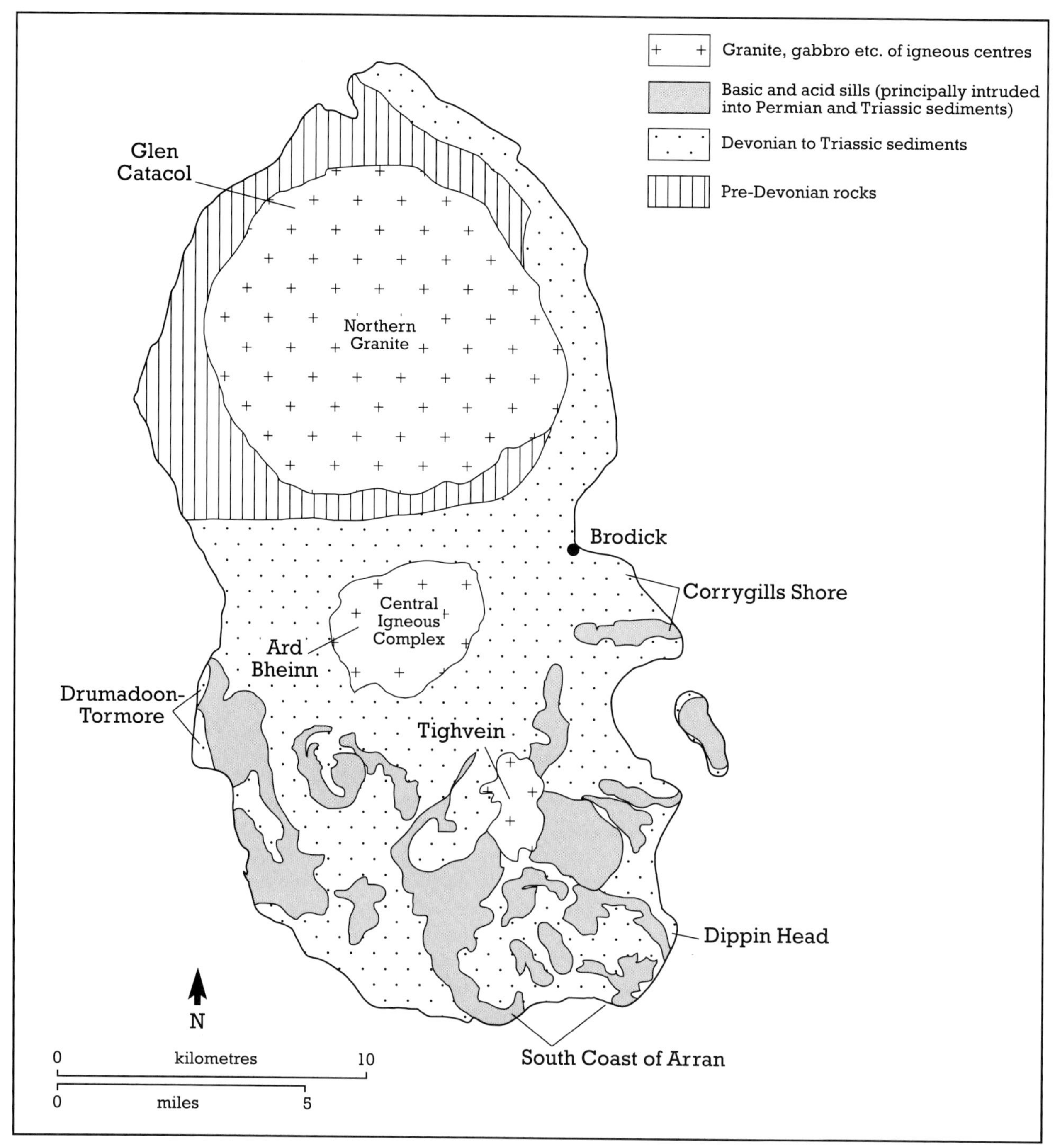

Figure 6.2 Map of the Isle of Arran, showing localities mentioned in the text.

(1955) who demonstrated that the varied rock types were grouped round four discrete foci, each interpreted as the eroded remains of a volcanic cone. These volcanoes built up on the floor of a caldera. Within this caldera there are large, subsided masses, the remains of Palaeocene basalt and Mesozoic sediments by which the complex is relatively dated. The complex thus preserves features of a comparatively high level of volcanicity that may be lacking, or ill exposed, in the other central complexes of the Province. The 'volcanic' part of the complex is more or less encircled by granites, sparse gabbros and some remarkable gabbro–granite hybrid rocks. The strata the complex intrudes at the present erosion level range from Devonian to Permian.

These sediments were already domed around the Northern Granites and were subsequently cut and tilted by the Central Ring Complex, which thus establishes the Northern Mountains granites as an early event in the Tertiary intrusive sequence of Arran. The site chosen to represent the features of the Central Ring Complex is the Creag an Fheidh–Beinn na h-Uaimh–Ard Bheinn area, simply denoted as the Ard Bheinn site (see below).

Despite the interest generated by these major intrusions, it is perhaps the minor intrusions of Arran which lend it most significance in the Province. The major Arran dyke swarm is magnificently exposed on the southern coast of the island, which cuts across the NNW–SSE-trending basaltic dykes from Bennan Head eastwards to Kildonan and beyond. This shore section, the South Coast site, has been chosen to represent this feature of the BTVP.

Dykes also figure prominently in the west coast site at Drumadoon and Tormore. These dykes are composite (dolerite + felsite + pitchstone) intrusions which, although present elsewhere in the Province, are especially prevalent and well exposed in Arran; having been described in the classic works of Judd (for example, 1893), they are usually referred to as 'Judd's Dykes' as a tribute to his acute observations. Analyses of several of the pitchstone dykes and sheets and their phenocrysts have been published by Carmichael (1960a, 1960b, 1962, 1963) and Carmichael and McDonald (1961). Sills of various types crop out over much of the southern part of the island and composite (dolerite + quartz porphyry) examples occur within both the Drumadoon and South Coast sites. These sills encompass a wide range of the compositions and structures associated with such bodies (for example, Tyrrell, 1928; Rao, 1958, 1959; Rogers and Gibson, 1977; Kanaris-Sotiriou and Gibb, 1985) and greatly enhance the significance of these sites. Basic sills are also a prominent feature and are either olivine-analcite dolerites (crinanites) or quartz dolerites. They form important members of the Dippin Head and Corrygills Shore sites. Recent studies on the Dippin Head sill (Henderson and Gibb, 1977; Gibb and Henderson, 1978a, 1978b) have added valuable information on the petrogenesis of the mildly alkaline basalt magma type which characterizes much of the Province. The Corrygills Shore site, in addition to exposing a crinanite sill which forms part of a possible cone-sheet system around Lamlash Bay (Tomkeieff, 1961), contains numerous other minor intrusions including the well-known Corrygills Pitchstone. The Tertiary igneous succession on the Isle of Arran is summarized in Table 6.1.

The Tertiary igneous rocks of the island have been the subject of geophysical studies and radiometric age determinations. The Northern Granite and Central Igneous Complex are the site of a positive Bouguer gravity anomaly, indicating that both centres are probably underlain by a considerable body of dense, mafic rock (McQuillin and Tuson, 1963). Palaeomagnetism investigations of the Arran dykes by Dagley *et al.* (1978) showed the majority to have reversed magnetization, although a substantial number of the dykes on the north-east coast showed normal magnetization; some of the sills and the Northern Granite sites also yielded normal magnetization. Age determinations on the Northern Granite range from *c.* 60 Ma (Dickin *et al.*, 1981) to *c.* 58 Ma (Evans *et al.*, 1973); a Palaeocene age (*c.* 58 Ma) was also obtained from the Central Igneous Complex (Evans *et al.*, 1973). Geochemical investigations by Dickin *et al.* (1981) included isotopic determinations on a variety of the Tertiary igneous rocks. From these the authors concluded that the isotopic variation is attributable to crustal contamination of mantle-derived, basic magma after differentiation.

The educational value of Arran is reflected in the number of geological guides to the island which have appeared over the years. Those now in common use are published by the Geological Society of Glasgow (MacDonald and Herriot, 1983) and the Geologists' Association (McKerrow and Atkins, 1985).

ARD BHEINN

Highlights

This site exposes some of the best evidence for the eroded remains of a caldera in the British Tertiary Volcanic Province. Collapsed blocks of basalt lavas and fossiliferous Mesozoic rocks have been preserved within the caldera; the mapping of lavas and volcaniclastic rocks has shown that several volcanic cones grew on the caldera floor.

Table 6.1 Tertiary igneous succession in the Isle of Arran (after Hodgson *et al.*, 1990, figure 8)

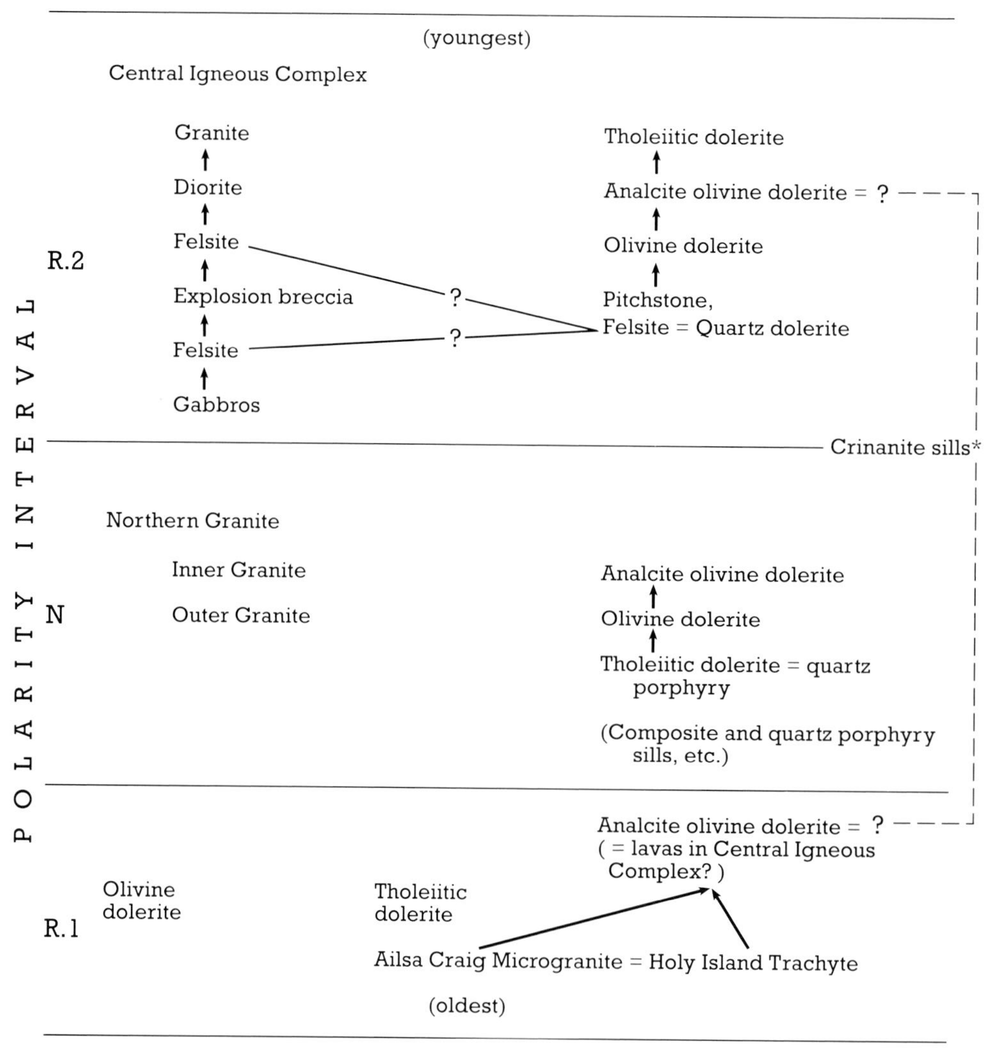

R.1, R.2 = reversed polarity intervals, N = normal polarity interval.
*Crinanite sills could be in either R.1 or R.2.
Arrows indicate probable sequence.

Introduction

The Ard Bheinn site demonstrates the characteristic features of the Central Igneous Complex of Arran. A large variety of rocks is exposed in the complex which are of both intrusive and effusive origin; they are compositionally highly variable. These include agglomerates, gabbros, felsites and granites which crop out as arcuate masses. In addition, blocks of Mesozoic sediments and Tertiary lavas are preserved within the complex. Volcanic rocks of intermediate and acidic composition are associated with several eruptive centres which developed on the caldera floor.

The foundations for our understanding of the geology of this complex area can be attributed to the Geological Survey (Peach *et al.*, 1901; Gunn, 1903; Tyrrell, 1928). These workers recognized the complex, volcanic nature of the area and demonstrated a Tertiary age by the identification of *remanié* blocks of Mesozoic sediments and similarities to ring intrusions found elsewhere in the BTVP. However, credit for our present detailed knowledge of the site is owed to King (1955) who mapped the area in considerable detail. King demonstrated that there were four, independent but overlapping, centres of volcanic activity in the Ard Bheinn area situated within a subsidence caldera (1955, Figs 1 and 2).

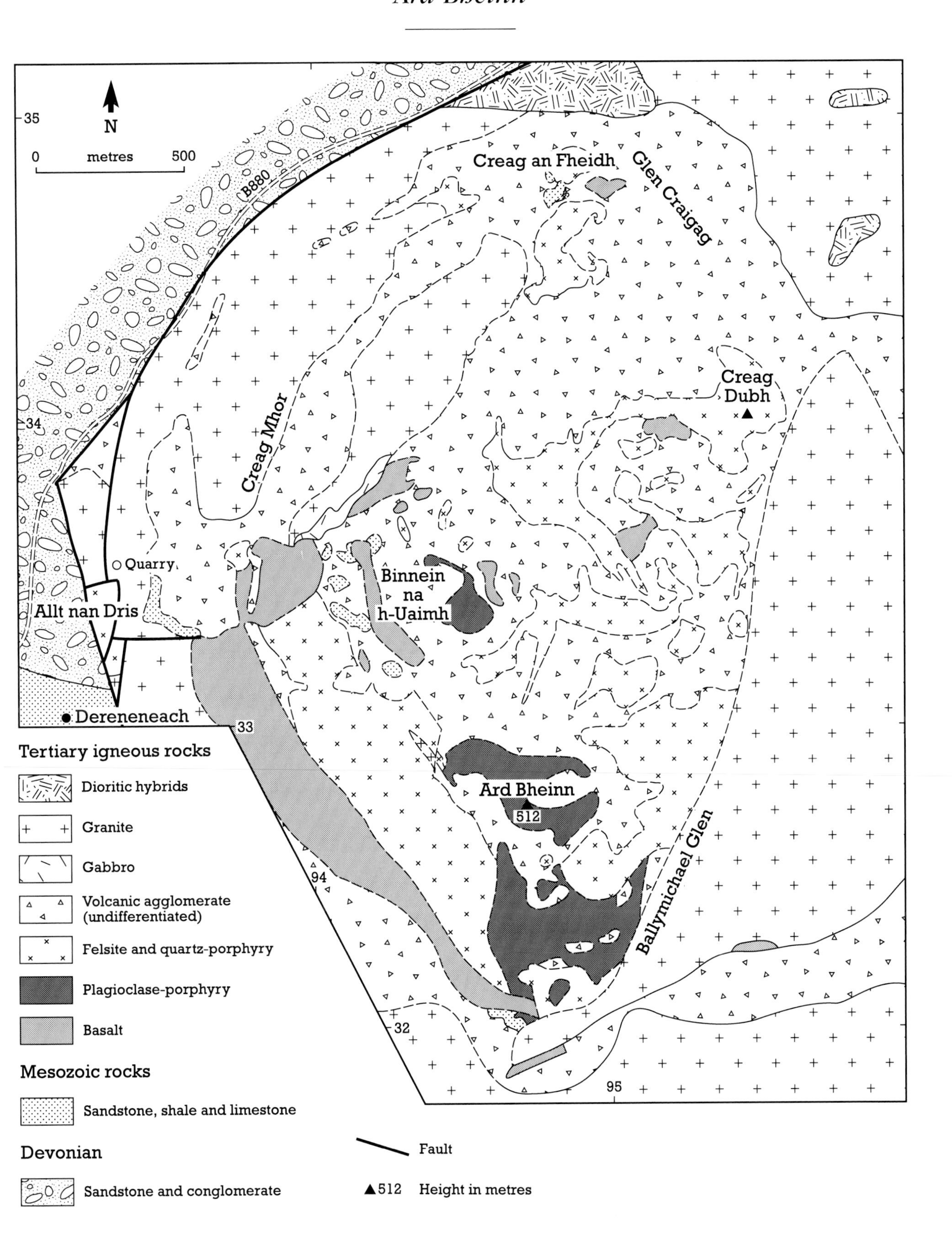

Figure 6.3 Geological map of the Ard Bheinn site (adapted from King, 1955, plate XVI).

Description

The site contains part of the poorly exposed Central Igneous Complex of Arran which has the overall structure of a major caldera (Fig. 6.3). The evolution of the centre is discussed in detail by King (1955) and the rocks in the site can be grouped accordingly:

(a) Pre-caldera rocks preserved as remanié *blocks.*

Sedimentary and basaltic pre-caldera *remanié* masses occur at various localities within the complex. One of the key localities is in the Allt nan Dris (NR 935 333) where Rhaetian and other Upper Triassic strata crop out. Another is on the southern slopes of Ard Bheinn (NR 947 323) where small areas of fossiliferous Lower Lias sediments occur, as they also do close to the Permian and Triassic rocks in Allt nan Dris. Probable Upper Cretaceous limestones, now silicified and baked, are found in two localities, one forming a small cave marking the site of an old limestone quarry at Creag an Fheidh (NR 948 348), the other in a cave (Pigeon Cave) north-west of Binnean na h-Uaimh (NR 941 335). A large *remanié* mass of basaltic lava, about 100 m in thickness, occurs west of Binnein na h-Uaimh and Ard Bheinn; at the northern end of this mass; the basalt is seen to overlie altered chalk at Pigeon Cave. Many smaller masses exist elsewhere throughout the complex.

(b) Arcuate masses of volcanic breccias, granites, gabbros and felsites formed during the early evolution of the complex.

The principal members of the igneous complex are confocally arranged, arcuate masses within the caldera. These comprise large outer masses of vent agglomerate, a peripheral belt of granite, gabbroic strips north-west of Binnein na h-Uaimh (NR 943 335), a broad belt of felsite outcropping as arcuate masses with boss- and dyke-like characteristics and a lenticular granite mass.

The relative ages of the various pyroclastic formations within the complex are virtually impossible to determine. Distinct types occur in the peripheral zones and in the volcanic centres distinguished by King (1955). The peripheral caldera volcanic breccias are relatively uniform, the dominant inclusions being sedimentary and low-grade metamorphic fragments of Dalradian rocks and include quartzite, schistose grit and quartz–mica schists and vein-quartz. They are frequently well-rounded pebbles and cobbles, and are almost certainly derived from Devonian conglomerates, since fragments of these conglomerates are also common. Post-Devonian sediments are much more sporadic in their occurrence. Igneous fragments are irregularly distributed, but seldom predominate over the sedimentary clasts, and include basaltic lavas and felsites. Basalt is ubiquitous, but mugearite and feldspar-phyric basalts are also found in marked contrast to their very minor occurrence in the *remanié* masses of early, pre-complex lavas. Rare andesitic, gabbro, dolerite and granite clasts are also present. Some of the basalt fragments are very well-rounded and up to several metres in diameter. They are possibly the result of prolonged attrition in a vent (cf. Richey and Thomas, 1930; Reynolds, 1954). This suggestion is supported by steeply inclined, 'bedded' structures and intricate, sinuous, 'intrusive' contacts between agglomerates of varying character.

The peripheral granites occur as an arcuate belt from Dereneneach and Creag Mhor to lower Glen Craigag and a mass terminating to the north of Ballymichael Burn (King, 1955). These are medium- to fine-grained, pinkish biotite granites containing abundant orthoclase (commonly perthitic), plagioclase, quartz and biotite with epidote as a common secondary mineral. The granites are in contact with the volcanic breccias against which they are found to be chilled, although the junction is often tectonized and the granite finely comminuted. A further lenticular mass of granite associated with the ring complex occurs to the north of Binnean na h-Uaimh. This is a more melanocratic rock characterized by strongly zoned plagioclases and green hornblende; it may be of hybrid origin. The thermal metamorphic effects caused by the emplacement of this granite are particularly pronounced along its north-eastern and south-eastern boundaries, where adjacent basalts and gabbros are acidified and have developed clouded feldspars. Hybrid, dioritic rocks occur in the stream at Glenloig Bridge (NR 946 351).

(c) Lavas and pyroclastics erupted from discrete volcanic centres developed within the confines of the caldera.

The central, higher part of the area contains at

least four separate volcanic centres superimposed upon the earlier steeply inclined, arcuate intrusions and volcanic breccias. Each volcanic centre is identified by groups of gently inclined lavas and pyroclastic accumulations. These centres are briefly summarized below (King, 1955).

Ard Bheinn Centre (NR 945 329)

This is the largest and least eroded of the centres and contains abundant basaltic andesite, andesite and plagioclase porphyry lavas and pyroclastics. The andesites dip to the south and south-east at low angles, presenting a confocal pattern. They are truncated by later felsites and felsitic agglomerates.

Binnein na h–Uaimh Centre (NR 943 335)

Concentrically arranged basalts and andesitic pyroclastic deposits appear to be arranged about a single focus of activity. Agglomerates associated with the basalts are exceptionally varied and comprise a succession of layers with varying amounts of coarse sedimentary, basaltic, plagioclase porphyry and felsitic debris in the form of gently north-westerly inclined sheets.

Creag Dubh Centre (NR 952 338)

Basalts, andesites and associated breccias and porphyritic felsites are well represented at this centre as at Ard Bheinn. The arcuate form of the outcrops suggests several intersecting and overlapping lavas and pyroclastic layers attributable to a number of minor volcanic vents.

Creag an Fheidh Centre (NR 948 347)

The disruption of peripheral structures of the complex indicate that a volcanic centre was probably present here. The lithology is varied and similar to that encountered in the other centres with basalts, plagioclase porphyries and felsite being dominant, although no andesites have been found. Two foci are suggested by the arcuate disposition of the felsites.

(d) Late granite mass.

The broad belt of granite extending from the source area of the Ballymichael Burn to mid-Glen Craigag is later than all of the other members of the complex. It is part of a late central granite intrusion located east of the Ring Complex and now occupying a large arcuate area mostly outside the limits of the Ard Bheinn site. The petrography of the granite is similar to that of the earlier peripheral granites. Small areas of fine-grained gabbro occur on the north-west margin of the Binnein na h-Uamha area. These are older than the granite and are comminuted where in contact with volcanic breccias. Well-developed flow layering in the gabbro has an arcuate pattern which conforms to the general structure of the complex (King, 1955, p. 331) but is cut across by the contacts; King suggested that this gabbro is a remnant of a much larger intrusion.

Interpretation

The Ard Bheinn area of the Central Igneous Complex contains probably some of the best-preserved evidence for an eroded caldera structure in the British Tertiary Volcanic Province. The arcuate outlines of the complex are clearly defined by later granitic intrusions and the evidence for central subsidence is unequivocal: the numerous, occasionally fossiliferous, *remanié* masses of Mesozoic sediments and closely associated lavas, similar to plateau lavas elsewhere in the BTVP, do not occur elsewhere on Arran, the country rocks adjoining the complex range from Devonian to Permian in age. Thus, King (1955) estimates that subsidence of *c.* 1000 m may have occurred. The nearest (on shore) exposures of similar fossiliferous rocks are on the North Antrim coast *c.* 60 km distant and even further afield in the Western Isles (cf. Peach *et al.*, 1901). The inference drawn from their presence, together with the Tertiary lavas, is that Arran and the surrounding areas were covered by Rhaetian, Lower Jurassic and Cretaceous sediments and Palaeocene plateau basalts prior to post-Eocene and Pleistocene erosion. Despite their preservation within a caldera and the presence of numerous later intrusions, the sedimentary rocks generally show little alteration beyond induration, although skarn minerals have been recorded from the contact between metamorphosed chalk and agglomerates at Creagh an Fheidh (Cressey, 1987).

Ard Bheinn is a convincing example of a volcanic cone within the caldera postulated by King (1955). Here, and at Binnein na h-Uamha, there are well-defined flows of andesite and dacite, and associated pyroclastic rocks. The high proportion of rocks derived from magmas of intermediate compositions is unusual in the BTVP, where compositions are generally distinctly basaltic or granitic. From the evidence of hybrid dioritic rocks at Glenloig Bridge and of mixing of basic and acid magmas (as at 'Hybrid Hill', *c.* NS 978 980–outside the site), a hybrid origin for these rocks is a possibility but this would imply very thorough homogenization of the contrasted, mixed magmas prior to eruption. The origin of these rocks must remain open, for no detailed geochemical work has been published on them since King's account (1955); they were not included in the radiochemical and geochemical investigations made by Dickin *et al.* (1981).

The coarser-grained granitic rocks of the complex appear to have been intruded along the margins of the caldera, possibly in association with the caldera collapse, although King's (1955) identification of crushing along the margins indicates post-consolidation movements. Emplacement of the granitic rocks may also have resulted in doming of the country rocks; this is particularly clear from the Survey's mapping to the south and south-east of the site, whereas the relationships along the northern and north-western margins of the complex suggest that the intrusions cut across the domed structure around the earlier Northern Granite.

The particular value of this site lies in the preservation of features of surface and near-surface volcanicity within a caldera collapse structure. Elsewhere in the BTVP, the present level of erosion is between 1–2 km (or even more) below what was the Palaeocene land surface, so that we are usually dealing with the eroded roots of the volcanoes. In this site much of the volcanic superstructure is preserved, and through the observations made here it is becoming increasingly recognized that some of the relicts of surface volcanism found in other central complexes (for example, Beinn na Caillich–Kilchrist on Skye) may also owe their preservation to subsidence within calderas. A positive gravity anomaly over the area indicates that it is underlain by mafic rocks at depth (McQuillin and Tuson, 1963).

The sequence of events within the Central Igneous Complex, and in particular this site, may be summarized as follows: after emplacement of the Northern Granite, a major igneous centre developed to the south. The initial activity involved formation of a caldera with central block subsidence of about 1 km and a diameter of at least 5 km. In the north-west, wall rocks collapsing into the caldera included olivine basalts from Tertiary lavas, which probably covered Arran and surrounding areas in the Palaeocene, and masses of fossiliferous Rhaetian and Lower Jurassic sedimentary rocks, together with Permian and Cretaceous rocks. At least four volcanic cones became established on the caldera floor and built up through the effusion of intermediate and acid lavas and pyroclastic deposits. Subsequently, granite and dioritic rocks were intruded along the ring fracture around the margins of the caldera. The status of gabbroic rocks in the north-west of the site, near Binnein na h-Uamha, is uncertain, but it may be a relict of a larger, pre-granite mass; the presence of the positive Bouguer gravity anomaly over the complex indicates that there are appreciable amounts of dense gabbroic rocks beneath the area.

Conclusions

The Ard Bheinn area contains a wide range of intrusive and extrusive igneous rocks with which are associated masses of fossiliferous and non-fossiliferous sediments, ranging in age from the Permian to the Cretaceous. This perplexing mixture of rock types is the result of surface and near-surface rocks being downfaulted within a major volcanic collapse structure, or caldera, which was subsequently intruded along its margins by granitic and other ring-dykes.

GLEN CATACOL

Highlights

The Northern Granite intrusions of Arran include a spectacular example of a diapir, uniquely well preserved in the British Tertiary Volcanic Province. The Dalradian country rocks have undergone extensive ductile deformation, with concentric folds formed around the granite. Excellent sharp intrusive contacts of the latter with the Dalradian are common within the site.

Introduction

The north-western part of the Northern Granite and adjoining updomed and metamorphosed

Dalradian metasediments lie within this site, which includes the glacially eroded and deepened Glen Catacol and much of the higher ground of the Bheinn Bharrain-Meall nan Damh ridge. Two distinct biotite granites comprise the Northern Granite, a coarse-grained Outer Granite often in chilled, intrusive contact with the Dalradian, and a younger fine-grained Inner Granite which has sharp intrusive contacts towards the Outer Granite. The Outer Granite forms the spectacular mountainous scenery of northern Arran (Fig. 6.1), whereas the Inner Granite is less resistant to weathering and is characterized by lower, more rounded hills. Granite–granite and granite–metasediment contacts are generally steep, a feature well displayed within the site which has over 700 m of relief. Post- and pre-granite basaltic dykes and post-granite felsitic dykes also occur throughout the site.

Early investigations on the Northern Granite (MacCulloch, 1819; Ramsay, 1841; Zirkel, 1871; Smith, 1896) were succeeded by those of the Geological Survey with the publication by Gunn (1903) and Tyrrell (1928) of the standard works on the geology of Arran. The form of the Northern Granite was considered by Bailey (1926) and subsequent work by Flett (1942) detailed the nature of the contact between the coarse- and fine-grained granites. More recently, the structures associated with the intrusion of the granite have received detailed attention from Woodcock and Underhill (1987) who have produced a new model for the high-level emplacement of the pluton; this topic has been further pursued by England (1988, 1990).

Description

The Outer Granite is a coarse-grained rock dominated by quartz and alkali feldspar accompanied by minor amounts of plagioclase and biotite. It is well exposed in the mountainous areas of the site and in the lower reaches of the River Catacol. Coarse granite, with characteristic rough weathering and widely spaced, subhorizontal and subvertical joint sets, is well-exposed on Meall nan Leac Sleamhuinn (NR 916 479), Madadh Lounie (NR 925 487) and Creag na h-Iolaire (NR 926 475) in close proximity to Dalradian country rocks (Fig. 6.4). The strong jointing, which often parallels the valley floors and sides, has been attributed to the unloading effects during the weathering of the granite. The Inner Granite has a similar mineralogy to the Outer Granite but is much finer grained and is frequently conspicuously drusy; the small cavities become especially numerous near the contacts with the earlier Outer Granite. The Inner Granite is occasionally granophyric, as at the confluence of the Allt nan Calman (NR 918 454). The intrusive relationship of the Inner Granite towards the Outer Granite has been well documented by Flett (1942). Intrusive contacts are exposed close to the confluence of the Diomhan and Catacol rivers (*c.* NR 923 469) on the south-east side of Meall nan Damh, in the col between that hill and Mheall Bhig, where thin sheets and veins of fine-grained granite cut the Outer Granite, and some 3 km south of Glas Choirein. The contacts are generally steeply dipping, with the Inner Granite passing under the Outer Granite.

Accessory minerals are fairly rare in both granites; they include magnetite, apatite, zircon and fluorite. However, both granites are noted for the variety of well-crystallized minerals lining the drusy cavities, which include feldspar, dark mica, quartz (clear, amethystine and Cairngorm varieties), pale blue to yellowish topaz and blue-green beryl.

Immediately to the north of Loch Tanna, fine-grained Inner Granite exposed in the river bed is slightly variable in grain size. Thin aplite veins and sheets are common in both granites and examples may be seen cutting the Outer Granite to the north-east of Meall nan Leac-Sleamhuinn (NR 915 478).

Dalradian metasediments crop out in the northern and western parts of the site in contact with the Outer Granite and provide important evidence that the granite was emplaced diapirically. The metasediments vary from coarse conglomerates to greenish-grey phyllites, arenaceous schists and dark slates. An impure schistose limestone associated with dark-grey graphitic slates is intermittently exposed between Pirnmill and Catacol. Details of the stratigraphy of these rocks are given by Anderson (1945), a summary of which appears in Macgregor (1965). The metasediments have been subdivided into:

a. North Sannox Grits (6 subdivisions); and
b. Loch Ranza Slates.

The Loch Ranza Slates occupy the core of the Loch Ranza Anticline, which is thought to be coincident with the Aberfoyle Anticline (also see

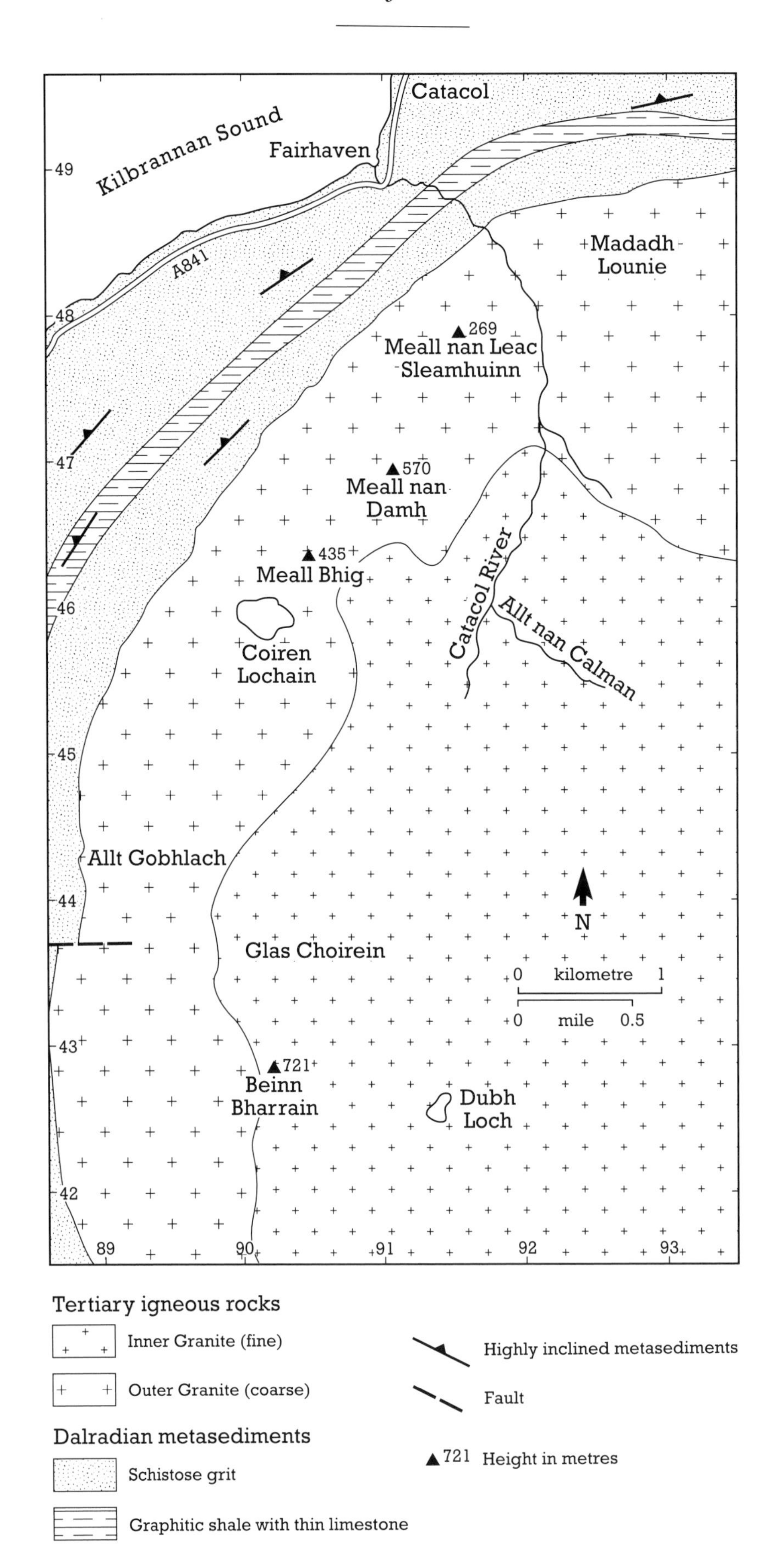

Catacol
Kilbrannan Sound
Fairhaven
A841
Madadh Lounie
269
Meall nan Leac Sleamhuinn
570
Meall nan Damh
435
Meall Bhig
Catacol River
Allt nan Calman
Coiren Lochain
Allt Gobhlach
N
Glas Choirein
0 kilometre 1
0 mile 0.5
721
Beinn Bharrain
Dubh Loch
49
48
47
46
45
44
43
42
89
90
91
92
93
Tertiary igneous rocks
Inner Granite (fine)
Outer Granite (coarse)
Dalradian metasediments
Schistose grit
Graphitic shale with thin limestone
Highly inclined metasediments
Fault
721 Height in metres

Johnston, *in* Craig, 1965) and lies within the Barrovian chlorite zone. The Catacol Synform to the north-east, on the other hand, has resulted from the deformation of the Dalradian rocks by the forcible emplacement of the granite and the steeply inclined Dalradian metasediments are frequently deflected into parallelism with the granite contacts.

Within 50–100 m of the granite, the Dalradian rocks become almost vertical in attitude and are often seen to have been considerably metamorphosed thermally, with resultant overprinting of their cleavage. At Fairhaven, Madadh Lounie, on the Allt Gobhlach and along the coast to the west of the site, the schists provide evidence for the updoming effects of the granite and dip north-westwards away from the granite, with minor folding parallel to strike of the contact. On the east side of Glen Catacol at Madadh Lounie, fine-grained Outer Granite is in sharp contact with schists in the prominent spurs along the cliff face. Dark schists are thermally altered in a zone of about 75 m at the contact and they take on a bluish hue. Occasionally, fine-grained veins and offshoots of granite penetrate the country rock and blocks of Dalradian, probably detached by stoping, are sometimes completely engulfed by granite (McKerrow and Atkins, 1985). In the Allt Gobhlach (at NR 888 439) granite is in steep contact with indurated quartzose schists and the metasediments within a few metres of the granite have developed epidote, biotite, cordierite and andalusite. West of Glen Catacol, the contact of granite and schist is well exposed on the northern side of Meall nan Leac Sleamhuinn (NR 913 483) where the marginal granite contains several metasedimentary xenoliths.

Minor intrusions, widely scattered throughout the site, include basic dykes which were intruded before and after the emplacement of the Northern Granite. A good example of a pre-granite dyke is seen near to the granite contact in the stream issuing from Lochan a'Mhill (NR 908 482). It has been substantially baked and the original clinopyroxene has been replaced by hornblende and accessory biotite (Gunn, 1903). Basaltic dykes intrusive into both varieties of granite trend NW–SE and can be observed in the Catacol River, some of its tributaries and along the better-exposed parts of Beinn Bharrain (NR 898 423). Felsitic dykes are less common but do intrude both granites, and examples occur near to the basic dyke at Lochan a'Mhill and within the granite of Glen Catacol. A small pitchstone sheet cuts the Inner Granite 500 m north-west of Dubh Loch (NR 908 428). It appears that the majority of acid dykes within the granite are younger than the basic dykes.

Figure 6.4 Geological map of the Glen Catacol site (adapted from the British Geological Survey 1:50 000 Special District Sheet, Arran).

Interpretation

The Northern Granite is the most spectacular pre-central complex granite pluton in the British Tertiary Volcanic Province and is thus a very significant feature in the geology of Arran. The site is chosen in preference to the more scenically spectacular eastern part of the mass because both granites, numerous minor intrusions and granite contacts with upturned Dalradian metasediments, are exposed to advantage here. The area around Catacol provides a clear demonstration of the local tectonics associated with the diapiric emplacement of the pluton, such as the pronounced updoming of the Dalradian country rock and accompanying annular Catacol Synform. The intrusion has a cylindrical, plug-like form with outwardly dipping steep margins. Its margins have unusual characteristics for an intrusion of this size (McKerrow and Atkins, 1985): features such as the limited, sometimes very narrow thermal aureole, the poorly developed or absent chilled margin, a general absence of veins of late-stage residua in the country rock and the total lack of mineralization in the country rock. These features and the presence of many faults on the eastern side of the intrusion which downthrow away from it, suggest that the final emplacement of the diapir occurred when the granite was nearly solid. The diapir therefore caused brittle and ductile deformation in the country rock and was too cool and too 'dry' to effect substantial thermal metamorphism and mineralization in the country rock. However, the contacts in the Catacol area differ from this overall interpretation in that the junction is highly irregular and the veining of the country rock, together with evidence for stoping, suggest that here at least the granite was still at, or near to, liquidus temperatures when it was emplaced. An alternative model for the emplacement of the pluton has been suggested by Woodcock and Underhill (1987), based on detailed investigations of the deformation of the country rock. A two-stage ballooning model is suggested, following the arrest of the uprising pluton at the Dalradian–New Red Sandstone unconformity (exposed on the eastern side of the pluton). Initially the arrested pluton steadily inflated, causing radial stretching and normal faulting. This was followed by continued ballooning of the pluton, resulting in updoming of the country rocks, with further upwards and outwards movement following the intrusion of the Inner Granite. This model, however, does not explain the unusual contact features discussed above, in particular the irregular veining of the Dalradian rocks and the evidence for stoping. England (1990) contends that the Outer Granite has most of the characteristics of a diapir and some of a ballooning pluton.

Walker (1975) considered that the emplacement of granite diapirs preceded the development of all of the BTVP central complexes, that of Arran being the only one which is fully exposed for study. He envisaged a sequence of events, starting with emplacement of hot, basaltic magma into the base of the crust. This gave rise to low-density acid magma which rose as a diapir, closely followed by basalt channelled into a magma trap below the diapir. The diapir was augmented by new magma formed by fractionation of the basalt, by partial melting of crustal rocks or, most probably, a combination of the two (see 'Introduction', Chapter 1). In the brittle, upper crust, the country rocks were deformed by folding and faulting and later events included the formation of cone-sheet swarms and 'confluent cone-sheet type' bodies of gabbro beneath the diapirs (Walker, 1975, Figs 5 and 6). Basalt eventually broke through the diapir to form ring-dykes of gabbro which may be so numerous as to obliterate much of the evidence for the original diapir (for example, Mull, Skye). In north Arran the later stages did not develop, and in this area there are relatively few basaltic intrusions which post-date the granites, although the gravity data suggest that there is underlying, dense (gabbroic) rock (Tuson, 1959).

The sharp, steep contacts of the Inner Granite with completely undeformed coarse Outer Granite suggest that the later Inner Granite was emplaced in a permissive manner by ring-faulting and block subsidence.

Conclusions

The Outer Granite of the Northern Granite of Arran is the most spectacular diapiric intrusion of granite in the British Tertiary Volcanic Province. Intrusion of the Outer Granite has deformed the country rocks, causing them to develop steep, radial dips off the intrusions and also to become folded parallel to the granite contact. The granite probably moved upwards as a nearly solid body in the final stages, although some mobile granite magma did remain and formed veins in the country rocks. The Inner Granite was intruded without any deformation of the earlier, coarse-grained, Outer Granite.

DRUMADOON–TORMORE

Highlights

Five major pitchstone dykes ('Judd's Dykes') are excellently exposed in coastal outcrops. These dykes and other sills and sheets are exceptionally clear examples of composite (basic–acid) intrusions, where coexisting acid and basic magmas have interacted to produce hybrid rocks. Petrographic and chemical evidence indicates that the basic rocks were also hybridized before they were intruded.

Introduction

The coastal section between Drumadoon Point and Tormore provides good sections through a number of dykes and sills of quartz–feldspar porphyry, felsite and pitchstone. Many of these intrusions are composite and contain tholeiitic basaltic components. The 4-km-long shore section has two principal interests – the composite dykes in the King's Cave (NR 884 309) to Leacan Ruadh area (NR 887 322) and composite and acid sills at Drumadoon.

The composite dykes around Tormore were first systematically described by Judd (1893) in the now classic publication on these intrusions. Further work on these dykes and the sills at Tormore was carried out by the Survey (Tyrrell, 1928). McKerrow and Atkins's guide (1985) contains useful discussions on the origin of the Drumadoon intrusions, and a recent study on the hybridization and petrogenesis of the composite dykes, in particular the one at An Cumhann, has

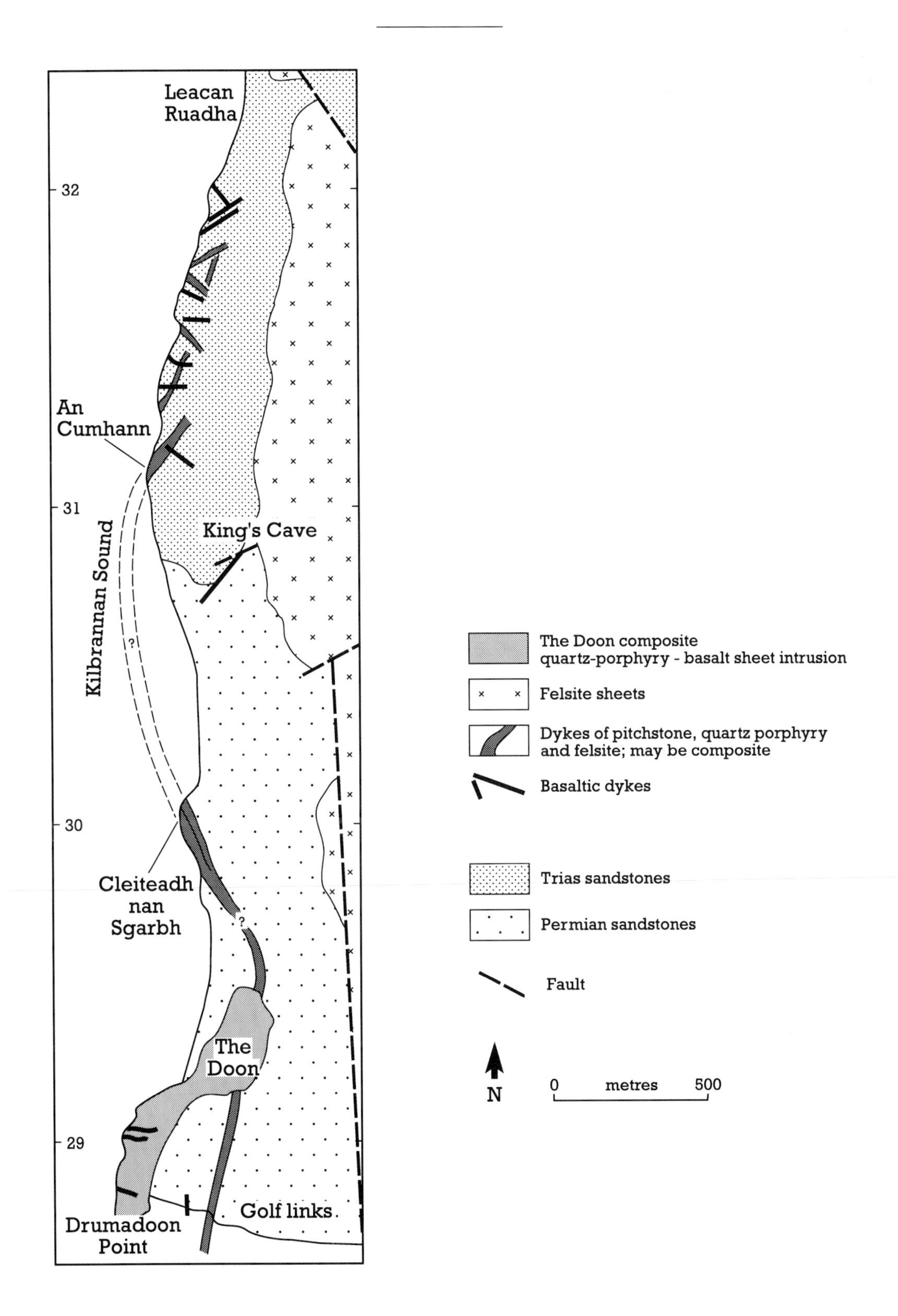

Figure 6.5 Geological map of the Drumadoon–Tormore site (adapted from the British Geological Survey 1:50 000 Special District Sheet, Arran, with additional information from McKerrow and Atkins, 1985, figure 11a).

been published by Kanaris-Sotiriou and Gibb (1985).

Description

Five major composite dykes on the shore some 2 km south-west of Tormore, between Kings Cave and Leacan Ruadha, are commonly referred to as Judd's Dykes I–V (Fig. 6.5). They are composed of quartz–feldspar porphyry, pitchstones and tholeiitic dolerite intruded into Permo-Triassic sediments.

The dykes are exposed on the wave-cut platform and also in the raised beach cliffs where the pitchstone and felsite dykes and sills in the soft sediments are frequently highly devitrified and spherulitic.

At An Cumhann (NR 884 312), a 30-m-thick NNE-trending composite intrusion (Judd IV) forms a low cliff (see Kanaris-Sotiriou and Gibb, 1985). It consists of quartz-feldspar porphyry flanked, and also intruded centrally by tholeiitic basalt which contains rounded xenocrysts of alkali feldspar and quartz. Locally, the basic rocks show marginal shearing and both components are cut by later, thin basaltic dykes and sheets. Bleached, indurated sandstones form a marginal ridge about a metre in width, a characteristic feature in all the intrusions of this section. An irregular hybrid zone occurs between the felsite and the basaltic members. To the north, the next dyke is Judd I, consisting of a central 5-m-wide greenish pitchstone with banded spherulitic felsite and tholeiitic margins. This dyke can be traced NNE for nearly 200 m along the shore and into the cliffs. Judd III is a 15 m wide, NW-trending dyke with a central 1–2-m-thick pitchstone with an olivine dolerite in its northern edge. Judd III, which assumes a horizontal, almost sill-like, aspect in places, has provided many of the beautifully flow-banded, dark greyish-green pitchstone boulders found on the wave-cut platform hereabouts. Towards the north end of the section, nearer Tormore, Judd II is present as a 10 m thick quartz-phyric felsite dyke bordered by spheroidally weathered dolerite; Judd V is exposed nearby, being a conspicuous pitchstone dyke on the shore. Both Judd II and V trend east–west and cannot be traced far inland. The felsites contain phenocrysts identical to those in the pitchstones and are most probably their altered, devitrified equivalents. Fresh pitchstones contain phenocrysts of quartz, andesine, ferroaugite, fayalitic olivine, hypersthene and Fe–Ti oxides. Running subparallel to this section, devitrified pitchstone and spherulitic felsite form at least two sills in the cliffs to the east and inland at Torr Righ Mor (NR 888 300), appearing at various levels due to minor faulting. The upper felsite is a continuation of the extensive Blackwaterfoot Felsite to the south and is not cut by any basic dykes.

The composite quartz-feldspar porphyry and tholeiitic basalt sill at Drumadoon is situated at the south end of the site. This 25–30-m-thick sill forms The Doon (NR 885 293), a spectacular columnar feature (Fig. 6.6) and extends south-west to Drumadoon Point, where the transgressive sill changes attitude to a more dyke-like form. The main part of the intrusion is formed by a quartz–feldspar porphyry with conspicuous feldspars up to 20 mm long, and smaller, glassy quartz crystals set in a pale felsitic matrix which becomes more basic next to the tholeiite margins. The lower part of the porphyry is sometimes crowded with rounded, lobate basic inclusions. The marginal tholeiitic sheets are 1–1.25 m thick and have sharply defined contacts with the acid component. Both the lower tholeiite and the rather poorly exposed upper tholeiite contain conspicuous quartz and alkali-feldspar xenocrysts, as do the lobate basic inclusions in the porphyry member.

About 400 m NNW of the northern end of Drumadoon, the small headland of Cleithadh nan Sgarbh consists of a thick porphyritic felsite with prominent flow-banding, a composite quartz porphyry–-dolerite sheet and a later, cross-cutting tholeiitic dyke. Xenocryst-bearing dolerite at the edge of the composite intrusion intrudes the flow-banded porphyritic felsite. Within the composite intrusion there is a thin zone of acid hybrid developed between the basic margin and the porphyry; the latter carries xenocryst-bearing, rounded and lobate basic xenoliths. The composite intrusion has virtually the same petrological characteristics as Judd IV at An Cumhann and may be its southerly extension (see Kanaris-Sotiriou and Gibb, 1985). It can be traced south-eastwards into sea cliffs at the edge of the raised beach and may link with the porphyritic felsite exposed on the beach south of Blackwaterfoot golf course, about 400 m east of Drumadoon Point. The xenocryst-free dolerite dyke is a multiple intrusion cutting the composite sheet and the flow-banded felsite. The dyke splits into several sheets in the felsite where it appears to

Figure 6.6 Columnar jointing in the composite sill, The Doon. Drumadoon–Tormore site, Arran. (Photo: A.P. McKirdy.)

have caused localized melting of the acid rock.

Minor intrusions fail to cut the (presumably later) flow-banded felsite at Cleitheadh nan Sgarbh and are absent within the felsite sheets in the raised beach cliffs. However, the numerous basic dykes observed in this section cut sediments and composite intrusions alike.

Interpretation

The Drumadoon to Tormore coastal section provides classic, well-exposed and easily accessible examples of the composite acid–basic intrusions which commonly occur on Arran. The importance of the site has been realized since the early studies of Judd (1893) and, as a consequence, the section is frequently visited and has unfortunately suffered heavy damage through indiscriminate hammering of the attractive glassy pitchstones.

The site clearly demonstrates that acid and basic magmas existed in this area at the same time, an important feature frequently noted in the BTVP (for example, Skye and Mull). A model for the petrogenesis of the composite intrusions in this area based upon a study of the dyke at An Cumhann, a possible feeder to the Drumadoon Sill, has been proposed by Kanaris-Sotiriou and Gibb (1985):

a. A differentiated, partially crystallized body of acid magma, bearing quartz, alkali feldspar and plagioclase phenocrysts existed at depth in the area.
b. Rising basic magma encountered the acid magma and passed through while generally retaining its identity; mixing was inhibited by the contrasting viscosity, composition and temperature properties of the two magmas. The basic magma, however, was partly hybridized by the incorporation of matrix and phenocrysts from the acid magma as xenocrysts.
c. The basic hybrid magma rose to higher levels and intruded Triassic sediments as dykes and sills of dolerite. Xenocrysts (especially alkali

feldspar) were resorbed. Flow differentiation in the dykes/sills resulted in the concentration of xenocrysts towards the centre.

d. Acid magma was subsequently emplaced along planes of weakness into the unconsolidated centres of the hybrid basic dykes/sills. Contamination of the acid porphyritic magma occurred at the contacts with the dolerite, by *in situ* assimilation of the basic rock and the incorporation of the basic rock as xenoliths.

e. The acid magma continued to fill the centres of the intrusion and became progressively more acidic as more evolved magma was tapped from the reservoir.

Kanaris-Sotiriou and Gibb (1985) suggested that all of the composite intrusions in this area formed in this way and all are related to a common source. They point out that composite sheets with cores of quartz–feldspar porphyry and dolerite margins are also developed in the immediate vicinity of other major granite bodies in the BTVP and cite Skye and the Mourne Mountains as examples. The inescapable conclusion seems to be that the composite sheets are formed by a single fundamental mechanism rather than by the chance association of partly crystallized acid magma and basalt magma. They do not consider that the acid magma was derived from the basic magma by differentiation. Possibly it was generated by remelting, or partial melting, of an earlier Tertiary granite.

Conclusions

The site contains excellent examples of glassy, flow-banded pitchstones some of which form the central parts of composite acid–basic dykes. The composite dykes, sheets and sills provide clear evidence that basaltic and granitic liquids were in existence at the same time in this area and were intruded in quick succession (basaltic magma, followed by acid magma). The crystals found as phenocrysts in the pitchstones, and particularly in the quartz–feldspar porphyries, are also present as xenocrysts in the enclosing, earlier dolerites. This indicates that the basalt magma mingled with partially crystallized acid magma before it was intruded; the basaltic and doleritic margins thus have some of the features of hybrid (mixed magma) rocks.

DIPPIN HEAD

Highlights

This site has exposures of the large alkaline Dippin Sill which contains crinanite, teschenite and pegmatite components and baked Triassic country rock. Primary nepheline, present in some of the components, indicates the undersaturated nature of some of the margins.

Introduction

The Dippin Head site represents an important locality for the exposure of the Dippin Sill – a basic, compositionally variable intrusion. It is an important member of the suite of minor intrusions seen in south-east Arran. The sill lies within baked Triassic marls and is intruded at this locality by a large doleritic dyke (Fig. 6.7).

The petrology of the sill has been described in a detailed study by Gibb and Henderson (1978a; 1978b), who proposed a model for the petrogenesis of the magma which fed the intrusion. An earlier description of the sill is also contained in the Arran Memoir (Tyrrell, 1928).

Description

The Dippin Sill crops out at Dippin Head (NS 050 222) and extends beyond the limits of the site between Cnoc na Comhairle (NS 036 240) and Cnocan Biorach (NS 034 222) and beneath much of the ground to the west. At Dippin Head, the sill attains a thickness of approximately 36 m and overlies highly baked Triassic marls. Here the sill is intruded by a thick, sparsely feldspar-phyric dolerite dyke with conspicuous tachylitic margins. Both intrusions display columnar jointing, which is vertical in the sill and horizontal in the dyke. The sill has the structure of a slightly transgressive sheet dipping to the south-east and thinning to the south-west, west and north. A thickness of 42.8 m was recorded from a borehole, in a stream at NS 043 228 (Gibb and Henderson, 1978a), which intersected both the roof and the floor of the intrusion.

Gibb and Henderson (1978a) described four main rock types within the sill, each variety having sharp, but unchilled, internal contacts, suggesting that the sill was probably intruded as a single event. These varieties and their petrologi-

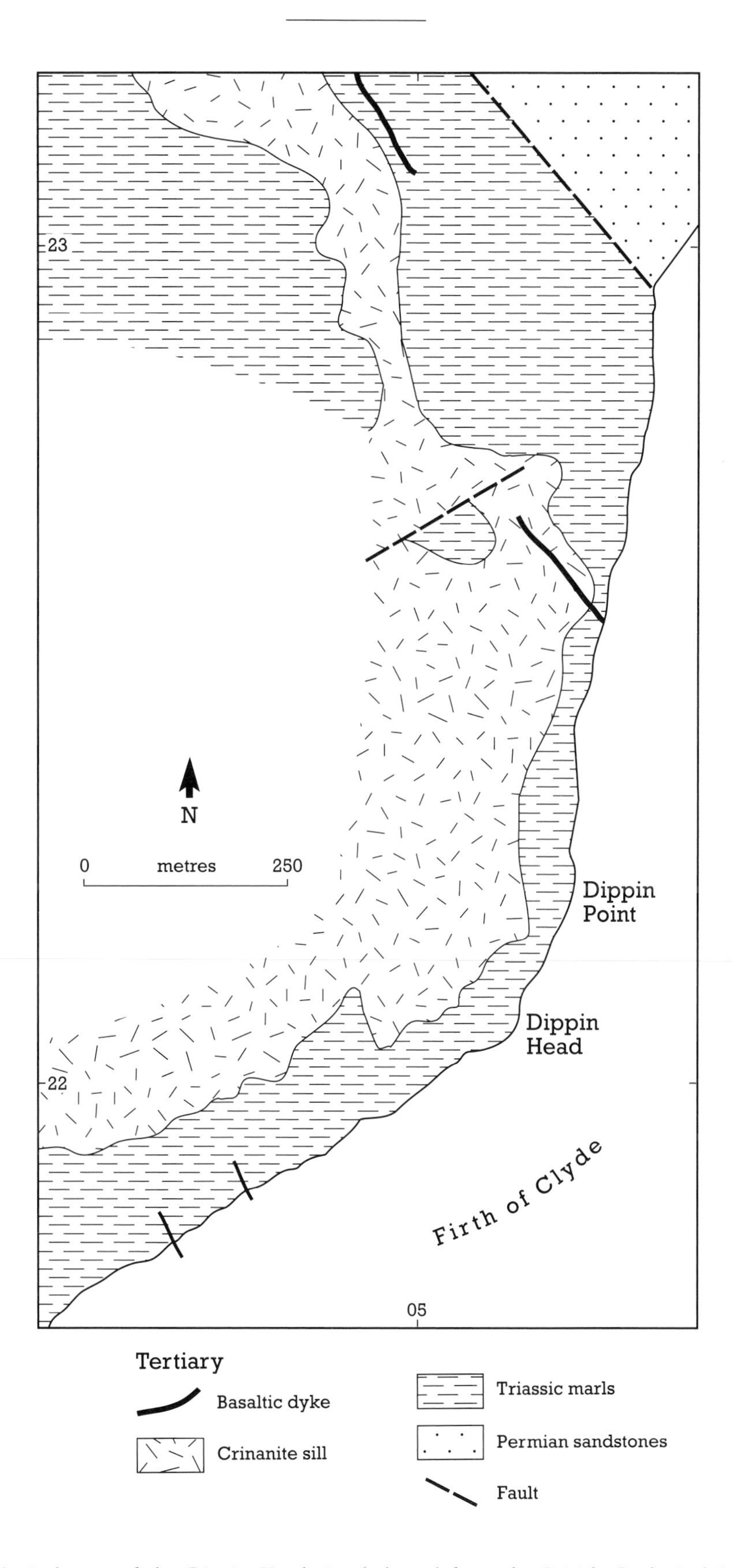

Figure 6.7 Geological map of the Dippin Head site (adapted from the British Geological Survey 1:50 000 Special District Sheet, Arran).

Table 6.2 Petrological variation within the Dippin Sill (based on Gibb and Henderson, 1978b, figure 4)

Rock type	Position within sill	Petrological features
(a) Crinanite	Central = forms the bulk of the intrusion	Plagioclase, analcite, olivine, ophitic Al-, Ti-rich augite. Zeolites. Analcite, secondary after nepheline and of hydrothermal origin. Olivine up to 12 vol.% about 10–15 m above base.
(b) Teschenite	Marginal facies = fine-grained margins showing quench textures	Lacks fresh olivine, substantial amounts of analcite, zeolites and calcite. Margins have skeletal Ti-augites.
(c) Augite teschenite	Patches within crinanites, especially towards base.	Augite, plagioclase, analcite. Alignment of augite suggests cumulate texture. Fe–Ti oxides more abundant than in crinanite.
(d) Pegmatite(i)	At several horizons throughout sill, centimetres to metres in thickness	Brown augite with emerald-green rims (Na-rich), plagioclase, analcite, Fe-oxides, apatite, rare blue riebeckitic amphibole and rare olivine pseudomorphs. Variant of augite teschenite.
(e) Pegmatite (ii)	As pegmatite (i)	Mineralogically as (i) but has less pyroxene and is much coarser grained. Skeletal magnetite and ophitic augite, rather than euhedral as in (i).

cal characteristics are listed in Table 6.2. Analcite, common in this and other similar intrusions, occurs in part after primary nepheline.

The large dolerite dyke at the north end of the cliff near the coast has good tachylitic margins against the sill rocks. It has developed a strong, flat-lying, columnar jointing.

Interpretation

Large basic alkaline sheets occur in Arran and elsewhere in the BTVP (Rubha Hunish, Skye, Shiant Isles). The one which is admirably exposed within this site, the Dippin Sill, is an excellent example of the characteristic petrological features of these intrusions.

Gibb and Henderson's (1978a; 1978b) detailed mineralogical and geochemical investigations were carried out on continuous drill-core samples obtained from a locality inland from the site. They showed that the distinctive rock types were generally sharply defined against one another, but there were no chilled contacts apart from those against the sediments; consequently, the sill must have built up from the injection of a series of compositionally contrasted pulses of magma which followed each other in rapid succession. This suggests that the source magma chamber was probably stratified at the time of sill injection. The sequence of events envisaged by Gibb and Henderson was as follows:

1. Alkali olivine basalt magma in a deep crustal reservoir underwent crystallization. Olivines settled towards the base of the magma under gravity and lighter, residual and more fractionated magma accumulated towards the top of the reservoir, which thus became compositionally zoned.
2. Fractionated magma from the top of the body was released and intruded at a high crustal level to form a sill. Reaction with the sediments modified the first-injected magma.
3. Successively less-fractionated, increasingly olivine-rich magmas were subsequently in-

jected into the central parts of the sill, where some *in situ* fractionation and flowage differentiation occurred, giving rise to further compositional variation in the high-level sill intrusion. The pegmatitic areas formed by segregation of residual magma after emplacement.

The sill is therefore important because it provides evidence for magmatic fractionation both prior to intrusion and also within the sill, after the injection of the pulses of closely-related, but compositionally differing, magmas. Henderson and Gibb's (1977) petrographic and mineralogical studies clearly show that some of the analcite in the sill is secondary after original nepheline. Analcite is very common in alkaline olivine dolerites throughout the BTVP, where it has been considered as a primary phase (for example, Harker, 1904). This is demonstrably not so in the Dippin Sill and it is likely that at least some analcite in other similar sills is of replacement origin, after original nepheline (as well as, for example, plagioclase).

Conclusions

The Dippin Sill shows considerable compositional variation from essentially olivine-free dolerites at the margins to olivine-enriched dolerites in the central parts. It was fed in a series of pulses from a compositionally zoned magma chamber at depth and, after emplacement, some of the magma underwent further segregation, for example, forming small pegmatitic patches. Nepheline crystallized from the magma and some of this has survived replacement by analcite; preservation of original nepheline is unusual in alkali-olivine dolerites in the BTVP.

SOUTH COAST OF ARRAN

Highlights

The site contains the best-exposed dyke swarm in the British Tertiary Volcanic Province and arguably one of the best examples in the world. The nearby Bennan Head composite sill has an exceptionally thick lower margin, and both its margins contain excellent examples of xenocrysts derived from granitic magma.

Introduction

The site encompasses the littoral shore zone from Cleitheadh Buidh east for about 5 km to Kildonan Castle, and the cliff at and immediately west of Bennan Head. The section cuts across the south Arran dyke swarm (consisting here of about 200 basalt and dolerite dykes), forming one of the best-exposed sections through a Tertiary dyke swarm (Figs 6.8 and 6.9). The adjoining Bennan Head composite sill is extremely well exposed around the base of the headland and in the Struey Water to the west; the upper basic member is highly xenocrystic and the lower part of the central quartz porphyry shows signs of hybridization. Baked Triassic sandstones are the country rocks in this area.

The dyke swarm of southern Arran has been recognized for many years, one of the earliest studies being that of Necker de Saussure (1840). Knapp (1973) has studied the form and structure of the dykes and Halsall (1978) has investigated the emplacement and compositional variation of the dykes within this site along the Kildonan shore. A palaeomagnetic study of the swarm has been carried out by Dagley *et al.* (1978).

Description

The site comprises a 10-km-long coastal section from Cleitheadh Buidh (NR 956 208) to Kildonan (NS 037 208), including the sea cliffs near Bennan Head. The exposures of the dyke swarm are principally on the shore (Figs 6.8 and 6.9); the composite sill is seen in the cliffs at Bennan Head and in small quarries to the north.

The southern Arran dyke swarm in this area has a dominant NW–SE trend, although north–south dykes are also abundant and there are subsidiary NE–SW-orientated dykes. Generally, the dykes are subvertical, but some have inclinations of between 60° and 70° to the north-east. Thicknesses vary from a few centimetres up to 30 m. The entire section contains about 200 dykes, representing a crustal dilation of about 10%. Some of the dykes bifurcate and, in several instances, dykes intersect. The shore section reveals clear, 'textbook' examples of features associated with dyke intrusion, such as prominent chilled margins, bedding offsets across dykes, flow-banding, vesicles and amygdales, jointing and dyke offsets.

Various rock types are represented by the dyke swarm, the commonest being: transitional alka-

Figure 6.8 Dyke swarm on the foreshore at Kildonan. The dykes weather out to form reefs; the softer Triassic sandstone in between has been eroded back. South Coast of Arran site, Arran. (Photo: C.H. Emeleus.)

Figure 6.9 Dolerite dykes forming part of the Arran dyke swarm on the shore below Kildonan Castle (NS 037 209). South coast of Arran site, Arran. (Photo: C.J. MacFadyen.)

line olivine dolerite, alkali olivine dolerite (crinanite and teschenite), tholeiitic olivine dolerite and quartz dolerite.

Red and white Triassic sandstones and coarser clastic rocks are the country rocks; they are often bleached and even occasionally metamorphosed to form tough, grey hornfelses which may contain quartz paramorphs after tridymite. The baked margins sometimes stand up as low walls along the dyke margins.

The major, thick composite sill of Brennan Head, with a central quartz–feldspar porphyry component, is flanked above and below by tholeiite. The sill is similar in many respects to the Drumadoon Sill. The base of the intrusion contains some large xenoliths, or rafts, of sandstone and, in the lower part of the porphyry, dark xenolithic bands of dolerite occur near to a small waterfall above the main Struey Falls (NR 993 203). The sections in and near Struey Burn (NR 994 205) show the acidic component to be more melanocratic than normal, and there is evidence for hybridization both by magma mixing and assimilation of xenolithic material. The marginal tholeiitic members, particularly the upper one, carry abundant quartz and alkali-feldspar xenocrysts (cf. Drumadoon Sill) and also scattered hypersthene crystals; the lower marginal tholeiite is exceptionally thick (>10 m).

Interpretation

The site contains the best example of a dyke swarm in the BTVP, including many of the 'textbook' features of dykes and a considerable range in compositions. Consequently, it is a classic section, with international status, and is frequently visited for educational purposes.

Halsall (1978) distinguished ten episodes of dyke and sill emplacement, based on successive, cross-cutting phases of intrusion. He showed that there was a general trend from alkali-basaltic to tholeiitic magmas with time, possibly attributable to increased partial melting of a mantle source as the swarm was intruded. However, superimposed on this simple pattern there was also random intrusion of olivine dolerite, tholeiitic olivine dolerite and tholeiitic dolerite dykes throughout the entire period, suggesting crustal magmatic plumbing similar to that envisaged for the Mull and Skye lavas (Chapter 1).

A palaeomagnetic study of the dykes (Dagley *et al.*, 1978) showed the majority to have reversed magnetization (>75%) and that the proportion showing reversed polarity increased from west to east. There is no simple correlation between polarity and petrography and neither is there a significant difference between mean pole directions of the alkali-basalt and tholeiitic intrusions. Overall, it appears that the dykes span three or four magnetic polarity episodes, reversed–normal–reversed or (more probably) reversed–normal–reversed–normal (Dagley *et al.*, 1978), during a period of about three million years, or four if four polarity periods were involved. Reliable radiometric age determinations are not as yet available for the dykes, but they pre-date the Holy Island trachyte which has been dated at about 58.5 Ma (Macintyre, 1973).

The high percentage of crustal dilation represented by the dykes indicates that Arran lay on a zone of crustal extension and possible thinning during the Palaeocene. Speight *et al.* (1982) consider that the dyke swarms formed above ridge-like, basaltic, magma chambers situated in the lower crust and extending into the mantle. They regard the ridges as fundamental structures in the BTVP, forming in the crust beneath the region in response to shearing stresses associated with the opening of the North Atlantic, and they compare the pattern of dyke swarms in the BTVP with similar, but smaller-scale patterns formed by quartz-filled tension gashes (cf. Speight *et al.*, 1982, Fig. 33.5).

The Bennan Head composite sill intrusion has many features in common with the composite intrusions of the Drumadoon–Tormore site. The principal differences are the thick, lower basaltic and doleritic member at Bennan and the occurrence of fairly numerous baked sedimentary xenoliths near the floor of the intrusion. Intrusion of the mafic rocks appears to have disturbed the country rocks to a much greater extent than at Drumadoon–Tormore. Although no quantitative modal data are available, the Bennan Head sill contains many more obvious examples of quartz and feldspar xenocrysts in its mafic members than are found at Drumadoon–Tormore, particularly towards the top, suggesting mixing with greater amounts of porphyritic acid magma.

Conclusions

The large number of well-exposed dykes and the wealth of intrusive features which they show, make the dyke swarm exposed on the south

Arran shore the best example to be found within the Province. The site was thus selected as the type section of a dyke swarm. The intrusion of the dykes has extended the crust by about 10%: this was achieved during the course of as many as ten phases of dyke injection.

Measurements of the remanent magnetization of the dykes show that the swarm was probably emplaced over a period of between three and four million years, spanning three or four episodes of magnetic polarity.

The Bennan Head composite sill provides additional evidence from Arran for the coexistence of basic and acid magmas and that mingling between these occurred both before intrusion and after emplacement of the intrusion.

CORRYGILLS SHORE

Highlights

The early, thick Clauchlands crinanite (analcite–olivine dolerite) sill forms part of a possible cone-sheet system focused on Lamlash Bay. The Corrygills Pitchstone is an outstanding example of a natural glass, representing a quenched granitic magma.

Introduction

The site comprises the coast and adjoining cliffs for about 1.5 km north from Clauchlands Point and extends as far inland as the hill of Dun Dubh. Excellent exposures of variably textured crinanite in a thick sill form Clauchlands Point, to the north, the Corrygills pitchstone sill crops out. The site also contains good examples of dolerite, basalt and felsite dykes. Permian sediments, including coarse breccias, conglomerates and dune-bedded sandstones, are the country rock in this area (Fig. 6.10).

Description

The 30–35 m thick Clauchlands Sill is well exposed on the scarp below Dun Fionn (NS 047 338) and at Clauchlands Point. The rock is a coarse-grained analcite olivine dolerite (crinanite) which is generally deeply weathered to give a spheroidal surface. The sill is traversed by veins of fine-grained basalt and contains segregations of pegmatite containing titaniferous augite crystals up to several centimetres in length. The rock characteristically has a speckled appearance caused by the presence of areas of altered analcite in the groundmass: this is a feature of all facies of the sill. Olivine-rich segregations are present and apatite is a common accessory mineral.

The sill is in the form of a sheet dipping to the west and south-west, towards Lamlash Bay. Its upper margin is transgressive towards the Permian sediments, but the lower contact is partly conformable with the sandstones which dip at between 30° and 50° SW or WSW. At Clauchlands Point there is a steep, almost vertical contact with the sediments and the sill has developed a 0.3-m-thick marginal zone of vertically flow-banded basalt. The steep margin at this point may be due to intrusion along an earlier fault.

Some way below the Clauchlands Sill there are 3- to 4-m-thick exposures of the Corrygills pitchstone sill which intrudes Permian sediments, and dips at about 30° SSW (Fig. 6.11). The pitchstone is a beautiful, dark-green, glassy rock with pale streaks and bands which define flow lines. It is nearly phenocryst free, the glassy base containing microlites and delicate branching growths of crystallites (probably pyroxene and/or plagioclase) which have figured in textbooks since the early days of microscopic petrography (for example, Teall, 1888; plate 34, fig. 4). The rock is closely comparable in composition with the more silica-rich granites of the BTVP (Carmichael, 1962).

Between Clauchlands Point and Corrygills Point the wave-cut sandstone platform is traversed by several dykes and sills which are only fully visible at low tide. The Mile Dyke is a prominent dolerite dyke exposed on the platform. It is between 1 and 2 m wide, trends in a NW–SE direction, and is bordered by indurated and bleached sandstone. Another pitchstone sheet, sometimes termed the Small Corrygills Pitchstone to distinguish it from the thicker one described above, crops out at the base of a thick felsite sheet exposed about 0.5 km SE of Corrygills Burn (NS 043 348). The pitchstone is about one metre in outcrop width and both it and the overlying felsite show well-defined flow-banding; the pitchstone is glassy and bottle-green in colour, whereas the felsite is distinctly spherulitic, greenish-grey rock. The sheet crosses the shore in a WNW direction and dips between 25° and 35° SSE.

Along the Corrygills shore section, which is

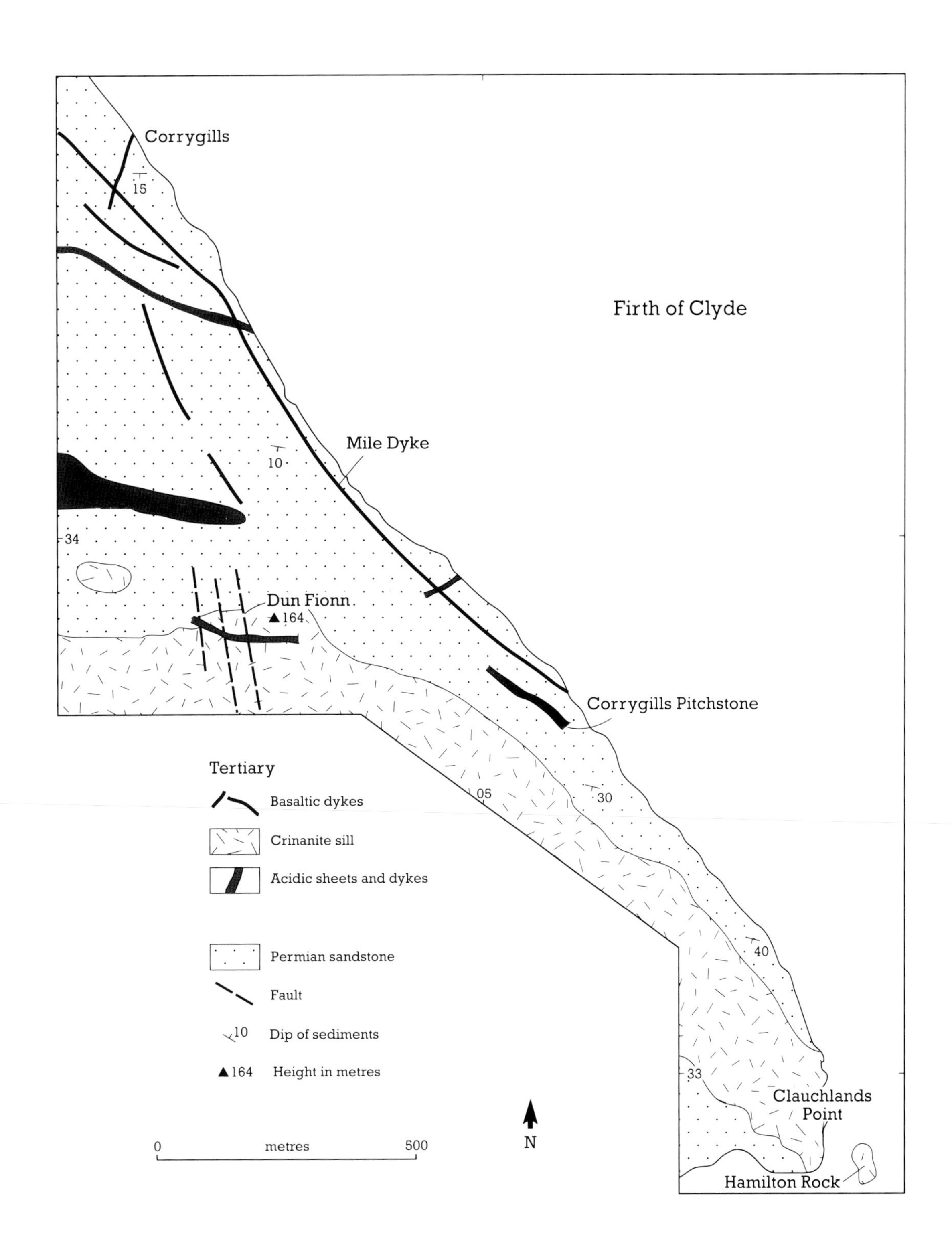

Figure 6.10 Geological map of the Corrygills Shore site (adapted from the British Geological Survey 1:50,000 Special District Sheet, Arran).

Figure 6.11 The Corrygills pitchstone sill in the cliff below the Clauchlands Sill (crinanite) at NS 050 337. Corrygills Shore site, Arran. (Photo: C.J. MacFadyen.)

liberally strewn with boulders of crinanite and glassy pitchstone, several other minor intrusions of basalt and quartz porphyry occur as small dykes.

Interpretation

The petrology of the Clauchlands Sill is similar to the Dippin Head Sill (see below) and both are probably part of the same sill complex, together with the sills at Monamore (*c.* NS 010 298) and Kingscross Point (NS 056 284). The sills have alkali-basalt compositions and are closely similar to alkali-basalt lavas (plateau basalts) in the BTVP (Tyrrell, 1928). Tyrrell suggested that the sills were in fact the hypabyssal facies of the plateau lavas, if this correlation is correct they could be equivalent to the large, subsided masses of basalt present within the Central Igneous Complex (Ard Bheinn) and could therefore be early members of the extensive suite of Palaeocene intrusions on Arran. An early date is supported by the manner in which they are freely cut by other minor intrusions (Tyrrell, 1928, pp. 112–3). The Clauchlands sill is cut by a pitchstone dyke near Dun Dubh and the equivalent crinanite sheet on the southern end of Holy Island (NS 068 288) is cut by dolerite dykes of the NW swarm, corroborating Tyrrell's suggestion. The Clauchlands Sill is one of a group of sills and sheets which appear to dip towards a focus under the north-west part of Lamlash Bay. In addition to the crinanites at Clauchlands, Monamore and Kingscross, this group of minor intrusions includes the Corrygills Pitchstone and numerous felsite sheets, some of which are intermittently exposed on the northern slopes of the Clauchland Hills (NS 035 335), immediately west of the site. Tomkeieff (1969) tentatively correlated the sheets around Lamlash Bay and suggested that they formed part of a cone-sheet system which focused on a 'magmatic hearth' (that is, an igneous centre) beneath Lamlash Bay (Tomkeieff, 1969, figs 3 and 4). He also suggested that the Holy Island riebeckite trachyte (NS 060 300) and the small quartz-porphyry intrusion at Dun Dubh (NS 038 343) were embryonic ring-dykes associated with the postulated Lamlash Bay igneous centre.

The essentially aphyric Corrygills Pitchstone is

one of numerous pitchstones in Arran. Tyrrell (1928) divided these into four groups on the basis of mineralogy and chemistry, the Corrygills Pitchstone being one type. It is the most siliceous of the four types, has a high K_2O/Na_2O ratio compared with the others, and is compositionally similar to granophyre in the Arran Central Igneous Complex. It is distinguished from the Tormore–Glen Shurig type of pitchstone by its lack of plagioclase, pyroxene and fayalite phenocrysts (cf. Tormore). It could be an extreme differentiate from a tholeiitic basalt magma and must have been completely liquid when intruded and quenched against the sediments. The pitchstone magma appears to have received little in the way of crustal contributions: it has a fairly low $^{87}Sr/^{86}Sr$ ratio (0.70855 at 59 Ma, Dickin *et al.*, 1981) and also only minor amounts of lead of crustal origin (Dickin *et al.*, 1981, table 2 and fig. 7). In common with most of the Arran Palaeocene intrusions there appears to have been little alteration due to circulating heated meteoric waters (Dickin *et al.*, 1981). The principal exception to this is the altered facies of the crinanite sills, thus providing a further indication of the early position of these intrusions in the igneous sequence.

Conclusions

The Clauchlands Sill was intruded at an early stage in the Palaeocene igneous activity in Arran and may have a similar age to large masses of basalt caught up in the Arran Central Igneous Complex. Together with other alkali-dolerite (crinanite) sills near Lamlash Bay and the numerous felsite and pitchstone sheets in this area, it may form part of a cone-sheet system with a focus beneath the bay. The Corrygills Pitchstone was intruded as a nearly crystal-free magma which quenched against Permian sedimentary rocks. It probably originated as an extreme differentiate of tholeiitic basaltic magma.

Chapter 7

Other Tertiary sites

INTRODUCTION

The British Tertiary Volcanic Province (BTVP) contains only a small proportion of the total igneous activity that occurred during the Palaeocene–Eocene opening of the North Atlantic. Within the Province, attention has been focused on the spectacular remnants of that activity in and around the central complexes and the majority of the sites described in this volume come from these areas and their immediate surroundings. Nevertheless, a number of important aspects of the igneous geology of the Province occur elsewhere and the sites described in this chapter cover them.

The NW- to NNW-oriented dyke swarms are a spectacular feature of the Province and examples of dense concentrations of dykes are described or noted close to the central complexes, for example, Kildonnan–Bennan Head, Arran and Loch Bà, Mull. The Cleveland Dyke of North Yorkshire is a far-flung, recently studied representative of the Mull swarm; this is described in the Langbaurgh Ridge–Cliff Ridge site. Mesozoic sediments filling basins adjoining the central complexes are often intruded by dolerite sill complexes. The Shiant Isles site is an excellently exposed, well-described example which exhibits a wide range of rock types attributed to multiple injection and magmatic differentiation. This site augments the information on differentiated sills from Dippin Head, Arran and Rubha Hunish, Skye. Two further sites, St Kilda and Rockall, lie in the North Atlantic to the west of Scotland (Fig. 1.1). St Kilda is formed by a central complex comparable in complexity with the other centres described; it consists of a succession of gabbroic and doleritic intrusions where basic and acid magmas frequently mingled, a later major granitic intrusion and many dykes and cone-sheets. The complex relationships of these bodies are magnificently exposed on the wave-washed cliffs of the islands in this archipelago. Further west, the isolated mass of Rockall provides one of the few examples in the BTVP of a peralkaline granitic intrusion, with a suite of rare minerals.

ROCKALL

Highlights

This is one of the few occurrences of alkali granite in the British Tertiary Volcanic Province and it is notable for its unusual mineralogy, including elpidite, leucophosphite and a Ba–Zr mineral 'bazirite'.

Introduction

The geological interest of Rockall lies in its relevance to the development of the North Atlantic Ocean and in its geochemistry and mineralogy which distinguishes it from the majority of intrusions in the BTVP. The island rises about 20 m above sea-level and consists entirely of peralkaline granite. Specimens were collected from Rockall early in the eighteenth century but the earliest detailed petrological and chemical descriptions were by Judd (1897) and Washington (1914). The geology has been described by Sabine (1960) and in accounts of expeditions to the islet in 1971 and 1972 (Harrison, 1975).

Description

Rockall is composed of moderately coarse-grained, aegirine/acmite-riebeckite granite with small segregations of finer-grained, peralkaline microgranite and xenoliths. Modal analyses of the granite gave the following mean (volume) percentage values: quartz 22%, feldspar 53%, ferromagnesian minerals 23% and accessory minerals 2%. Both felsic and mafic variants occur within the main body of granite; the variant with the highest modal percentage of ferromagnesian minerals (68%) has been termed rockallite (cf. Sabine, 1960). The granite contains drusy cavities which are lined by a variety of minerals including elpidite, leucophosphite and a barium–zirconium silicate (bazirite). Monazite may be embedded in the elpidite and apatite has been observed. Fine-grained, ovoid dark-coloured xenoliths up to 1 m by 0.3 m frequently carry quartz and feldspar megacrysts but otherwise have virtually identical mineralogy to the main mass of the granite. Sharp boundaries exist between the inclusions and the granite but no definite chilled contacts are present. The granite has been dated at *c.* 52 Ma (Harrison, 1975).

The Rockall Granite is apparently emplaced into the lavas and microgabbros of probable Cretaceous age which form Helen's Reef (Roberts *et al.*, 1974). The presence of continental crust is indicated by granulite-facies Precambrian rocks recovered from drill cores and bottom dredging

(Roberts *et al.*, 1972, 1973); the metamorphic rocks are similar to granulites exposed in the Outer Hebrides and north-west Scotland. The Helen's Reef microgabbro is associated with troctolites and they are probably responsible for a major negative magnetic anomaly in this area.

Interpretation

The Rockall Granite and the nearby microgabbro of Helen's Reef may be part of a central complex intruded into Cretaceous igneous rocks and Precambrian granulites. Although the Helen's Reef rocks have yielded a Cretaceous age in comparison with the Eocene age obtained from the Rockall Granite, it has been argued that the microgabbro could be of early Cenozoic age (Beckinsale, in discussion of Durant *et al.*, 1976). Harrison (1982) compares the composition of Helen's Reef microgabbro with (tholeiitic) non-porphyritic central magma-type rocks of the Inner Hebrides (for example, those of Centre 2, Mull), pointing out that the composition is unlike that of oceanic tholeiites; he also argues that there is no petrographic or chemical affinity with the Rockall Granite. However, this would not preclude both intrusions belonging to the same central complex, since both tholeiitic and (mildly) alkaline intrusions are well-documented from other centres, as, for example, on Skye (compare the mafic rocks of the Cuillin Centre with the mildly alkaline granite of Strath na Creitheach).

The Rockall Granite, with its suite of rare minerals, is compositionally comparable with the earlier, alkali-microgranite boss of Ailsa Craig (62 Ma; Harrison *et al.*, 1987) and with some members of the Eocene and younger alkaline complexes present along much of the East Greenland coastal belt (Nielsen, 1987; Upton, 1988). The Rockall Granite may have been formed by a small amount of partial melting in the upper mantle during the waning stages of igneous activity on the margins of a major hot-spot which existed over the area of the North Atlantic during the Tertiary (White, 1988).

Conclusions

Rockall is a small intrusion of granite which contains a suite of uncommon minerals. It appears to intrude lavas and metamorphic rocks which form the Rockall microcontinent. It was probably formed at a late stage in Tertiary igneous activity within the British Tertiary Volcanic Province and the North Atlantic, at a time when the upper mantle was cooling off after the main phase of magmatism during the Palaeocene.

THE SHIANT ISLES

Highlights

The Shiant Isles sill complex provides spectacular evidence for the differentiation of alkali-olivine basalt magma: basal olivine-rich picrites formed by accumulation under gravity outcrop on Eilean an Tighe–Garbh Eilean and analcite syenites, representing late differentiates, are present on Eilean Mhuire.

Introduction

The rocky Shiant Isles lie about 20 km north of Skye and 8 km south-east of Lewis. They consist of sheets of dolerite intruded into Jurassic sediments and form the exposed part of a sheet complex in the Little Minch. Their geological interest lies in the considerable textural, compositional and mineralogical variation exhibited by the dolerite which is magnificently exposed in the cliff sections (Fig. 7.1) and has been further explored by drilling.

Early accounts of the geology of the Shiant Isles were given by MacCulloch (1819), Judd (1878, 1885), Heddle (1884) and Geikie (1897), and

Figure 7.1 The north-west corner of Garbh Eilean, showing the main sill (left) and the lower sill (with natural arch), Shiant Isles. (Photo: F.G.F. Gibb.)

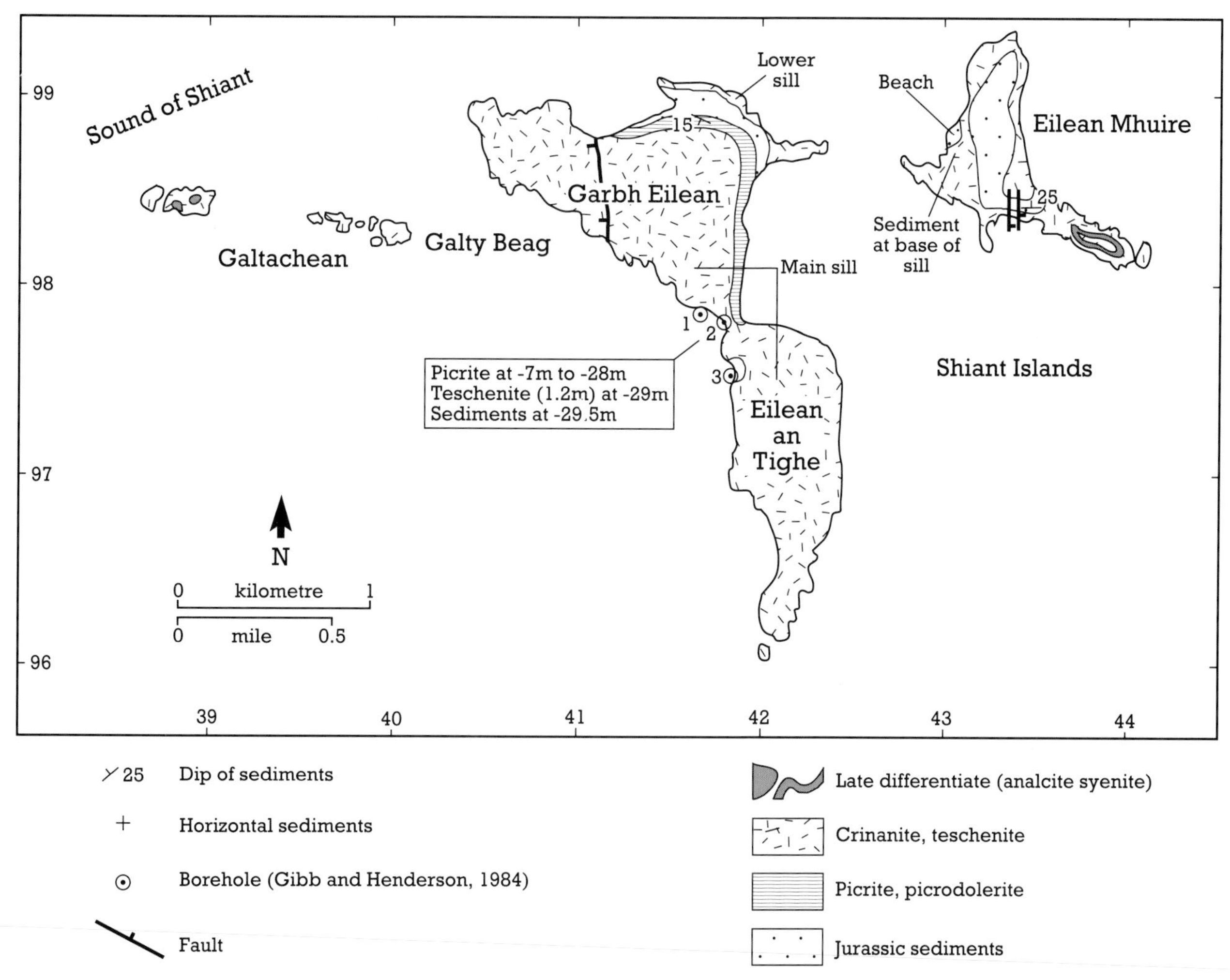

Figure 7.2 Geological map of the Shiant Isles (after Walker, 1930, plate 36, with additions from Gibb and Henderson, 1984, figure 1).

the picrites were figured in the classic *British Petrography* by Teall (1888, plate 3). The first detailed account was by Walker (1930) and subsequently, the mineralogy and petrology were investigated by Johnston (1953), Murray (1954), Drever (1953, 1957), Drever and Johnston (1958, 1965) and Gibb (1973). Reassessments of the structure of the sill complex has been provided by Gibb and Henderson (1984, 1989).

Description

The sills of the Shiant Isles are members of a major alkali-dolerite sheet complex intruding Jurassic sediments of the Little Minch Basin (Chesher *et al.*, 1983, Fig. 3) and probably represent an extension of the Trotternish sills of northern Skye (Rubha Hunish). Thick sheets of alkaline-olivine dolerite form the islands of Eilean Mhuire, Garbh Eilean, Eilean an Tighe and the Galtachean islets to the west. Jurassic (Lias) sedimentary rocks crop out on all three large islands although the outcrop on Eilean an Tighe is limited to a small coastal exposure. Fairly flat-lying contacts between sediments and dolerites, and the steep, massive columnar jointing of the dolerite (cf. Walker, 1930, plates 31, 32 and 33), indicate that the intrusions form sheets more or less conformable to the bedding of the sediments. The main body of sediments on Eilean Mhuire forms a major raft in the dolerite and the base of the sill is probably exposed on the west of the island, where Gibb and Henderson (1984) found sediments at sea-level. On Garbh Eilean, a lower sill on the north coast is separated by sediments from the main sill which is at least 130 m thick. The southern continuation of the main Garbh

Eilean sill is on Eilean an Tighe and the base of this sill was proved about 30 m below sea-level in a drill hole near the north end of the isthmus connecting the two islands. Walker (1930) considered that the Shiant Isles consisted of three separate sills:

1. the main sill of Garbh Eilean–Eilean an Tighe plus the Galtachean outcrops;
2. the lower sill of Garbh Eilean plus, the upper sill of Eilean Mhuire; and
3. the lower sill of Eilean Mhuire.

Drever (1957) argued that there were only two, subsequently reduced to a single sill (for example, Drever and Johnston, 1965; Wager and Brown, 1968, p. 532). A re-examination of the three main islands by Gibb and Henderson (1984) provided convincing evidence that there are at least three, and probably four, sills:

1. the main sill of Gharb Eilean–Eilean an Tighe;
2. the lower sill of Garbh Eilean;
3. the Eilean Mhuire sill with a large raft of sediment, and perhaps also a separate sill;
4. forming the Galtachean islets.

Analcite-olivine dolerite (crinanite) forms the majority of exposures, but towards the base of the Garbh Eilean main sill there is a noteworthy increase in modal olivine giving rise to over 45 m thickness of picrodolerite; this is in sharp contact with an underlying 23 m of picrite (modal olivine $>40\%$). This in turn is separated from underlying sediments by a thin layer of fine-grained, olivine-free analcite dolerite (teschenite) (Drever and Johnston, 1965; Gibb and Henderson, 1984, Fig. 2). A thin picrite (*c.* 2 m) occurs in sharp contact with underlying crinanite at the top of the main sill (Gibb and Henderson, 1978a, 1984, Fig. 2; Fig. 7.3). Vesicular crinanite high on Eilean an Tighe, estimated to be from near the top of the sill, contains poikilitic nepheline. The upper part of the Eilean Mhuire sill, exposed towards the south-east end of the island, contains a layer about 20 m thick much enriched in alkali feldspar and alkali pyroxene; this is an analcite syenite with essexite segregations (Walker, 1930). It is therefore apparent that both the main sill of Garbh Eilean–Eilean an Tighe and the Eilean Mhuire sill show vertical variations suggestive of crystal fractionation and differentiation. Variation in modal mineralogy and texture, for example, is accompanied by variation in the mineral compositions (cryptic variation); clinopyroxenes range upwards in the sills from diopsidic augite to hedenbergite, with marked soda-enrichment in late pegmatitic facies. The pyroxenes follow the typical flattish alkaline trend when plotted in the pyroxene quadrilateral (Gibb, 1973, Figs 3 and 4). Olivines in the picrite and picrodolerite maintain core compositions of about Fo_{80}, but at higher levels in the sills become increasingly zoned, with margins of more iron-rich olivine. Intense zoning of the olivine also occurs in the crinanites (Johnston, 1953, Fig. 2). The olivine-rich rocks contain zoned plagioclases, again the core compositions remain fairly constant at about An_{80} but zoning to sodic margins becomes more pronounced upwards and is strongly developed in the crinanites. Pegmatitic veins cut the picrites and pegmatite veins and bands occur in the higher parts of sills and, according to Walker (1930), are common in the Galtachean crinanites. The veins tend to be rich in analcite and natrolite, after original feldspar, and some carry small proportions of nepheline, alkali amphibole, aegirine and biotite.

Interpretation

Correlation of the isolated outcrops of the Shiant Isles Sill Complex is difficult simply because the outcrops are on well-separated islands. However, the presence of roof and floor relationships on Eilean Mhuire, the occurrence of an obvious floor to the sheet forming much of Eilean an Tighe and Garbh Eilean, and a roof-like contact below sediments on the north of Garbh Eilean, make it fairly certain that we are dealing with a sheet complex with at least three and possibly four leaves.

The Shiant Isles sills were interpreted by Walker (1930) to provide a particularly clear example of magmatic differentiation through the *in situ* gravity settling of early-formed crystals within cooling magma: the picrites and picrodolerites represent rocks enriched through the settling of early-formed olivine, while the later-crystallizing and generally lower-temperature phases were concentrated towards the upper parts of the sill where the analcite syenite (syenoteschenite, of Gibb and Henderson, 1984, p. 29) and the alkaline segregations represent the extreme products of fractionation. The discovery by Drever and Johnston (1965) that the picrite had a sharp, undulating contact with the overlying picrodolerite rather than a gradational relationship, necessitated some reassessment of Walker's interpretation.

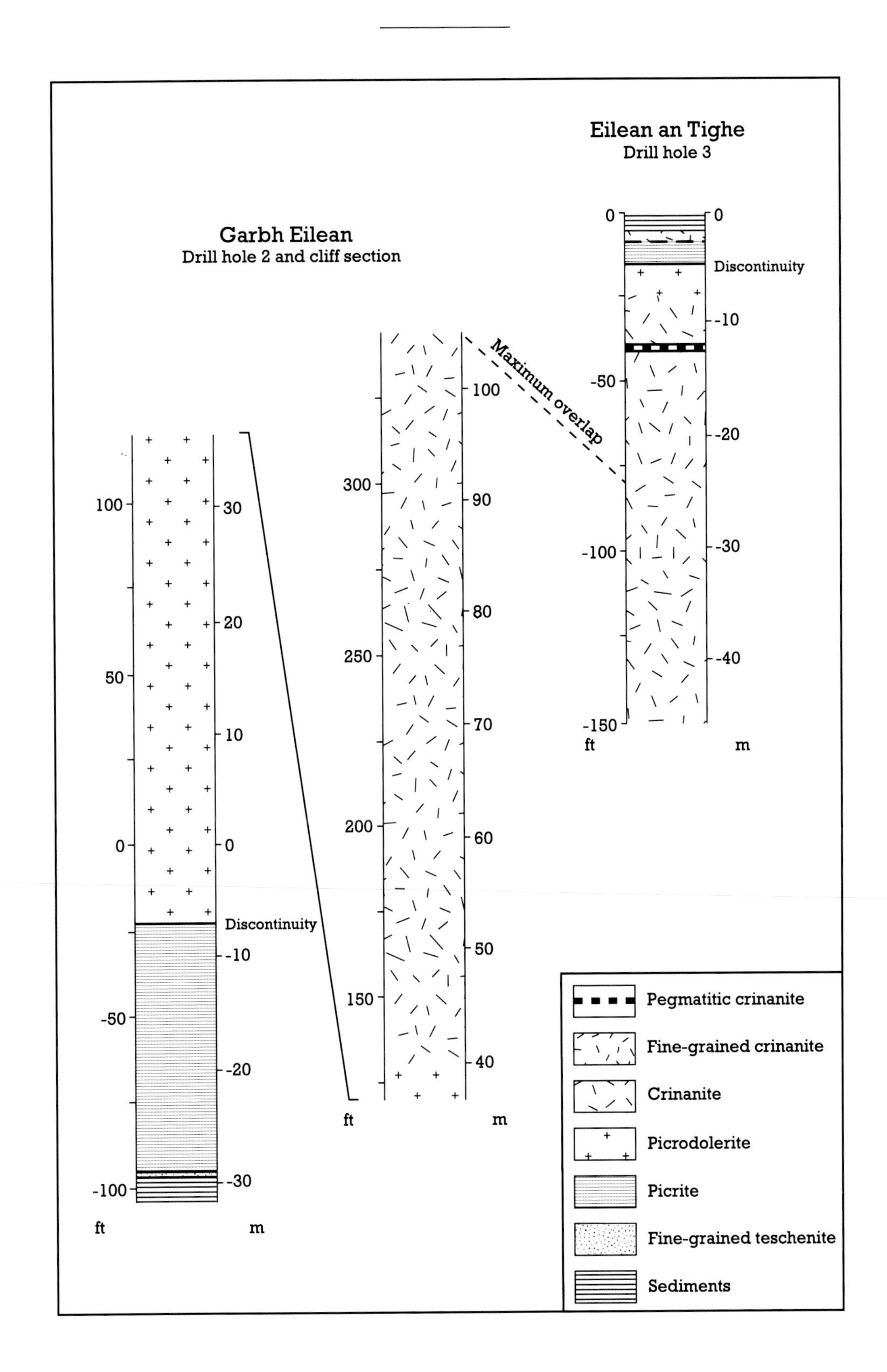

Figure 7.3 Simplified vertical sections through the main Garbh Eilean–Eilean an Tighe sill, Shiant Isles (after Gibb and Henderson, 1984, figure 2).

Conclusions

The Shiant Isles sill is a multiple intrusion in which an initial injection of picrite was followed by a large volume of analcite-olivine basalt magma. The sheets provide spectacular evidence for gravitational settling of olivine to form basal picrodolerites, with the concentration of residual differentiates to give segregations of analcite syenite in their higher parts. The contrasted rock types demonstrate the wide range of products which may be derived from an alkali-olivine basalt magma.

It is possible to demonstrate from the field-relationships that the dolerites form a series of separate sheets rather than a single sill.

ST KILDA

Highlights

The exposures on St Kilda show hybrid rocks in which chilled, lobate, pillow-like masses of basalt in acid matrices provide convincing demonstrations of the coexistence of acid and basic magmas. At Glen Bay, a gabbro body has been most unusually chilled to a glass at its margin. Cliffs in east Hirta show some of the most spectacularly exposed cone-sheets in the BTVP.

Introduction

Relict fragments of a Tertiary central complex form the St Kilda group of rocky islets about 80 km west of Harris, Outer Hebrides. Profound erosion has left only a small proportion of the complex above sea-level and the exposed rocks are entirely intrusive. It is likely, however, that Lewisian gneisses are present offshore. A group of early, layered gabbros dominates the site; they are cut by a mixed (acid and basic) magma complex which is in turn intruded by a granophyre, forming the last major intrusion. A number of dolerite and felsite dykes and cone-sheets cut the plutonic rocks.

The islands were visited by MacCulloch (1819) who recognized the presence of basic and acid rocks. Ross (1884) showed that acid rocks veined the basic intrusions and therefore considered them to be younger. Geikie (1897) described the rock types and compared them with those of the Hebridean Tertiary central complexes. Cockburn (1935) published the first detailed map and account of the geology of St Kilda and recognized several subdivisions to the mafic rocks. Wager described relationships which he interpreted as indicating that basic magma had chilled against acid magma (Wager and Bailey, 1953; Fig. 7.4). The most detailed modern investigations of the geology have been those of Harding (1966, 1967), who made a particular study of the relationships between the acid and basic rocks which display a number of unusual features better represented on St Kilda than elsewhere. In 1979 and 1980 the islands were remapped by the British Geological Survey and the results of this research, and the accompanying 1:25 000 map, have been published (Harding *et al.*, 1984). The igneous sequence is summarized in Table 7.1.

Description

The islands of the St Kilda group comprise St Kilda or Hirta, Dun, Soay, Borery and surrounding stacks, and Levenish (Fig. 7.5). Of these, all but Hirta and Dun are virtually inaccessible and appear to be made up of breccias of gabbro and dolerite cut by a few composite felsite–dolerite sheets. The geology of Hirta and Dun is, however, more varied (Table 7.1). The earliest intrusion was the Western Gabbro, which is a layered intrusion forming the western edges of the two islands. The layering, which is a reflection of varying modal proportions of mafic minerals (principally olivine Fo_{65} and diopsidic augite; Harding *et al.*, 1984) and plagioclase, dips towards a focus *c.* 2 km ENE of Hirta. The gabbro also contains minor orthopyroxene, amphibole, spinel and iron–titanium oxides. This gabbro appears to have been sheared and crushed and then veined by numerous sheets of basalt and dolerite on the west cliffs of the Cambir (*c.* NF 075 101). Similar breccias of gabbro in finer-grained basic sheets form the outlying islands and north Hirta and all have been grouped as a separate unit (Harding *et al.*, 1984). The Glen Bay Gabbro intrudes and is chilled against the basic breccias. The chilled contact on the east of Glen Bay is most unusual since there is complete textural gradation from a 10 mm border zone of splintery, glassy basalt to gabbro exposed on the east side of the bay. Fine, vertical banding occurs

Figure 7.4 'Pillows' of basic rock in a granitic matrix. Probably due to the simultaneous intrusion of granitic and basaltic liquids; an example of 'mixed magmas'. South end of Hirta, near Dun, St Kilda. (Photo: H. Armstrong.)

Table 7.1 Geological succession in the St Kilda archipelago (adapted from the British Geological Survey 1:25 000 Special Sheet, St Kilda)

Pleistocene glaciation
Palaeocene igneous activity
Basaltic and composite (acid and basic) inclined sheets and dykes
Conachair Granite
Mullach Sgar Complex (mixed magma (basic–acid) intrusions)
Glen Bay Granite
Glen Bay Gabbro
Breccias of gabbro and dolerite
Western Gabbro (layered in places)

No pre-Palaeocene rocks are exposed, but the complex is thought to be intruded into Lewisian gneisses.

in the marginal zone parallel to the contact and the effects of chilling are estimated to extend for 100–120 m into the gabbro (Harding *et al.*, 1984, p. 12). Gabbro on the west of the bay is much sheared and granulated and is separated from the eastern outcrops by the oldest granite on Hirta, the Glen Bay Granite. This granite is chilled against the earlier gabbro and, like the gabbro, shows signs of fragmentation. The next intrusive phase involved four pulses of mixed basic and acid magmas which formed the Mullach Sgar Complex between Glen Bay and Village Bay (Fig. 7.5). This group of rocks includes dolerite, microdiorite, microgranite and rocks of hybrid (mixed dolerite and granite) aspect. Angular and lobate masses of marginally chilled basic rocks occur in more acid matrices and are veined by felsic material (cf. Fig. 7.4); a large amount of shattering of the dolerite and basalt has occurred, giving areas of complex net-veining (cf. Ardnamurchan Point to Sanna). Although extremely complex in detail, Harding *et al.* (1984) suggest that initial intrusions of basaltic magma were followed successively by granitic magma and further basalt. The final major intrusion is the Conachair Granite which forms the high ground north-east of Village Bay. This granite intrudes the Mullach Sgar Complex without notable chilling along the contact. The granite typically contains quartz, turbid, perthitic alkali feldspar and albitic plagioclase, with minor amounts of biotite and amphibole. Other accessory minerals include zircon, sphene, rutile, anatase, Ti-magnetite, fluorite and needle-like, deep-brown crystals of the rare-earth-bearing silicate chevkinite. The Conachair Granite characteristically has a microgranitic texture often with a considerable content of granophyrically intergrown quartz and feldspar. Some of the larger quartz crystals are interpreted as corroded, inverted high-temperature quartz. Radiometric age determinations on this granite give a date of *c.* 55 Ma, indicating a Palaeocene age.

Several generations of minor intrusions with compositions ranging from basalt to rhyolite have been recognized (Cockburn, 1935; Harding *et al.*, 1984). Frequently these are inclined sheets (Harding *et al.*, 1984, Figs 24c and 25c) whose disposition suggests that they once formed a classic cone-sheet complex. Many of the inclined sheets cut the Conachair Granite and are therefore the latest intrusions in St Kilda.

Interpretation

St Kilda is formed from the remains of a central complex of Palaeocene age which is situated towards the margin of the European continental shelf where it is probably emplaced into Lewisian gneisses. The earliest intrusions were coarse, layered gabbros which subsequently became crushed and shattered and were then intruded by a multiplicity of dolerite and basalt sheets and veins. Some of these must have been emplaced prior to complete solidification of the earlier, very variable-textured gabbros. The distinctly later Glen Bay Gabbro is most unusual among the gabbroic intrusions of the BTVP in possessing a glassy, chilled-margin which gradually grades into normal gabbro. Presumably the intrusion was emplaced into cold, solidified, earlier gabbro which itself had a high-melting point and was possibly effectively anhydrous. Normally, BTVP gabbros have complex contacts with earlier, relatively low-melting point acid rocks (cf. Rum, Harris Bay; Skye, Coire Uaigneich), or are not conspicuously chilled against other mafic bodies. This occurrence would appear to be unique in the Province.

The Mullach Sgar Complex provides a superb example of the coexistence of acid and basic magmas and their near simultaneous intrusion. Evidence for mixed basic and acid magmas occurs elsewhere in the Province (cf. Ardnamurchan Point to Sanna; Arran, Ard Bheinn and Drumadoon–Tormore; Skye, Marsco and Mheall a' Mhaoil, Kilchrist and Rubha' an Eirannaich; Mull, Cruach Choireadail, Allt Molach–Beinn Chàisgidle and Loch Bà–Ben More), but the pillowed exposures of chilled basaltic rocks in unchilled felsic matrices are exceptionally fine. The early recognition of their significance by Wager and Bailey (1953) has been crucial in elucidating some of the more puzzling field relationships within the Province, particularly where limited outcrops suggest that acid rocks veining and brecciating dolerite or gabbro are significantly younger than the mafic rocks, yet the broader relationships clearly show that this is not the case (cf. Rum, Harris Bay). The pervasive shattering of many of the St Kilda gabbros and dolerites is a striking feature of the complex and suggests that explosive release of water may have occurred towards the end of their solidification, followed by rapid injection of quickly cooled basaltic magma. It is also possible that the highly unusual

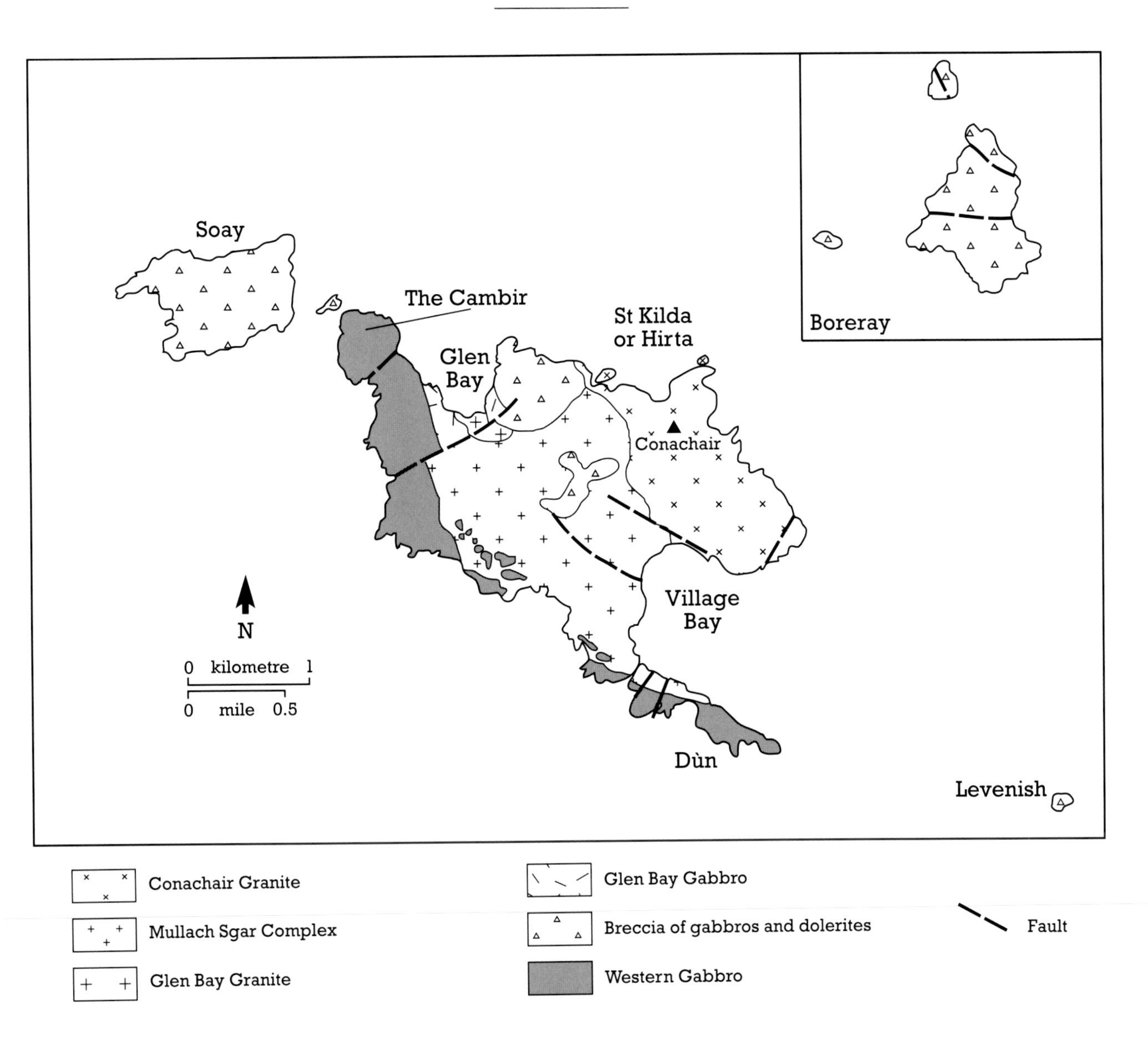

Figure 7.5 Geological map of the St Kilda archipelago (adapted from the British Geological Survey 1:25 000 Special Sheet, St Kilda).

glassy, quenched contact of the Glen Bay Gabbro may owe its origin to high-temperature, hydrothermal quenching.

Conclusions

An early layered gabbro intrusion forms much of the islands of the St Kilda archipelago. On Hirta and Dun, and to a lesser extent elsewhere, it underwent penetrative shattering and brecciation which may have been caused by explosive release of water as it completed crystallization. A further gabbro intrusion was quenched to a glassy rock against the breccias and itself intruded by granite. Basaltic and granitic magmas coexisted at this stage and the next intrusion consisted of several pulses of mixed basic and acid magmas. The last major intrusion, following soon after the mixed magma bodies, was a major body of granite in the east of Hirta. A final phase of basalt intrusion gave rise to a suite of cone-sheets which focuses to the north-east of Hirta.

LANGBAURGH RIDGE AND CLIFF RIDGE

Highlights

The dyke is the most laterally extensive and best exposed member of the Tertiary swarm of north-east England. It shows remarkable chemical uniformity and is compositionally identical with rocks in Mull. Calculations suggest that it was emplaced by rapid lateral flow over a period of as little as 1–5 days, from a source beneath Mull.

Introduction

The Tertiary Cleveland Dyke cuts and indurates Jurassic sediments in the Langbaurgh and Cliff Ridges close to the village of Great Ayton, on the border of Cleveland and North Yorkshire (Fig.

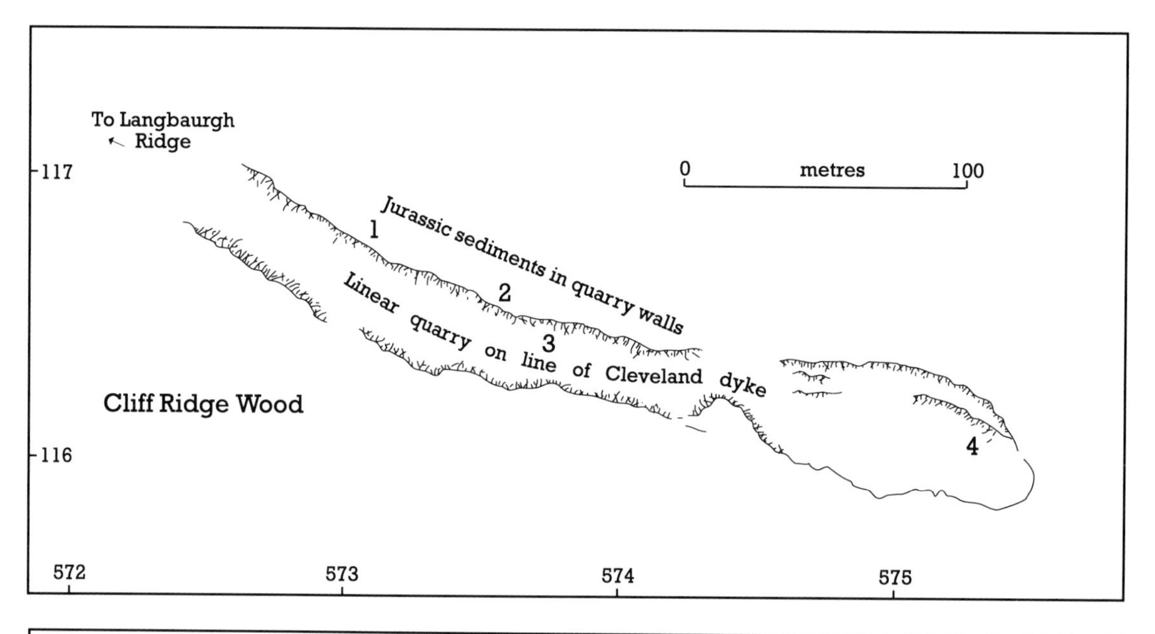

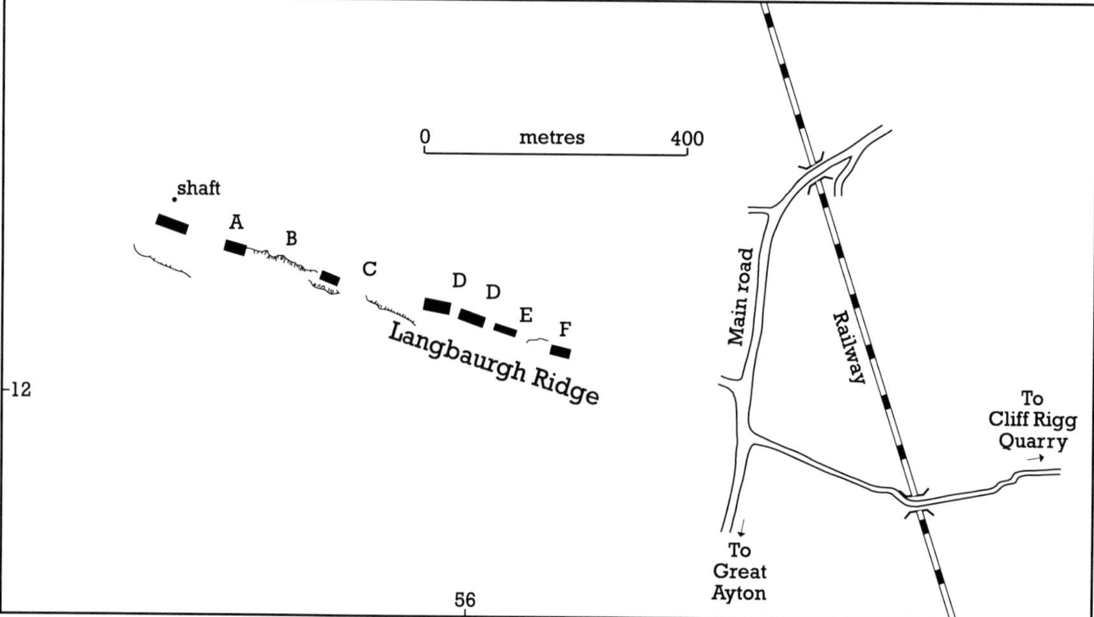

Figure 7.6 Sketch maps showing outcrops of the Cleveland Dyke near Great Ayton, North Yorkshire: (lower) Langbaurgh Ridge. Localities A–F refer to points where the north margin of the dyke has been preserved. (upper) Upper part of Cliff Rigg Quarry. For explanation of localities 1–4 see text.

7.6). The basaltic dyke is a member of the Mull dyke swarm, the site is about 370 km from the focus of the swarm in the Mull centre (see Chapter 5).

Early descriptions of the dyke include those of Tate and Blake (1876) and Barrow (1888) who also described the petrography. Petrographic descriptions were also given by Teall (1884, who cited an analysis by Stock) and by Holmes and Harwood (1929). Numerous chemical analyses were made by Hornung *et al.* (1966). MacDonald *et al.* (1988) have modelled emplacement of the dyke.

Description

The Cleveland Dyke is the southernmost member of a swarm of east–west to WNW–ESE-trending tholeiite (quartz dolerite) dykes in north-east England which focus on the Mull Tertiary central complex (Fig. 1.1). It is the most laterally extensive and best-exposed member of this swarm. The dyke consists of a fine-grained, porphyritic basaltic andesite which has up to 13% normative quartz. Phenocrysts of plagioclase are conspicuous in hand specimen as are small rounded vesicles up to 10 mm diameter. The plagioclase phenocrysts, which may be aggregated, are generally of labradorite (An_{50-60}) or, rarely, of anorthite (An_{90}) composition. They are accompanied by microphenocrysts of pigeonite which sometimes mantle hypersthene cores. The groundmass plagioclase, augite and opaque oxides may enclose areas of quartz and alkali feldspar or, in the chilled margins, there may be intersertal areas of clear brown glass. Cognate basaltic inclusions up to 5 mm in diameter and derived from the margins are common. The vesicles contain quartz, calcite, chlorite and clay minerals (which may expand on exposure to the atmosphere causing disintegration of the vesicles' contents), rare epidote, pyrite, pectolite and mesolite (cf. Barrow, 1888). The basalt at the contacts with highly fossiliferous Middle Lias sandstones and ironstones is distinctly finer grained than the centre of the dyke, but tachylitic rock is not found. Subhorizontal, columnar jointing is developed in the marginal dolerite and has given rise to good examples of spheroidal weathering.

The dyke is up to 25 m in width and appears to have produced little alteration of the sediments apart from discoloration and induration for a distance of about 2 m from the contact. Its course across country is readily observed since its *en échelon* segments form elongate, low ridges; the effect is well seen when Langbaurgh Ridge is viewed from Cliff Ridge (Locality 3, Fig. 7.6a). Since the majority of the rocks of North Yorkshire and Cleveland are soft and crumbling, the dyke has been extensively quarried and even mined (as under Langbaurgh Quarry) for setts and aggregate; hence, both parts of this site and many other 'outcrops' resemble railway cuttings. In the disused Langbaurgh Quarry five segments of the north wall of the dyke have been preserved. In the upper Cliff Rigg Quarry a cross-section of the dyke occurs at the extreme east end (4), where the dyke may become 'headed'. A contact of baked sediments against marginal dyke occurs 100 m into the site from its west end (2) and on the same side a longitudinal section of the dyke is preserved near the west end (1).

Interpretation

The dyke was the subject of early attempts to date rocks using radiometric techniques when Dubey and Holmes (1929) demonstrated that it was younger than the (Stephanian) Whin dolerites of north-east England. Subsequently, the Palaeocene age of the Cleveland Dyke was confirmed when a date of *c.* 58 Ma was obtained by the K–Ar method (Evans *et al.*, 1973) and by its magnetic properties (for example, Dagley, 1969). It is demonstrably post-Jurassic and the clear connection with the Mull central complex also supports a Palaeocene age.

There has been debate whether the dyke was actually fed from a source in Mull, or rose vertically from subjacent mantle along a fracture system propagated from Mull. The considerable distance from Mull (*c.* 370 km) and the absence of any systematic increase in thermal metamorphism around the dyke when traced towards Mull have perhaps supported the suggestion that the magma rose vertically, but this does require a laterally extensive magma source beneath the dyke over its entire extent, for which there is little evidence (MacDonald *et al.*, 1988). MacDonald *et al.* have made a detailed examination of the petrology of the dyke which substantiates the earlier claims that it is extremely similar compositionally to some of the Mull non-porphyritic central magma-type intrusions. They have also carried out numerical modelling of the flow of magma through a dyke of this size from which they conclude that the dyke could have been fed by lateral flow from a large magma chamber beneath Mull, and that its emplacement could have taken place in the very short time of 1–5 days.

Conclusions

The Cleveland Dyke is a compositionally uniform quartz dolerite which closely resembles basaltic and doleritic rocks in the Mull central complex. It is a compact, fine-grained rock with scattered, small, plagioclase crystals and small vesicles. As it is the only durable rock in North Yorkshire and Teeside it has been extensively quarried and mined for aggregate.

Recent research has shown that the dyke was probably intruded laterally from a source beneath the Isle of Mull in western Scotland, and that lateral emplacement may have taken place in a few days.

References

Agrell, S.O. (1965) Polythermal metamorphism of limestones at Kilchoan, Ardnamurchan. *Mineralogical Magazine*, **34** (Tilley Volume), 1–15.

Allwright, A.E. (1980) The structure and petrology of the Tertiary volcanic rocks of Eigg, Muck and Canna, N.W. Scotland. Unpublished M.Sc. thesis, University of Durham.

Allwright, A.E. and Hudson, J.D. (1982) The Sgùrr of Eigg. *Journal of the Geological Society of London*, **139**, p. 215 (abstract).

Almond, D.C. (1960) The Tertiary igneous geology of Strathaird, Skye. Unpublished Ph.D. Thesis, University of Durham.

Almond, D.C. (1964) Metamorphism of Tertiary lavas in Strathaird, Skye. *Transactions of the Royal Society of Edinburgh*, **65**, 413–34.

Anderson, E.M. (1936) Dynamics of formation of cone-sheets, ring-dykes and cauldron-subsidences. *Proceedings of the Royal Society of Edinburgh*, **61**, 128–57.

Anderson, F.W. and Dunham, K.C. (1966) *The Geology of Northern Skye*. Memoir of the Geological Survey of Great Britain, HMSO, Edinburgh.

Anderson, J.G.C. (1945) The Dalradian rocks of Arran. *Transactions of the Geological Society of Glasgow*, **20**, 264–86.

Argyll, Duke of (1851) On Tertiary leaf-beds in the Isle of Mull. With a note on the vegetable remains from Ardtun Head by Prof. E. Forbes. *Quarterly Journal of the Geological Society of London*, **7**, 89–103.

Bailey, E.B. (1914) The Sgùrr of Eigg. *Geological Magazine (Decade 6)*, **1**, 296–305.

Bailey, E.B. (1926) Domes in Scotland and South Africa: Arran and Vredefort. *Geological Magazine*, **63**, 481–95.

Bailey, E.B. (1945) Tertiary igneous tectonics of Rhum (Inner Hebrides). *Quarterly Journal of the Geological Society of London*, **100** (for 1944), 165–91.

Bailey, E.B. (1954) Relations of Torridonian to Durness Limestone in the Broadford–Strollamus district of Skye. *Geological Magazine*, **91**, 73–8.

Bailey, E.B. (1956) Hebridean notes: Rhum and Skye. *Liverpool and Manchester Geological Journal*, **1**, 420–6.

Bailey, E.B. and Anderson, E.M. (1925) *The Geology of Staffa, Iona and western Mull*. Memoir of the Geological Survey of Great Britain, HMSO, Edinburgh.

Bailey, E.B., Clough, C.T., Wright, W.B. *et al.* (1924) *Tertiary and Post-Tertiary Geology of Mull, Loch Aline and Oban*. Memoir of the Geological Survey of Great Britain, HMSO, Edinburgh.

Barrow, G. (1888) *The Geology of North Cleveland*. Memoir of the Geological Survey of Great Britain, HMSO, London.

Beckinsale, R.D. (1974) Rb–Sr and K–Ar age determinations and oxygen isotope data for the Glen Cannel granophyre, Isle of Mull, Argyllshire, Scotland. *Earth and Planetary Science Letters*, **22**, 267–74.

Beckinsale, R.D., Pankhurst, R.J., Skelhorn, R.R. *et al.* (1978) Geochemistry and petrogenesis of the early Tertiary lava pile of the Isle of Mull, Scotland. *Contributions to Mineralogy and Petrology*, **66**, 415–27.

Bédard, J.H., Sparks, R.S.J., Renner, R. *et al.* (1988) Peridotite sills and metasomatic gabbros in the Eastern Layered Series of the

Rhum complex. *Journal of the Geological Society of London*, **145**, 207–24.

Bell, B.R. (1982) The evolution of the Eastern Red Hills Tertiary igneous centre, Skye, Scotland. Unpublished Ph.D. Thesis, University of London.

Bell, B.R. (1983) Significance of ferrodioritic liquids in magma mixing processes. *Nature*, **306**, 323–7.

Bell, B.R. (1984a) The basic lavas of the Eastern Red Hills district, Isle of Skye. *Scottish Journal of Geology*, **20**, 73–86.

Bell, B.R. (1984b) The geochemistry of Lower Tertiary basic dykes in the Eastern Red Hills district, Isle of Skye, and their significance for the proposed magmatic evolution of the Skye Centre. *Mineralogical Magazine*, **48**, 365–72.

Bell, B.R. (1985) The pyroclastic rocks and rhyolitic lavas of the Eastern Red Hills district, Isle of Skye. *Scottish Journal of Geology*, **21**, 57–70.

Bell, B.R. and Emeleus, C.H. (1988) A review of the silicic pyroclastic rocks in the British Tertiary Volcanic Province. In *Early Tertiary Volcanism and the Opening of the NE Atlantic*, (eds A.C. Morton and L.M. Parson), Geological Society Special Publication, No. 39, pp. 365–80.

Bell, B.R. and Harris, J.W. (1986) *An Excursion Guide to the Geology of the Isle of Skye*. Geological Society of Glasgow, 317 pp.

Bell, J.D. (1966) Granites and associated rocks of the eastern part of the Western Red Hills Complex, Isle of Skye. *Transactions of the Royal Society of Edinburgh*, **66**, 307–43.

Bell, J.D. (1976) The Tertiary intrusive complex on the Isle of Skye. *Proceedings of the Geologists' Association*, **87**, 247–71.

Berggren, W.A., Kent, D.V. and Flynn, J.J. (1985) Jurassic to Palaeogene: Part 2 Palaeogene geochronology and chronostratigraphy. In *The geochronology of the geological record*, (ed. N.J. Snelling). Memoir of the Geological Society of London, **10**, 141–95.

Bhattacharji, S. and Smith, C.H. (1964) Flowage differentiation. *Science*, **145**, 150–3.

Binns, P.E., McQuillin, R. and Kenolty, N. (1974) *The Geology of the Sea of the Hebrides*. Report of the Institute of Geological Sciences, No. 73/14, 43 pp.

Black, G.P. (1952a) The age relationship of the granophyre and basalt of Orval, Isle of Rhum. *Geological Magazine*, **89**, 106–12.

Black, G.P. (1952b) The Tertiary volcanic succession of the Isle of Rhum, Inverness-shire. *Transactions of the Edinburgh Geological Society*, **15**, 39–51.

Black, G.P. (1954) The acid rocks of western Rhum. *Geological Magazine*, **91**, 257–72.

Black, G.P. (1955) The junction between Jurassic sandstones and Tertiary granophyre near Dunan, Isle of Skye: a re-interpretation. *Transactions of the Edinburgh Geological Society*, **16**, 217–22.

Black, G.P. (1974) Appendix 4: The geology of Rhum. In *Isle of Rhum Nature Reserve: the Reserve Handbook*. Nature Conservancy Council, Scotland.

Black, G.P. and Welsh, W. (1961) The Torridonian succession of the Isle of Rhum. *Geological Magazine*, **98**, 265–76.

Blake, D.H., Elwell, R.W.D., Gibson, I.L. *et al.* (1965) Some relationships resulting from the intimate association of acid and basic magmas. *Quarterly Journal of the Geological Society of London*, **121**, 31–49.

Bott, M.H.P. and Tantrigoda, D.A. (1987) Interpretation of the gravity and magnetic anomalies over the Mull Tertiary intrusive complex, NW Scotland. *Journal of the Geological Society of London*, **144**, 17–28.

Bott, M.H.P. and Tuson, J. (1973) Deep structure beneath the Tertiary volcanic regions of Skye, Mull and Ardnamurchan, north-west Scotland. *Nature, Physical Sciences*, **242**, 114–16.

Boué, Ami (1820) Short comparison of the volcanic rocks of France with those of a similar nature found in Scotland. *Edinburgh Philosophical Journal*, **2**, p. 326.

Bowen, N.L. (1928) *The Evolution of the Igneous Rocks*. Princeton University Press, Princeton, New Jersey, 334 pp.

Boyd, W.W. (1974) A geochemical investigation of composite bodies involving intermediate members of the alkali—basalt—trachyte suite. Unpublished Ph.D. Thesis, University of Edinburgh.

Bradshaw, N. (1961) The mineralogy and petrology of the eucrites of the Centre 3 igneous complex, Ardnamurchan, Scotland. Unpublished Ph.D. Thesis, University of Manchester.

Brown, G.C. and Mussett, A.E. (1976) Evidence for two discrete centres in Skye. *Nature, London*, **261**, 218–20.

Brown, G.M. (1954) A suggested igneous origin for the banded granular hornfelses within the hypersthene-gabbro of Ardnamurchan, Argyllshire. *Mineralogical Magazine*, **30**, 529–33.

References

Brown, G.M. (1956) The layered ultrabasic rocks of Rhum, Inner Hebrides. *Philosophical Transactions of the Royal Society*, **B240**, 1–53.

Brown, G.M. (1963) Melting relations of Tertiary granitic rocks in Skye and Rhum. *Mineralogical Magazine*, **33**, 533–62.

Brown, G.M. (1969) *The Tertiary Igneous Geology of the Isle of Skye*. Geologists' Association Guide, **13**, 37 pp.

Buist, D.S. (1959) The composite sill of Rudh' an Eireannaich, Skye. *Geological Magazine*, **96**, 247–52.

Butcher, A.R. (1985) Channelled metasomatism in Rhum layered cumulates – evidence from late-stage veins. *Geological Magazine*, **122**, 503–18.

Butcher, A.R., Young, I.M. and Faithfull, J.W. (1985) Finger structures in the Rhum Complex. *Geological Magazine*, **122**, 491–502.

Butchins, C.S. (1973) An extension of the granophyric quartz-dolerite intrusion of Centre 2, Ardnamurchan, Argyllshire. *Geological Magazine*, **110**, 473–5.

Cann, J.R. (1965) The metamorphism of amygdales at 'S Airde Beinn, northern Mull. *Mineralogical Magazine*, **34**, 92–106.

Carmichael, I.S.E. (1960a) The feldspar phenocrysts of some Tertiary acid glasses. *Mineralogical Magazine*, **32**, 587–608.

Carmichael, I.S.E. (1960b) The pyroxenes and olivines from some Tertiary acid glasses. *Journal of Petrology*, **1**, 309–36.

Carmichael, I.S.E. (1962) A note on the composition of some natural acid glasses. *Geological Magazine*, **99**, 253–64.

Carmichael, I.S.E. (1963) The crystallization of feldspar in volcanic acid liquids. *Quarterly Journal of the Geological Society of London*, **119**, 95–131.

Carmichael, I.S.E. (1964) The petrology of Thingmuli, a Tertiary volcano in eastern Iceland. *Journal of Petrology*, **5**, 435–60.

Carmichael, I.S.E. and McDonald, A. (1961) The geochemistry of some natural acid glasses from the North Atlantic Tertiary Volcanic Province. *Geochimica et Cosmochimica Acta*, **25**, 189–222.

Carr, J.M. (1952) An investigation of the Sgùrr na Stri–Druim Hain sector of the basic igneous complex of the Cuillin Hills, Isle of Skye. Unpublished D. Phil. Thesis, University of Oxford.

Cheeney, R.F. (1962) Early Tertiary fold movements in Mull. *Geological Magazine*, **99**, 227–32.

Chesher, J.A., Smythe, D.K. and Bishop, P. (1983) *The geology of the Minches, Inner Sound and Sound of Raasay*. Report of the Institute of Geological Sciences, No. 83/6, 29 pp.

Cockburn, A.M. (1935) The geology of St Kilda. *Transactions of the Royal Society of Edinburgh*, **58**, 511–47.

Cooper, J. (1979) Lower Tertiary fresh-water Mollusca from Mull, Argyllshire. *Tertiary Research*, **2**, 69–74.

Craig, G.Y. (ed.) (1965) *The Geology of Scotland*, 1st edn, Oliver and Boyd, Edinburgh, 556 pp.

Craig, G.Y. (ed.) (1983) *Geology of Scotland*. 2nd edn, Scottish Academic Press, Edinburgh, 472 pp.

Cressey, G. (1987) Skarn formation between metachalk and agglomerate in the Central Ring Complex, Isle of Arran, Scotland. *Mineralogical Magazine*, **51**, 231–46.

Dagley, P. (1969) Palaeomagnetic results from some British Tertiary dykes. *Earth and Planetary Science Letters*, **6**, 349–54.

Dagley, P. and Mussett, A.E. (1981) Palaeomagnetism of the British Tertiary Igneous Province: Rhum and Canna. *Geophysical Journal of the Royal Astronomical Society*, **65**, 475–91.

Dagley, P. and Mussett, A.E. (1986) Palaeomagnetism and radiometric dating of the British Tertiary Volcanic Province: Muck and Eigg. *Geophysical Journal of the Royal Astronomical Society*, **85**, 221–42.

Dagley, P., Mussett, A.E. and Skelhorn, R.R. (1987) Polarity, stratigraphy and duration of the Tertiary igneous activity of Mull, Scotland. *Journal of the Geological Society of London*, **144**, 985–96.

Dagley, P., Mussett, A.E., Wilson, R.L. *et al.* (1978) The British Tertiary igneous province: palaeomagnetism of the Arran dykes. *Geophysical Journal of the Royal Astronomical Society*, **54**, 75–91.

Day, S.J. (1989) The geology of the Hypersthene Gabbro of Ardnamurchan Point and implications for its evolution and as upper crustal basic magma chamber. Unpublished Ph.D. Thesis, University of Durham.

Deer, W.A. (1969) *Field Excursion Guide to the Tertiary Volcanic Rocks of Ardnamurchan*. International Association of Volcanology and Chemistry of the Earth's Interior. Symposium on volcanoes and their roots. Oxford, England, 7–13 September 1969.

Dickin, A.P. (1981) Isotope geochemistry of Tertiary igneous rocks from the Isle of Skye, NW Scotland. *Journal of Petrology*, **22**, 155–89.

Dickin, A.P. and Exley, R.A. (1981) Isotopic and geochemical evidence for magma mixing in the petrogenesis of the Coire Uaigneich granophyre, Isle of Skye, N.W. Scotland. *Contributions to Mineralogy and Petrology*, **76**, 98–108.

Dickin, A.P. and Jones, N.W. (1983) Isotopic evidence for the age and origin of pitchstones and felsites, Isle of Eigg, N.W. Scotland. *Journal of the Geological Society of London*, **140**, 691–700.

Dickin, A.P., Moorbath, S. and Welke, H.J. (1981) Isotope, trace element and major element geochemistry of Tertiary igneous rocks, Isle of Arran, Scotland. *Transactions of the Royal Society of Edinburgh: Earth Sciences*, **72**, 159–70.

Dickin, A.P., Brown, J.L., Thompson, R.N. *et al.* (1984) Crustal contamination and the granite problem in the British Tertiary Volcanic Province. *Philosophical Transactions of the Royal Society of London*, **A310**, 755–80.

Donaldson, C.H. (1974) Olivine crystal types in harrisitic rocks of the Rhum pluton and in Archaean spinifex rocks. *Bulletin of the Geological Society of America*, **85**, 1721–6.

Donaldson, C.H. (1975) Ultrabasic breccias in layered intrusions – the Rhum Complex. *Journal of Geology*, **83**, 33–45.

Donaldson, C.H. (1977a) Petrology of anorthite—bearing gabbroic anorthosite dykes in north-west Skye. *Journal of Petrology*, **18**, 595–620.

Donaldson, C.H. (1977b) Laboratory duplication of comb-layering in the Rhum pluton. *Mineralogical Magazine*, **41**, 323–36.

Donaldson, C.H. (1982) Origin of some of the Rhum harrisite by segregation of intercumulus liquid. *Mineralogical Magazine*, **45**, 201–9.

Donaldson, C.H., Drever, H.I. and Johnston, R. (1973) Crystallization of poikilo-macro-spherulitic feldspar in a Rhum peridotite. *Nature, Physical Sciences*, **243**, 69–70.

Drever, H.I. (1953) A note on the field relations of the Shiant Isles picrite. *Geological Magazine*, **90**, 159–60.

Drever, H.I. (1957) A note on the occurrence of rhythmic layering in the Eilean Mhuire Sill, Shiant Isles. *Geological Magazine*, **94**, 277–80.

Drever, H.I. and Johnston, R. (1958) The petrology of picritic rocks in minor intrusions – a Hebridean group. *Transactions of the Royal Society of Edinburgh*, **63**, 459–99.

Drever, H.I. and Johnston, R. (1965) New petrographical data on the Shiant Isles picrite. *Mineralogical Magazine*, **34**, 194–203.

Drever, H.I. and Johnston, R. (1966) A natural high—lime silicate liquid more basic than basalt. *Journal of Petrology*, **7**, 414–20.

Dubey, S. and Holmes, A. (1929) Estimates of the ages of the Whin Sill and Cleveland Dyke by the helium method. *Nature*, **123**, 794–5.

Dunham, A.C. (1962) The petrology and structure of the northern edge of the Tertiary igneous complex of Rhum. Unpublished D.Phil. Thesis, University of Oxford.

Dunham, A.C. (1964) A petrographic and geochemical study of back-veining and hybridization at a gabbro–felsite contact in Coire Dubh, Rhum, Inverness-shire. *Mineralogical Magazine*, **33**, 887–902.

Dunham, A.C. (1965a) The nature and origin of the groundmass textures in felsites and granophyres from Rhum, Inverness-shire. *Geological Magazine*, **102**, 8–23.

Dunham, A.C. (1965b) A new type of banding in ultrabasic rocks from central Rhum, Inverness-shire. *American Mineralogist*, **50**, 1410–20.

Dunham, A.C. (1968) The felsites, granophyre, explosion breccias and tuffisites of the north-eastern margin of the Tertiary igneous complex of Rhum, Inverness-shire. *Quarterly Journal of the Geological Society of London*, **123** (for 1967), 327–52.

Dunham, A.C. and Emeleus, C.H. (1967) The Tertiary geology of Rhum, Inner Hebrides. *Proceedings of the Geologists' Association*, **78**, 391–418.

Dunham, A.C. and Wadsworth, W.J. (1978) Cryptic variation in the Rhum layered intrusion. *Mineralogical Magazine*, **42**, 347–56.

Durant, G.P., Dobson, M.R. and Kokelaar, B.P. *et al.* (1976) Preliminary report on the nature and age of the Blackstones Bank Igneous Centre, western Scotland. *Journal of the Geological Society of London*, **132**, 319–26.

Durant, G.P., Kokelaar, B.P. and Whittington, R.J. (1982) *The Blackstones Bank Igneous Centre, western Scotland.* Proceedings of the 6th Symposium of the Confederation Mondiale des Activites Subaquatique, Heriot-Watt University, September 1980. Natural Environment Research Council, London, 297–308.

References

Durrance, E.M. (1967) Photoelastic stress studies and their application to a mechanical analysis of the Tertiary ring-complex of Ardnamurchan, Argyllshire. *Proceedings of the Geologists' Association*, **78**, 289–318.

Emeleus, C.H. (1973) Granophyre pebbles in Tertiary conglomerate on the Isle of Canna, Inverness-shire. *Scottish Journal of Geology*, **9**, 157–9.

Emeleus, C.H. (1980) Rhum: Solid geology map, 1:20 000. Nature Conservancy Council, Scotland.

Emeleus, C.H. (1982) The central complexes. In *Igneous Rocks of the British Isles* (ed. D.S. Sutherland), Wiley, Chichester, pp. 369–414.

Emeleus, C.H. (1983) Tertiary igneous activity. In *Geology of Scotland* (ed. G.Y. Craig), 2nd edn, Scottish Academic Press, Edinburgh, pp. 357–98.

Emeleus, C.H. (1985) The Tertiary lavas and sediments of north-west Rhum, Inner Hebrides. *Geological Magazine*, **122**, 419–37.

Emeleus, C.H. (1987) The Rhum layered complex, Inner Hebrides, Scotland. In *Origins of igneous layering* (ed. I. Parsons), NATO ASI Series, Series C: Mathematical and Physical Sciences, **196**, 263–86, Reidel, Dordrecht.

Emeleus, C.H. (in preparation) *The geology of Rum and adjoining islands*. Memoir of the British Geological Survey.

Emeleus, C.H. and Forster, R.M. (1979) *Field guide to the Tertiary igneous rocks of Rhum*. Nature Conservancy Council, London, 44 pp.

Emeleus, C.H., Dunham, A.C. and Thompson, R.N. (1971) Iron-rich pigeonite from acid rocks in the Tertiary Igneous Province, Scotland. *American Mineralogist*, **56**, 940–51.

Emeleus, C.H., Wadsworth, W.J. and Smith, N.J. (1985) The early igneous and tectonic history of the Rhum Tertiary Volcanic Centre. *Geological Magazine*, **122**, 451–7.

England, R.W. (1988) The ascent and emplacement of granitic magmas: the Northern Arran Granite. Unpublished Ph.D. Thesis, University of Durham.

England, R.W. (1990) The identification of granite diapirs. *Journal of the Geological Society of London*, **147**, 931–4.

Esson, J., Dunham, A.C. and Thompson, R.N. (1975) Low alkali, high calcium olivine-tholeiite lavas from the Isle of Skye, Scotland. *Journal of Petrology*, **16**, 488–97.

Evans, A.L. (1969) On dating the British Tertiary Igneous Province. Unpublished Ph.D. Thesis, University of Cambridge.

Evans, A.L., Fitch, F.J. and Miller, J.A. (1973) Potassium–argon age determinations on some British Tertiary igneous rocks. *Journal of the Geological Society of London*, **129**, 419–43.

Faithfull, J.W. (1985) The Lower Eastern Layered Series of Rhum. *Geological Magazine*, **122**, 459–68.

Fenner, C.N. (1937) A view of magmatic differentiation. *Journal of Geology*, **45**, 158–68.

Flett, W.R. (1942) The contact between the granites of North Arran. *Transactions of the Geological Society of Glasgow*, **20**, 180–204.

Forester, R.W. and Taylor, H.P., Jun. (1976) ^{18}O-depleted igneous rocks from the Tertiary complex of the Isle of Mull, Scotland. *Earth and Planetary Science Letters*, **32**, 11–17.

Forester, R.W. and Taylor, H.P., Jun. (1977) ^{18}O/^{16}O, D/H and ^{13}C/^{12}C studies of the Tertiary igneous complex of Skye, Scotland. *American Journal of Science*, **277**, 136–77.

Forster, R.M. (1980) A geochemical and petrological study of the Tertiary minor intrusions of Rhum, north-west Scotland. Unpublished Ph.D. Thesis, University of Durham.

Gardner, J.S. (1887) On the leaf-beds and gravels of Ardtun, Carsaig, etc. in Mull. *Quarterly Journal of the Geological Society of London*, **43**, 270–300.

Gass, I.G. and Thorpe, R.S. (1976) Igneous case study: the Tertiary igneous rocks of Skye, NW Scotland. In *Science: a third level course. Earth science topics and methods*, (ed. F. Aprahamian), Open University Press, Milton Keynes.

Geikie, A. (1888) The history of volcanic action during the Tertiary period in the British Isles. *Transactions of the Royal Society of Edinburgh*, **35**, 21–184.

Geikie, A. (1894) On the relations of the basic and acid rocks of the Tertiary volcanic series of the Inner Hebrides. *Quarterly Journal of the Geological Society of London*, **50**, 212–31.

Geikie, A. (1897) *The Ancient Volcanoes of Great Britain*. 2 vols, Macmillan, London.

Geikie, A. and Teall, J.J.H. (1894) On the banded structure of some Tertiary gabbros in the Isle of Skye. *Quarterly Journal of the Geological Society of London*, **50**, 645–60.

Gibb, F.G.F. (1973) The zoned clinopyroxenes of the Shiant Isles Sill, Scotland. *Journal of Petrology*, **14**, 203–30.

Gibb, F.G.F. (1976) Ultrabasic rocks of Rhum and

Skye: the nature of the parent magma. *Journal of the Geological Society of London*, **132**, 209–22.

Gibb, F.G.F. and Henderson, C.M.B. (1978a) The petrology of the Dippin Sill, Isle of Arran. *Scottish Journal of Geology*, **14**, 1–27.

Gibb, F.G.F. and Henderson, C.M.B. (1978b) Possible higher pressure relics within titaniferous augites in a basic sill. *Geological Magazine*, **115**, 55–62.

Gibb, F.G.F. and Henderson, C.M.B. (1984) The structure of the Shiant Isles Sill Complex, Outer Hebrides. *Scottish Journal of Geology*, **20**, 21–9.

Gibb, F.G.F. and Henderson, C.M.B. (1989) Discontinuities between picritic and crinanitic units in the Shiant Isles Sill: evidence of multiple intrusion. *Geological Magazine*, **126**, 127–37.

Gibson, S.A. (1988) The geochemistry, mineralogy and petrology of the Trotternish Sill Complex, northern Skye, Scotland. Unpublished Ph.D. Thesis, Kingston Polytechnic.

Gibson, S.A. (1990) The geochemistry of the Trotternish sills, Isle of Skye: crustal contamination in the British Tertiary Volcanic Province. *Journal of the Geological Society of London*, **147**, 1071–81.

Gibson, S.A. and Jones, A.P. (1990) Igneous stratigraphy and internal structure of the Little Minch Sill Complex, Trotternish Peninsula, northern Skye, Scotland. *Geological Magazine*, **128**, 51–66.

Green, J. and Wright, J.B. (1969) Ardnamurchan Centre 1 – does it need re-defining? *Geological Magazine*, **106**, 599–601.

Green, J. and Wright, J.B. (1974) Ardnamurchan, Centre 1 – new radiometric evidence. *Geological Magazine*, **111**, 163–4.

Greenwood, R.C. (1987) Geology and petrology of the margin of the Rhum ultrabasic intrusion, Inner Hebrides, Scotland. Unpublished Ph.D. Thesis, University of St Andrews.

Greenwood, R.C., Donaldson, C.H. and Emeleus, C.H. (1990) The contact zone of the Rhum ultrabasic intrusion: evidence of peridotite formation from magnesian magmas. *Journal of the Geological Society of London*, **147**, 209–12.

Gribble, C.D. (1974) The dolerites of Ardnamurchan. *Scottish Journal of Geology*, **10**, 71–89.

Gribble, C.D., Durrance, E.M. and Walsh, J.N. (1976) *Ardnamurchan: a Guide to Geological Excursions*. Edinburgh Geological Society, Edinburgh, 122 pp and map.

Gunn, W. (1900) On the old volcanic rocks of the Island of Arran. *Transactions of the Geological Society of Glasgow*, **11**, 174–91.

Gunn, W. (1903) *The Geology of North Arran, South Bute and the Cumbraes, with Parts of Ayrshire and Kintyre*. Memoir of the Geological Survey of Great Britain, HMSO, Edinburgh.

Halsall, T.J. (1978) The emplacement of the Tertiary dykes of the Kildonan shore, south Arran. *Journal of the Geological Society of London*, **135**, p. 462 (abstract).

Hampton, C.M. and Taylor, P.N. (1983) The age and nature of the basement of southern Britain: evidence from Sr and Pb isotopes in granites. *Journal of the Geological Society of London*, **140**, 499–509.

Harding, R.R. (1966) The Mullach Sgar Complex, St Kilda, Outer Hebrides. *Scottish Journal of Geology*, **2**, 165–78.

Harding, R.R. (1967) The major ultrabasic and basic intrusions of St Kilda, Outer Hebrides. *Transactions of the Royal Society of Edinburgh*, **66**, 419–44.

Harding, R.R., Merriman, R.J. and Nancarrow, P.H.A. (1984) *St Kilda: an illustrated account of the geology*. Report of the British Geological Survey, 16, 46 pp and 1:25 000 map.

Harker, A. (1904) *The Tertiary Igneous Rocks of Skye*. Memoir of the Geological Survey of Great Britain, HMSO, Edinburgh.

Harker, A. (1908) *The Geology of the Small Isles of Inverness-shire*. Memoir of the Geological Survey of Great Britain, HMSO, Edinburgh.

Harker, A. (1917) Some aspects of igneous action in Britain. *Proceedings of the Geological Society of London*, **73**, lxvii–xcvi.

Harland, W.B., Cox, A.V. and Llewellyn, P.G. *et al.* (1982) *A Geologic Time Scale*. Cambridge University Press, Cambridge, 131 pp.

Harris, J.P. and Hudson, J.D. (1980) Lithostratigraphy of the Great Estuarine Group (Middle Jurassic), Inner Hebrides. *Scottish Journal of Geology*, **16**, 231–50.

Harrison, R.K. (ed.) (1975) *Expeditions to Rockall 1971–1972*. Report of the Institute of Geological Sciences, No. 75/1, 72 pp.

Harrison, R.K. (1982) Mesozoic magmatism in the British Isles and adjacent areas. In *Igneous Rocks of the British Isles* (ed. D.S. Sutherland), Wiley, Chichester, pp. 333–41.

Harrison, R.K., Stone, P. and Cameron, I.B. *et al.* (1987) Geology, petrology and geochemistry of Ailsa Craig, Ayrshire. *Report of the British*

Geological Survey, **16/9**, 29 pp.

Heddle, M.F. (1884) On the geognosy and mineralogy of Scotland. *Mineralogical Magazine*, **5**, 41–71.

Henderson, C.M.B. and Gibb, F.G.F. (1977) Formation of analcime in the Dippin Sill, Isle of Arran. *Mineralogical Magazine*, **41**, 534–7.

Hodgson, B.D., Dagley, P. and Mussett, A.E. (1990) Magnetostratigraphy of the Tertiary igneous rocks of Arran. *Scottish Journal of Geology*, **26**, 99–118.

Hoersch, A.L. (1979) General structure of the Skye Tertiary igneous complex and detailed structure of the Beinn an Dubhaich Granite from magnetic evidence. *Scottish Journal of Geology*, **15**, 231–45.

Hoersch, A.L. (1981) Progressive metamorphism of the chert-bearing Durness Limestone in the Beinn an Dubhaich aureole, Isle of Skye, Scotland: a re-examination. *American Mineralogist*, **66**, 491–506.

Holland, J.G. and Brown, G.M. (1972) Hebridean tholeiitic magmas: a geochemical study of the Ardnamurchan cone sheets. *Contributions to Mineralogy and Petrology*, **37**, 139–60.

Holmes, A. (1936) The idea of contrasted differentiation. *Geological Magazine*, **73**, 228–38.

Holmes, A. and Harwood, H.F. (1929) The tholeiite dykes of the north of England. *Mineralogical Magazine*, **22**, 1–52.

Hornung, G., Al-Ani, A. and Stewart, R.M. (1966) The composition and emplacement of the Cleveland Dyke. *Transactions of the Leeds Geologists' Association*, **7**, 232–49.

Hughes, C.J. (1960a) The Southern Mountains Igneous Complex, Isle of Rhum. *Quarterly Journal of the Geological Society of London*, **116**, 111–38.

Hughes, C.J. (1960b) An occurrence of tilleyite-bearing limestones in the Isle of Rhum, Inner Hebrides. *Geological Magazine*, **97**, 384–8.

Hunter, R.H. (1987) Textural equilibrium in layered rocks. In *Origins of Igneous Layering*, (ed. I. Parsons), NATO ASI Series, Series C: Mathematical and Physical Sciences, **196**, 473–504, Reidel, Dordrecht.

Huppert, H.E. and Sparks, R.S.J. (1980) The fluid dynamics of a basaltic magma chamber replenished by influx of hot, dense, ultrabasic magma. *Contributions to Mineralogy and Petrology*, **75**, 279–89.

Hutchison, R. (1964) The Tertiary basic igneous rocks of the Western Cuillin, Isle of Skye. Unpublished Ph. D. Thesis, University of Glasgow.

Hutchison, R. (1966a) The age relationships between the Sgùrr Dubh ultrabasic laccolite and Cuillin gabbros. *Scottish Journal of Geology*, **2**, 227–8.

Hutchison, R. (1966b) Intrusive tholeiites of the western Cuillin, Isle of Skye. *Geological Magazine*, **103**, 352–63.

Hutchison, R. (1968) Origin of White Allivalite, western Cuillin, Isle of Skye. *Geological Magazine*, **105**, 338–47.

Hutchison, R. and Bevan, J.C. (1977) The Cuillin layered igneous complex – evidence for multiple intrusion and former presence of a picritic liquid. *Scottish Journal of Geology*, **13**, 197–209.

Hutton, J. (1795) *Theory of the Earth, with Proofs and Illustrations*. Edinburgh.

Irvine, T.N. (1987) Layering and related structures in the Duke Island and Skaergaard intrusions: similarities, differences and origins. In *Origins of Igneous Layering* (ed. I. Parsons), NATO ASI Series, Series C: Mathematical and Physical Sciences, **196**, 185–246, Reidel, Dordrecht.

Jassim, S.Z. and Gass, I.G. (1970) The Loch na Crèitheach volcanic vent, Isle of Skye. *Scottish Journal of Geology*, **6**, 285–94.

Johnston, R. (1953) The olivines of the Garbh Eilean Sill, Shiant Isles. *Geological Magazine*, **90**, 161–71.

Judd, J.W. (1874) The Secondary rocks of Scotland. Second Paper. On the ancient volcanoes of the Highlands and the relations of their products to the Mesozoic strata. *Quarterly Journal of the Geological Society of London*, **30**, 220–301.

Judd, J.W. (1878) The Secondary rocks of Scotland. Third Paper. The strata of the western coast and islands. *Quarterly Journal of the Geological Society of London*, **34**, 660–743.

Judd, J.W. (1885) On the Tertiary and older peridotites of Scotland. *Quarterly Journal of the Geological Society of London*, **41**, 354–418.

Judd, J.W. (1886) On the gabbros, dolerites and basalts of Tertiary age in Scotland and Ireland. *Quarterly Journal of the Geological Society of London*, **42**, 49–97.

Judd, J.W. (1889) The Tertiary volcanoes of the Western Isles of Scotland. *Quarterly Journal of the Geological Society of London*, **45**, 187–219.

Judd, J.W. (1890) The propylites of the Western Isles of Scotland, and their relation to the andesites and diorites of the district. *Quarterly Journal of the Geological Society of London*, **46**, 341–85.

Judd, J.W. (1893) On composite dykes in Arran. *Quarterly Journal of the Geological Society of London*, **49**, 536–65.

Judd, J.W. (1897) On the petrology of Rockall. *Transactions of the Royal Irish Academy*, **31**, 48–58.

Kanaris-Sotiriou, R. and Gibb, F.G.F. (1985) Hybridization and the petrogenesis of composite intrusions: the dyke at An Cumhann, Isle of Arran, Scotland. *Geological Magazine*, **122**, 361–72.

Kennedy, W.Q. (1931a) The parent magma of the British Tertiary Province. *Geological Survey of Great Britain, Summary of Progress* (for 1930), **II**, 61–73.

Kennedy, W.Q. (1931b) On composite lava flows. *Geological Magazine*, **68**, 166–81.

Kennedy, W.Q. (1933) Trends of differentiation in basaltic magmas. *American Journal of Science*, **25**, 239–56.

Kennedy, W.Q. and Anderson, E.M. (1938) Crustal layers and the origin of magmas. *Bulletin Volcanologique*, séries II, tome III, 23–82.

Kille, I.C., Thompson, R.N., Morrison, M.A. *et al.* (1986) Field evidence for turbulence during flow of basalt magma through conduits from south-west Mull. *Geological Magazine*, **123**, 693–7.

King, B.C. (1953) Structure and igneous activity in the Creag Strollamus area of Skye. *Transactions of the Royal Society of Edinburgh*, **62**, 357–402.

King, B.C. (1955) The Ard Bheinn area of the Central Igneous Complex of Arran. *Quarterly Journal of the Geological Society of London*, **110** (for 1954), 323–55.

King, B.C. (1960) The form of the Beinn an Dubhaich granite, Skye. *Geological Magazine*, **97**, 326–33.

King, P.M. (1977) The secondary minerals of the Tertiary lavas of northern and central Skye – zeolite zonation patterns, their origin and formation. Unpublished Ph.D. Thesis, University of Aberdeen.

Knapp, R.J. (1973) The form and structure of the Islay, Jura and Arran Tertiary basic dyke swarms. Unpublished Ph.D. Thesis, University of London.

Koomans, C. and Kuenen, Ph.H. (1938) On the differentiation of the Glen More ring-dyke, Mull. *Geological Magazine*, **75**, 145–60.

Kuenen, Ph.H. (1937) Intrusion of cone-sheets. *Geological Magazine*, **74**, 177–83.

Le Bas, M.J. (1959) The term eucrite. *Geological Magazine*, **96**, 497–502.

Le Bas, M.J. (1971) Cone-sheets as a mechanism of uplift. *Geological Magazine*, **108**, 373–6.

Lee, G.W. and Bailey, E.B. (1925) *The Pre-Tertiary Geology of Mull, Loch Aline and Oban*. Memoir of the Geological Survey of Great Britain, HMSO, Edinburgh.

Le Maitre, R.W. (1989) *A Classification of Igneous Rocks and Glossary of Terms*. Blackwell Scientific Publications, Oxford.

Lewis, J.D. (1968) Form and structure of the Loch Bà ring-dyke, Isle of Mull. *Proceedings of the Geological Society of London*, **1649**, 110–11.

Longman, C.D. and Coward, M.P. (1979) Deformation around the Beinn an Dubhaich granite, Skye. *Scottish Journal of Geology*, **15**, 301–11.

MacCulloch, J. (1819) *A Description of the Western Islands of Scotland, including the Isle of Man. Comprising an account of their Geological Structure; with Remarks on their Agriculture, Scenery, and Antiquities.* 3 vols, Hurst Robinson, London.

MacDonald, G.A. and Katsura, T. (1964) Chemical composition of Hawaiian lavas. *Journal of Petrology*, **5**, 82–133.

MacDonald, J.G. and Herriot, A. (1983) *Macgregor's excursion guide to the Geology of Arran*. 3rd edition (revised), Geological Society of Glasgow, Glasgow, 210 pp.

MacDonald, R., Wilson, L., Thorpe, R.S. and Martin, A. (1988) Emplacement of the Cleveland Dyke: evidence from geochemistry, mineralogy, and physical modelling. *Journal of Petrology*, **29**, 559–83.

MacGregor, A.G. (1931) Clouded feldspars and thermal metamorphism. *Mineralogical Magazine*, **22**, 524–38.

Macgregor, M. (1965) *Excursion Guide to the Geology of Arran*. Geological Society of Glasgow, Glasgow, 192 pp.

Macintyre, R.M. (1973) Lower Tertiary geochronology of the North Atlantic continental margins. In *Geochronology and Isotope Geology of Scotland*. Field guide and reference.

Marshall, L.A. and Sparks, R.S.J. (1984) Origins of some mixed-magma and net-veined ring intrusions. *Journal of the Geological Society of London*, **141**, 171–82.

Martin, J.H. (1969) A petrographic study of the allivalite dykes of north Skye. Unpublished Ph.D. Thesis, University of St Andrews.

Mattey, D.P., Gibson, I.L., Marriner, G.F. *et al.* (1977) The diagnostic geochemistry, relative abundance, and spatial distribution of high-calcium, low-alkali olivine tholeiite dykes in the Lower Tertiary regional swarm of the Isle of Skye, N.W. Scotland. *Mineralogical Magazine*, **41**, 273–85.

McBirney, A.R. (1975) Differentiation of the Skaergaard intrusion. *Nature*, **253**, 691–4.

McBirney, A.R. and Noyes, R.M. (1979) Crystallization and layering of the Skaergaard intrusion. *Journal of Petrology*, **20**, 487–554.

McClurg, J.E. (1982) Geology and structure of the northern part of the Rhum ultrabasic complex. Unpublished Ph.D. Thesis, University of Edinburgh.

McKerrow, W.S. and Atkins, F.B. (1985) *Isle of Arran: a Field Guide for Students of Geology*. Geologists' Association guide, 96 pp.

McQuillin, J. and Tuson, J. (1963) Gravity measurements over the Rhum Tertiary plutonic complex. *Nature*, **199**, 1276–7.

McQuillin, R., Bacon, M. and Binns, P.E. (1975) The Blackstones Tertiary igneous complex. *Scottish Journal of Geology*, **11**, 179–92.

Meighan, I.G., Hutchison, R. and Williamson, I.T. (1981) Geological evidence for the different relative ages of the Rhum and Skye Tertiary central complexes. *Geological Society of London Newsletter*, **10**, p. 12 (Abstract).

Meighan, I.G., McCormick, A.G., Gibson, D. *et al.* (1988) Rb–Sr isotopic determinations and the timing of Tertiary central complex magmatism in NE Ireland. In *Early Tertiary Volcanism and the Opening of the NE Atlantic*, (eds A.C. Morton and L.M. Parson), Geological Society Special Publication, No. 39, pp. 349–60.

Miller, J.A. and Brown, P.E. (1965) Potassium–argon age studies in Scotland. *Geological Magazine*, **102**, 106–34.

Mitchell, J.G. and Reen, K.P. (1973) Potassium–argon ages from the Tertiary ring complexes of the Ardnamurchan Peninsula, western Scotland. *Geological Magazine*, **110**, 331–40.

Moorbath, S. and Bell, J.D. (1965) Strontium isotope abundance studies and rubidium–strontium age determinations on Tertiary igneous rocks from the Isle of Skye, north-west Scotland. *Journal of Petrology*, **6**, 37–66.

Moorbath, S. and Thompson, R.N. (1980) Strontium isotope geochemistry and petrogenesis of the Early Tertiary lava pile of the Isle of Skye, Scotland, and other basic rocks of the British Tertiary Province: an example of magma–crust interaction. *Journal of Petrology*, **21**, 295–321.

Moorbath, S. and Welke, H. (1969) Lead isotope studies on igneous rocks from the Isle of Skye, north-west Scotland. *Earth and Planetary Science Letters*, **5**, 217–30.

Morrison, M.A. (1978) The use of 'immobile' trace elements to distinguish palaeotectonic affinities of metabasalts: applications to the Palaeocene basalts of Mull and Skye, NW Scotland. *Earth and Planetary Science Letters*, **39**, 407–16.

Morrison, M.A. (1979) Igneous and metamorphic geochemistry of Mull lavas. Unpublished Ph.D. Thesis, University of London.

Morrison, M.A., Thompson, R.N. and Dickin, A.P. (1985) Geochemical evidence for complex magmatic plumbing during development of a continental volcanic centre. *Geology*, **13**, 581–4.

Muir, I.D. and Tilley, C.E. (1961) Mugearites and their place in alkali igneous rock series. *Journal of Geology*, **69**, 186–203.

Murray, R.J. (1954) The clinopyroxenes of the Garbh Eilean Sill, Shiant Isles. *Geological Magazine*, **91**, 17–31.

Mussett, A.E. (1984) Time and duration of Tertiary igneous activity of Rhum and adjacent areas. *Scottish Journal of Geology*, **20**, 273–9.

Mussett, A.E. (1986) ^{40}Ar-^{39}Ar step-heating ages of the Tertiary igneous rocks of Mull, Scotland. *Journal of the Geological Society of London*, **143**, 887–96.

Mussett, A.E., Brown, G.C., Eckford, M. *et al.* (1973) The British Tertiary Igneous Province: K–Ar ages of some dykes and lavas from Mull, Scotland. *Geophysical Journal of the Royal Astronomical Society*, **30**, 405–14.

Mussett, A.E., Dagley, P. and Skelhorn, R.R. (1980) Magnetostratigraphy of the Tertiary igneous succession of Mull, Scotland. *Journal of the Geological Society of London*, **137**, 349–57.

Mussett, A.E., Dagley, P. and Skelhorn, R.R. (1988) Time and duration of activity in the British Tertiary Igneous Province. In *Early Tertiary Volcanism and the opening of the NE Atlantic*, (eds A.C. Morton and L.M.

Parson), Geological Society of London Special Publication, No. 39, 337–48.

Necker de Saussure, L.A. (1840) Documents sur les Dykes de Trap d'une Partie de l'île d'Arran. *Transactions of the Royal Society of Edinburgh*, **14**, 677–98.

Nicholson, R. (1970) A note on deformed igneous sheets in the Durness Limestone of the Strath district of Skye. *Geological Magazine*, **107**, 229–33.

Nicholson, R. (1985) The intrusion and deformation of Tertiary minor sheet intrusions, west Suardal, Isle of Skye, Scotland. *Geological Journal*, **20**, 53–72.

Nielsen, T.F.D. (1987) Tertiary alkaline magmatism in East Greenland: a review. In *Alkaline Igneous Rocks* (eds J.G. Fitton and B.G.J. Upton), Geological Society Special Publication, No. 30, 489–516.

Nockolds, S.R. and Allen, R. (1954) The geochemistry of some igneous rock series, Part II. *Geochimica et Cosmochimica Acta*, **5**, 245–85.

Paithankar, M.G. (1968) Petrological study and intrusion history of the granophyre of Grigadale and associated gabbros, Ardnamurchan, Argyllshire, Scotland. *Neues Jahrbuch für Mineralogie*, **110**, 1–23.

Palacz, Z.A. and Tait, S.R. (1985) Isotopic and geochemical investigation of unit 10 from the Eastern Layered Series of the Rhum intrusion, north-west Scotland. *Geological Magazine*, **122**, 485–90.

Pankhurst, R.J., Walsh, J.N., Beckinsale, R.D. *et al.* (1978) Isotopic and other geochemical evidence for the origin of the Loch Uisg granophyre, Isle of Mull, Scotland. *Earth and Planetary Science Letters*, **38**, 355–63.

Parsons, I. (ed.) (1987) *Origins of Igneous Layering*. NATO ASI Series, Series C: Mathematical and Physical Sciences, **196**, Reidel, Dordrecht.

Peach, B.N., Gunn, W. and Newton, E.T. (1901) On a remarkable volcanic vent of Tertiary age in the Island of Arran, enclosing Mesozoic fossiliferous rocks. *Quarterly Journal of the Geological Society of London*, **62**, 226–43.

Peach, B.N., Horne, J., Woodward, H.B. *et al.* (1910) *The Geology of Glenelg, Lochalsh and South-East Part of Skye*. Memoir of the Geological Survey of Great Britain, HMSO, Edinburgh.

Pennant, T. (1774) *A Tour in Scotland and Voyage to the Hebrides, 1772*. Chester (Arran, volume 1, pp. 168–89).

Phillips, W.J. (1974) The dynamic emplacement of cone-sheets. *Tectonophysics*, **24**, 69–84.

Preston, J. (1963) The dolerite plug at Slemish, Co. Antrim, Ireland. *Liverpool and Manchester Geological Journal*, **3**, 301–14.

Ramsay, A.C. (1841) *The Geology of the Island of Arran from Original Survey*. Glasgow, 78 pp.

Rao, M.S. (1958) Composite and multiple intrusions of Lamlash–Whiting Bay region, Arran. *Geological Magazine*, **95**, 265–80.

Rao, M.S. (1959) Minor acid intrusions and dykes of the Lamlash–Whiting Bay region, Arran. *Geological Magazine*, **96**, 237–46.

Rast, D.E. (1968) Age relationships and geometry of the Knock and Beinn a Ghraig granophyres, Isle of Mull. *Proceedings of the Geological Society of London*, **1649**, 114–15.

Rast, N., Diggens, J.N. and Rast, D.E. (1968) Triassic rocks of the Isle of Mull: their sedimentation, facies, structure, and relationship to the Great Glen Fault and the Mull caldera. *Proceedings of the Geological Society of London*, **1645**, 299–304.

Ray, P.S. (1960) Ignimbrite in the Kilchrist vent, Skye. *Geological Magazine*, **97**, 229–38.

Ray, P.S. (1962) A note on some acid breccias in the Kilchrist vent, Skye. *Geological Magazine*, **99**, 420–6.

Ray, P.S. (1964) On the association of an indurated basic tuff and a felsite intrusion in the Kilchrist vent, Skye. *Geological Magazine*, **101**, 289–301.

Ray, P.S. (1966) An association of rhyolite and ignimbrite in the Kilchrist vent, Skye. *Geological Magazine*, **103**, 8–18.

Ray, P.S. (1972) A rhyolitic injection-breccia in tuff near Allt Slapin, Strath, Skye. *Geological Magazine*, **109**, 427–34.

Raybould, J.G. (1973) The form of the Beinn an Dubhaich Granite, Skye, Scotland. *Geological Magazine*, **110**, 341–50.

Renner, R. and Palacz, Z. (1987) Basaltic replenishment of the Rhum magma chamber: evidence from unit 14. *Journal of the Geological Society of London*, **144**, 961–70.

Reynolds, D.L. (1951) The geology of Slieve Gullion, Foughill and Carrickcarnan: an actualistic interpretation of a Tertiary gabbro–granophyre complex. *Transactions of the Royal Society of Edinburgh*, **62**, 85–143.

Reynolds, D.L. (1954) Fluidization as a geological process, and its bearing on the problem of intrusive granites. *American Journal of Sci-*

ence, **252**, 577–613.

Richey, J.E. (1928) The structural relations of the Mourne granites (Northern Ireland). *Quarterly Journal of the Geological Society of London*, **83** (for 1927), 653–88.

Richey, J.E. (1932) Tertiary ring structures in Britain. *Transactions of the Geological Society of Glasgow*, **19**, 42–140.

Richey, J.E. (1933) Summary of the geology of Ardnamurchan. *Proceedings of the Geologists' Association*, **44**, 1–56.

Richey, J.E. (1937) Some features of Tertiary volcanicity in Scotland and Ireland. *Bulletin Volcanologique*, séries II, tome **I**, 13–34.

Richey, J.E. (1938) The rhythmic eruptions of Ben Hiant, Ardnamurchan, a Tertiary volcano. *Bulletin Volcanologique*, séries II, tome **III**, 2–21.

Richey, J.E. (1940) Association of explosive brecciation and plutonic intrusion in the British Tertiary Igneous Province. *Bulletin Volcanologique*, séries II, tome **VI**, 157–75.

Richey, J.E. and Thomas, H.H. (1930) *The Geology of Ardnamurchan, North-west Mull and Coll*. Memoir of the Geological Survey of Great Britain, HMSO, Edinburgh.

Richey, J.E., Stewart, F.H. and Wager, L.R. (1946) Age relations of certain granites and marscoite in Skye. *Geological Magazine*, **83**, p. 293.

Richey, J.E. (1961) *British Regional Geology, Scotland: the Tertiary Volcanic Districts*. 3rd edn, revised by A.G. MacGregor and F.W. Anderson, HMSO, Edinburgh, 120 pp.

Ridley, W.I. (1971) The petrology of some volcanic rocks from the British Tertiary Province: the islands of Rhum, Eigg, Canna and Muck. *Contributions to Mineralogy and Petrology*, **32**, 251–66.

Ridley, W.I. (1973) *The Petrology of Volcanic Rocks from the Small Isles of Inverness-shire*. Report of the Institute of Geological Sciences, No. 73/10, 55 pp.

Roberts, D.G., Ardus, D.A. and Dearnley, R. (1973) Precambrian rocks drilled on the Rockall Bank. *Nature, Physical Sciences*, **244**, 21–3.

Roberts, D.G., Matthews, D.H. and Eden, R.A. (1972) Metamorphic rocks from the southern end of the Rockall Bank. *Journal of the Geological Society of London*, **128**, 501–6.

Roberts, D.G., Flemming, N.C. and Harrison, R.K. *et al.* (1974) Helen's Reef: a microgabbroic intrusion in the Rockall Intrusive Centre, Rockall Bank. *Marine Geology*, **16**, M21–M30.

Rogers, N.W. and Gibson, I.L. (1977) The petrology and geochemistry of the Creag Dubh composite sill, Whiting Bay, Arran, Scotland. *Geological Magazine*, **114**, 1–8.

Ross, A. (1884) A visit to St Kilda. *Transactions of the Inverness Scientific Society*, **3**, 72–91.

Sabine, P.A. (1960) The geology of Rockall, North Atlantic. *Bulletin of the Geological Survey of Great Britain*, **16**, 156–78.

Simpson, J.B. (1961) The Tertiary pollen-flora of Mull and Ardnamurchan. *Transactions of the Royal Society of Edinburgh*, **64**, 421–68.

Simkin, T.E. (1965) The picritic sills of north Skye, Scotland. Unpublished Ph.D. Thesis, University of Princeton.

Simkin, T.E. (1967) Flow differentiation in the picritic sills of north Skye. In *Ultramafic and Related rocks*. (ed. P.J. Wyllie), Wiley, New York, pp. 64–9.

Skelhorn, R.R. (1969) *The Tertiary Igneous Geology of the Isle of Mull*. Geologists' Association Guide, No. 20.

Skelhorn, R.R. and Elwell, R.W.D. (1966) The structure and form of the granophyric quartz-dolerite intrusion, Centre II, Ardnamurchan, Argyllshire. *Transactions of the Royal Society of Edinburgh*, **66**, 285–306.

Skelhorn, R.R. and Elwell, R.W.D. (1971) Central subsidence in the layered hypersthene-gabbro of Centre II, Ardnamurchan, Argyllshire. *Journal of the Geological Society of London*, **127**, 535–51.

Smith, D.I. (1957) The structure and petrology of the third ring-dyke complex, Ardnamurchan. Unpublished Ph.D. Thesis, University of Edinburgh.

Smith, J. (1896) A new view of the Arran granite mountains. *Transactions of the Geological Society of Glasgow*, **10**, 216–56.

Smith, N.J. (1985) The age and structural setting of limestones and basalts on the Main Ring Fault in south-east Rhum. *Geological Magazine*, **122**, 439–45.

Smith, N.J. (1987) The age and structure of limestone and basalt on the Main Ring Fault of south-east Rhum, Inner Hebrides, Scotland. Unpublished M.Sc. Thesis, University of Durham.

Sparks, R.S.J. (1988) Petrology and geochemistry of the Loch Bà ring-dyke, Mull (NW Scotland): an example of the extreme differentiation of tholeiitic magmas. *Contributions to Mineralogy and Petrology*, **100**, 446–61.

Sparks, R.S.J., Huppert, H.E. and Turner, J.S.

(1984) The fluid dynamics of evolving magma chambers. *Philosophical Transactions of the Royal Society of London*, **A310**, 511–31.

Sparks, R.S.J., Huppert, H.E., Kerr, R.C. *et al.* (1985) Postcumulus processes in layered intrusions. *Geological Magazine*, **122**, 555–68.

Speight, J.M., Skelhorn, R.R., Sloan, T., *et al.* (1982) The dyke swarms of Scotland. In *Igneous Rocks of the British Isles.* (ed. D.S. Sutherland), Wiley, Chichester, pp. 449–59.

Stewart, F.H. (1965) Tertiary igneous activity. In *The Geology of Scotland.* (ed. G.Y. Craig), 1st edn, Oliver and Boyd, Edinburgh, pp. 417–65.

Streckeisen, A. (1978) IUGS Subcommission on the systematics of igneous rocks. Classification and nomenclature of volcanic rocks, lamprophyres, carbonatites and melilite rocks. Recommendations and suggestions. *Neues Jahrbuch für Mineralogie*, Stuttgart, Abhandlung, **143**, 1–14.

Sutherland, D.S. (ed.) (1982) *Igneous Rocks of the British Isles.* Wiley, Chichester, 645 pp.

Tait, S.R. (1985) Fluid dynamic and geochemical evolution of cyclic unit 10, Rhum, Eastern Layered Series. *Geological Magazine*, **122**, 469–84.

Tate, R. and Blake, J.F. (1876) *The Yorkshire Lias.* John van Voorst, London.

Taylor, H.P. Jun. and Forester, R.W. (1971) Low-^{18}O igneous rocks from the intrusive complexes of Skye, Mull and Ardnamurchan, western Scotland. *Journal of Petrology*, **12**, 465–97.

Teall, J.J.H. (1884) Petrological notes on some north-of-England dykes. *Quarterly Journal of the Geological Society of London*, **40**, 209–47.

Teall, J.J.H. (1888) *British Petrography.* Dulau, London, 469 pp.

Thirlwall, M.F. and Jones, N.W. (1983) Isotope geochemistry and contamination mechanics of Tertiary lavas from Skye, north-west Scotland. In *Continental Basalts and Mantle Xenoliths* (eds C.J. Hawkesworth and M.J. Norry), Shiva, Nantwich, 186–208.

Thomas, H.H. (1922) Xenolithic Tertiary minor intrusions in the Island of Mull. *Quarterly Journal of the Geological Society of London*, **78**, 229–60.

Thompson, R.N. (1969) Tertiary granites and associated rocks of the Marsco area, Isle of Skye. *Quarterly Journal of the Geological Society of London*, **124**, 349–85.

Thompson, R.N. (1982) Magmatism of the British Tertiary Volcanic Province. *Scottish Journal of Geology*, **18**, 49–107.

Thompson, R.N. and Morrison, M.A. (1988) Asthenospheric and lower lithospheric contributions to continental extensional magmatism: an example from the British Tertiary Province. *Journal of Geophysical Research*, **91**, 5985–97.

Thompson, R.N., Esson, J. and Dunham, A.C. (1972) Major element chemical variation in the Eocene lavas of the Isle of Skye, Scotland. *Journal of Petrology*, **13**, 219–53.

Thompson, R.N., Gibson, I.L., Marriner, G.F. *et al.* (1980) Trace-element evidence of multistage mantle fusion and polybaric fractional crystallization in the Palaeocene lavas of Skye, NW Scotland. *Journal of Petrology*, **21**, 265–93.

Thompson, R.N., Dickin, A.P. and Gibson, I.L. (1982) Elemental fingerprints of isotopic contamination of Hebridean Palaeocene mantle-derived magmas by Archaean sial. *Contributions to Mineralogy and Petrology*, **79**, 159–68.

Thompson, R.N., Morrison, M.A., Dickin, A.P. *et al.* (1986) Two contrasting styles of interaction between basic magmas and continental crust in the British Tertiary Volcanic Province. *Journal of Geophysical Research*, **91**, B6, 5985–97.

Tilley, C.E. (1947) The gabbro–limestone contact zone of Camas Mor, Muck, Inverness-shire. *Comptes Rendus de la Société geologique de Finlande*, No. **140**, 97–105.

Tilley, C.E. (1949) An alkali facies of granite at granite–dolomite contacts in Skye. *Geological Magazine*, **86**, 81–93.

Tilley, C.E. (1951) The zoned contact-skarns of the Broadford area, Skye: a study of boron–fluorine metasomatism in dolomites. *Mineralogical Magazine*, **29**, 621–66.

Tilley, C.E. (1952) Some trends of basaltic magma in limestone syntexis. *American Journal of Science* (Bowen Volume), 529–45.

Tilley, C.E. and Harwood, H.F. (1931) The dolerite–chalk contact of Scawt Hill, Co. Antrim. The production of basic alkali-rocks by the assimilation of limestone by basaltic magma. *Mineralogical Magazine*, **22**, 439–68.

Tilley, C.E. and Muir, I.D. (1962) The Hebridean plateau magma type. *Transactions of the Edinburgh Geological Society*, **19**, 208–15.

Tilley, C.E. and Muir, I.D. (1967) Tholeiite and tholeiitic series. *Geological Magazine*, **104**, 337–43.

Tomkeieff, S.I. (1942) The Tertiary lavas of Rum. *Geological Magazine*, **79**, 1–13.

Tomkeieff, S.I. (1961) *Isle of Arran*. Excursion Guide, No. 32, Geologists' Association.

Tomkeieff, S.I. (1969) *Isle of Arran*. Excursion Guide, No. 32 (revised edition), Geologists' Association, 35 pp.

Tuson, J. (1959) A geophysical investigation of the Tertiary volcanic districts of western Scotland. Unpublished Ph.D. Thesis, University of Durham.

Tuttle, O.F. and Bowen, N.L. (1958) Origin of granite in the light of experimental studies in the system $NaAlSi_3O_8$–$KAlSi_3O_8$–SiO_2–H_2O. *Geological Society of America Memoir*, **74**.

Tuttle, O.F. and Keith, M.L. (1954) The granite problem: evidence from the quartz and feldspar of a Tertiary granite. *Geological Magazine*, **91**, 61–72.

Tyrrell, G.W. (1928) *The Geology of Arran*. Memoir of the Geological Survey of Great Britain, HMSO, Edinburgh.

Upton, B.G.J. (1988) History of Tertiary igneous activity in the N. Atlantic borderlands. In *Early Tertiary Volcanism and the Opening of the NE Atlantic* (eds A.C. Morton and L.M. Parson), Geological Society Special Publication, No. 39, pp. 3–14.

Vann, I.R. (1978) The siting of Tertiary vulcanicity. In *Crustal Evolution in Northwestern Britain and Adjacent Regions* (eds D.R. Bowes and B.E. Leake), Geological Journal Special Issue, No. 10, 393–414.

Vogel, T.A. (1982) Magma mixing in the acidic–basic complex of Ardnamurchan: implications on the evolution of shallow magma chambers. *Contributions to Mineralogy and Petrology*, **79**, 411–23.

Vogel, T.A., Younker, L.W., Wilband, J.T. *et al.* (1984) Magma mixing: the Marsco Suite, Isle of Skye, Scotland. *Contributions to Mineralogy and Petrology*, **87**, 231–41.

Volker, J.A. (1983) The geology of the Trallval area, Rhum, Inner Hebrides. Unpublished Ph.D. Thesis, University of Edinburgh.

Volker, J.A. and Upton, B.G.J. (1990) The structure and petrogenesis of the Trallval and Ruinsival areas of the Rhum Ultrabasic Complex. *Transactions of the Royal Society of Edinburgh: Earth Sciences*, **81**, 69–88.

Wadsworth, W.J. (1961) The layered ultrabasic rocks of south-west Rhum, Inner Hebrides. *Philosophical Transactions of the Royal Society*, **B244**, 21–64.

Wadsworth, W.J. (1982) The major basic intrusions. In *Igneous rocks of the British Isles*. (ed. D.S. Sutherland), Wiley, Chichester, pp. 415–25.

Wager, L.R. (1956) A chemical definition of fractionation stages as a basis for comparison of Hawaiian, Hebridean and other basic lavas. *Geochimica et Cosmochimica Acta*, **9**, 217–48.

Wager, L.R. and Bailey, E.B. (1953) Basic magma chilled against acid magma. *Nature, London*, **172**, 68–9.

Wager, L.R. and Brown, G.M. (1951) A note on rhythmic layering in the ultrabasic rocks of Rhum. *Geological Magazine*, **88**, 166–8.

Wager, L.R. and Brown, G.M. (1968) *Layered Igneous Rocks*. Oliver and Boyd, Edinburgh, 588 pp.

Wager, L.R. and Deer, W.A. (1939) Geological investigations in East Greenland, Part III – The petrology of the Skaergaard intrusion, Kangerdlugssuaq, East Greenland. *Meddelelser om Grønland*, **105**, 1–352.

Wager, L.R. and Vincent, E.A. (1962) Ferrodiorite from the Isle of Skye. *Mineralogical Magazine*, **33**, 26–36.

Wager, L.R., Brown, G.M. and Wadsworth, W.J. (1960) Types of igneous cumulates. *Journal of Petrology*, **1**, 73–85.

Wager, L.R., Vincent, E.A., Brown, G.M. *et al.* (1965) Marscoite and related rocks from the Western Red Hills complex, Isle of Skye. *Philosophical Transactions of the Royal Society*, **A257**, 273–307.

Wager, L.R., Weedon, D.S. and Vincent, E.A. (1953) A granophyre from Coire Uaigneich, Isle of Skye, containing quartz paramorphs after tridymite. *Mineralogical Magazine*, **30**, 263–76.

Walker, F. (1930) The geology of the Shiant Isles (Hebrides). *Quarterly Journal of the Geological Society of London*, **86**, 355–98.

Walker, F. (1932) Differentiation in the sills of northern Trotternish (Skye). *Transactions of the Royal Society of Edinburgh*, **57**, 241–57.

Walker, G.P.L. (1959) Some observations on the Antrim basalts and associated dolerite intrusions. *Proceedings of the Geologists' Association*, **70**, 179–205.

Walker, G.P.L. (1960) Zeolite zones and dike distribution in relation to the structure of the basalts of eastern Iceland. *Journal of Geology*, **68**, 515–28.

Walker, G.P.L. (1971) The distribution of amygdale minerals in Mull and Morvern (western

Scotland). In *Studies in Earth Sciences, W.D. West Commemorative Volume.* (eds T.V.V.G.R.K. Murty and S.S. Rao), pp. 181–94.

Walker, G.P.L. (1975) A new concept of the evolution of the British Tertiary intrusive centres. *Journal of the Geological Society of London*, **131**, 121–41.

Walsh, J.N. (1971) The geochemistry and mineralogy of the Centre 3 igneous complex, Ardnamurchan. Unpublished Ph.D. Thesis, University of London.

Walsh, J.N. (1975) Clinopyroxenes and biotites from the Centre 3 igneous complex, Ardnamurchan, Argyllshire. *Mineralogical Magazine*, **40**, 335–45.

Walsh, J.N. and Henderson, P. (1977) Rare earth element patterns of rocks from the Centre 3 igneous complex, Ardnamurchan, Argyllshire. *Contributions to Mineralogy and Petrology*, **60**, 31–8.

Walsh, J.N., Beckinsale, R.D., Skelhorn, R.R. *et al.* (1979) Geochemistry and petrogenesis of Tertiary granitic rocks from the Island of Mull, north-west Scotland. *Contributions to Mineralogy and Petrology*, **71**, 99–116.

Washington, H.S. (1914) The composition of rockallite. *Quarterly Journal of the Geological Society of London*, **70**, 294–302.

Weedon, D.S. (1960) The Gars-bheinn ultrabasic sill, Isle of Skye. *Quarterly Journal of the Geological Society of London*, **116**, 37–54.

Weedon, D.S. (1961) Basic igneous rocks of the Southern Cuillin, Isle of Skye. *Transactions of the Geological Society of Glasgow*, **24**, 190–212.

Weedon, D.S. (1965) The layered ultrabasic rocks of Sgùrr Dubh, Isle of Skye. *Scottish Journal of Geology*, **1**, 41–68.

Wells, M.K. (1951) Sedimentary inclusions in the hypersthene-gabbro, Ardnamurchan, Argyllshire. *Mineralogical Magazine*, **29**, 715–36.

Wells, M.K. (1954a) The structure and petrology of the hypersthene–gabbro intrusion, Ardnamurchan, Argyllshire. *Quarterly Journal of the Geological Society of London*, **109** (for 1953), 367–97.

Wells, M.K. (1954b) The structure of the granophyric quartz–dolerite intrusion of Centre 2, Ardnamurchan, and the problem of net-veining. *Geological Magazine*, **91**, 293–307.

Wells, M.K. and McRae, D.G. (1969) Palaeomagnetism of the hypersthene–gabbro intrusion, Ardnamurchan. *Nature*, **223**, 608–9.

Whetton, J.T. and Myers, J.O. (1949) Geophysical survey of magnetite deposits in Strath, Isle of Skye. *Transactions of the Geological Society of Glasgow*, **21**, 263–77.

White, R.S. (1988) A hot-spot model for early Tertiary volcanism in the N. Atlantic. In *Early Tertiary Volcanism and the Opening of the NE Atlantic* (eds A.C. Morton and L.M. Parson), Geological Society Special Publication, No. 39, pp. 3–14.

Whitten, E.H.T. (1961) Modal variation and the form of the Beinn an Dubhaich Granite, Skye. *Geological Magazine*, **98**, 467–72.

Williams, P.J. (1985) Pyroclastic rocks in the Cnapan Breaca felsite, Rhum. *Geological Magazine*, **122**, 447–50.

Williamson, I.T. (1979) The petrology and structure of the Tertiary volcanic rocks of west-central Skye, N.W. Scotland. Unpublished Ph.D. Thesis, University of Durham.

Wills, K.J.A. (1970) The inner complex of Centre 3, Ardnamurchan. Unpublished B.Sc. Thesis, University of London (Royal School of Mines).

Wilson, G.V. (1937) In *Geological Survey of Great Britain, Summary of Progress* (for 1936), 77–9.

Woodcock, N.H. and Underhill, J.R. (1987) Emplacement-related fault patterns around the Northern Granite, Arran, Scotland. *Geological Society of America Bulletin*, 515–27.

Wyatt, M. (1952) The Camasunary (Skye) gabbro–limestone contact. Unpublished Ph.D. Thesis, University of Cambridge.

Wyllie, P.J. and Drever, H.I. (1963) The petrology of picritic rocks in minor intrusions – a picritic sill on the Island of Soay (Hebrides). *Transactions of the Royal Society of Edinburgh*, **65**, 155–77.

Young, I.M., Greenwood, R.C. and Donaldson, C.H. (1988) Formation of the Eastern Layered Series of the Rhum Complex, north-west Scotland. *Canadian Mineralogist*, **26**, 225–33.

Zinovieff, P. (1958) The basic layered intrusion and the associated igneous rocks of the central and eastern Cuillin Hills, Isle of Skye. Unpublished D.Phil. Thesis, University of Oxford.

Zirkel, F. (1871) Geologische Skizzen von der Westküste Schottlands. *Zeitskrifte Deutsches Geologisches Gesellschaft*, **23**, 1–124.

Glossary

This glossary contains simple explanations of a selection of the more important technical terms used in Chapter 1 and in the Introduction, Highlights and Conclusions sections of Chapters 2 to 7. These explanations do not pretend to be scientific definitions. Rock groups are usually explained in terms of their chemistry rather than by reference to their precise mineral content. Only major mineral groups are included. Stratigraphical terms are omitted as they are related to their contexts within the tables and figures. Bold face indicates a further glossary entry.

Throughout the glossary and the volume as a whole the following grain (i.e. crystal) sizes are assumed for the igneous rocks:

coarse-grained – grains over 3mm on average
medium-grained – grains between 1 and 3mm on average
fine-grained – grains under 1mm on average (including non-crystalline glass)

Acid: coarse- to fine-grained **igneous rocks** relatively enriched in silica (SiO_2 nominally over 66%) which was originally thought to reflect the proportion of 'silicic acid'. An alternative term is 'silicic'.

Agglomerate: a **volcaniclastic rock** composed of large, often angular rock and mineral fragments (**clasts**).

Alkali-feldspar: see **feldspar**.

Allivalite: a coarse-grained **ultrabasic igneous rock** composed largely of the minerals plagioclase **feldspar** and **olivine**.

Amygdale: a **vesicle** infilled by minerals.

Aphyric: **igneous rocks**, especially those which are generally fine-grained, which contain no particularly large crystals; = non-**porphyritic**.

Arkose: sandstone containing abundant fragmental grains (**clasts**) of alkali-**feldspar**.

Aureole: the **metamorphic rocks** adjacent to an **igneous** intrusion.

Basalt: a fine-grained, **basic igneous rock** consisting largely of the minerals plagioclase **feldspar**, **pyroxene(s)** +/− **Olivine**. Usually a lava or a **dyke**.

Basic: coarse- to fine-grained **igneous rocks** relatively enriched in the 'bases' of early chemistry i.e. MgO, FeO, FeO_2, FeO_3CaO etc; silica (SiO_2) relatively low (nominally 45–53%.

Benmoreite: a fine-grained **igneous rock**, usually a lava, consisting essentially of soda-rich alkali-**feldspar**; see also **hawaiite**.

Bole: the iron-rich (sub-)soil produced by the surface weathering of **basalts**.

Breccia: a **volcaniclastic**, sedimentary, or fault-related rock composed of very large, usually angular rock fragments (**clasts**).

Caldera: a very large, approximately circular, fault-bounded basin formed by the collapse of a volcano.

Clast: a fragment.

Cone-sheet: a cone-like **igneous** intrusion which dips towards its centre, i.e. which closes downwards on projection. (cf. **ring-dyke**).

Crinanite: an alkali-(especially soda-)rich **dolerite** (or **gabbro**).

Diapir: a body, e.g. of **igneous rock/magma** which has risen through other rocks in consequence of its lower density and/or greater plasticity.

Glossary

Diorite: a coarse-grained, lime-rich **intermediate igneous rock** containing plagioclase **feldspar** and various ferromagnesian silicate minerals.

Dolerite: a medium-grained, **basic igneous rock** containing plagioclase **feldspar** and **pyroxene(s)**; usually a **dyke** or **sill**.

Dyke: a sheet-like body of **igneous rock** which cross-cuts the structure of the rocks it intrudes; often steeply inclined and composed of **dolerite** or **basalt** (cf. **sill**).

Eucrite: a coarse-grained, **ultrabasic igneous rock** containing plagioclase **feldspar**, **pyroxene(s)** and **olivine**.

Eutaxitic: the 'streaky' fabric (i.e. gross texture) exhibited by an **ignimbrite**.

Feldspars: a series of alumino-silicate minerals between lime-soda-rich (plagioclase) and potash-soda-rich (alkali-feldspar) end-members; the most abundant minerals in the earth's crust.

Felsite: medium- to fine-grained, equigranular, **acid igneous rock** consisting largely of alkali-**feldspar** and **quartz**; often a **dyke** or **sill**.

Gabbro: a coarse-grained, **basic igneous rock** consisting largely of plagioclase **feldspar**, **pyroxene(s)** +/− **olivine**; usually in large intrusions.

Gneiss: a coarse-grained, often banded **metamorphic rock**.

Granite: a coarse-grained, **acid igneous rock** consisting largely of alkali-**feldspar** and **quartz**; usually in large intrusions.

Granophyre: a medium- to coarse-grained **acid igneous rock** which often displays a complicated, angular ('graphic') intergrowth between **quartz** and alkali-**feldspar**; usually in large intrusions.

Granulite: an even-grained granular **metamorphic rock**.

Harrisite: a coarse-grained, **ultrabasic igneous rock** which displays branching crystals of **olivine**.

Hawaiite: a variety of **trachybasalt** rich in soda. Hawaiite, **mugearite** and **benmoreite** form a compositional series between alkali-**basalt** and **trachyte**; all usually occur as lava flows.

Hornfels: a well-baked, hard **metamorphic rock**.

Hyaloclastite: **volcaniclastic rock** composed of quenched, glassy fragments (**clasts**) formed when magma cools and shatters on coming into contact with water.

Hydrothermal: to do with hot water.

Igneous Rocks: rocks which have solidified (usually crystallized) from molten rock (**magma**).

Ignimbrite: a lava-like sheet of **volcaniclastic rock** formed by the compaction and welding of an ash-flow (see also **tuff** and **eutaxitic**).

Intermediate: coarse- to fine-grained **igneous rocks** intermediate in compositions between **acid** and **basic**.

Laterite: a red (sub-)soil rich in iron and alumina.

Lherzolite: a **peridotite** consisting largely of **olivine** and **pyroxenes**; commonly exhibits a **metamorphic** texture indicating (re-) crystallization deep in the earth.

Mafic: see **basic**.

Magma: molten rock; referred to as lava when on the earth's surface.

Metamorphic Rocks: rocks whose texture and mineralogy have been changed in the solid, i.e. without melting, by heat and/or pressure.

Metasomatism: a group of processes by which rocks change their chemical composition in the solid, i.e. without melting.

Meteoric Water: water derived directly from the atmosphere.

Mullite: a very high-temperature alumino-silicate mineral.

Mugearite: see **hawaiite**.

Olivine: a silicate mineral enriched in magnesium (and/or iron).

Pegmatite: applied to very coarse-grained varieties of **igneous rocks**; however, 'pegmatite' normally implies very coarse-grained **granite**.

Peridotite: a coarse-grained, **ultrabasic igneous rock** consisting largely of **olivine** and **pyroxene(s)**.

Petrography: (the study of) the mineralogy and texture (fabric) of rocks.

Phyric: denotes the type(s) of the large crystals in a **porphyritic igneous rock** (usually a lava), e.g. **feldspar**-phyric.

Picrite: a lava of **ultrabasic** composition, particularly enriched in **olivine**.

Pitchstone: a **rhyolite** composed largely of volcanic glass.

Plagioclase: see **feldspar**.

Porphyrite: an **igneous rock**, often of **intermediate** composition and of medium grain-size, which displays **porphyritic** texture.

Porphyritic: the texture of those **igneous rocks**, often lavas, in which large crystals (megacrysts, phenocrysts) are surrounded by a matrix of smaller crystals and/or glass.

Pyroclastic: see **volcaniclastic**.

Pyroxenes: the most abundant ferromagnesian silicate minerals in the earth's crust, e.g. augite and hypersthene.

Quartz: a mineral composed entirely of silica (SiO_2)

Rheomorphism: the processes by which a rock is (re-)melted.

Ring-Dyke: a near-cylindrical **igneous** intrusion which tends to close upwards on projection (cf. **cone-sheet**).

Skarn: a rock containing iron-rich and other minerals sometimes found at the contact between an **igneous** intrusion and its **aureole**, especially where this contains limestones; **metasomatism** is often invoked.

Sill: a sheet-like body of **igneous rock** which, in general, does not cross-cut the structure of the rocks which it intrudes; often gently inclined, medium-grained and composed of **dolerite** or **basalt** (cf. **dyke**).

Syenite: a coarse-grained, **intermediate igneous rock** consisting largely of alkali-**feldspar** and various ferromagnesian silicate minerals; usually in large intrusions.

Teschenite: an alkali-rich **gabbro**.

Tholeiite: **basalt** relatively enriched in silica $(SiO)_2$ and deficient in the alkalis (NaO_2 and K_2O).

Trachybasalt: **basalt** containing both plagioclase and alkali-**feldspar**.

Trachyte: a fine-grained, **intermediate igneous rock** consisting largely of alkali-**feldspar** and various ferromagnesian silicate minerals; usually a lava or **dyke**.

Transitional Basalt: **basalt** transitional between alkali-basalt and **tholeiite**.

Tridymite: a high-temperature equivalent (paramorph) of **quartz**.

Troctolite: a coarse-grained **basic igneous rock** consisting largely of **olivine** and plagioclase **feldspar**.

Tuff: consolidated volcanic ash (see also **volcaniclastic rocks**).

Tuffisite: a **Tuff**-like rock usually formed by the **hydrothermal** breakdown of volcanic rocks close to a rock fracture.

Ultrabasic Rocks: coarse- to fine-grained **igneous rocks** which are particularly enriched in 'bases' (see also **basic**) and relatively deficient in silica (SiO_2 nominally under 45%).

Vesicles: gas bubble cavities in consolidated lavas.

Volcaniclastic Rocks: rocks made up of volcanic fragments (**clasts**); also known as pyroclastic rocks.

Xenocrysts/-liths: (fragments of) crystals and rocks that are foreign to the **igneous rock** in which they are found.

Zeolites: a group of low-temperature, hydrous, elumino-silicate minerals.

Index

Page numbers in *italic* refer to figures and those in **bold** refer to tables.

Index

Index

Index

Index

Index

Index

Index

Index

Index

Index

Index

Index